Software Metrics

A Rigorous and Practical Approach

THIRD EDITION

Chapman & Hall/CRC Innovations in Software Engineering and Software Development

Series Editor
Richard LeBlanc
Chair, Department of Computer Science and Software Engineering, Seattle University

AIMS AND SCOPE

This series covers all aspects of software engineering and software development. Books in the series will be innovative reference books, research monographs, and textbooks at the undergraduate and graduate level. Coverage will include traditional subject matter, cutting-edge research, and current industry practice, such as agile software development methods and service-oriented architectures. We also welcome proposals for books that capture the latest results on the domains and conditions in which practices are most effective.

PUBLISHED TITLES

Software Essentials: Design and Construction
Adair Dingle

Software Test Attacks to Break Mobile and Embedded Devices
Jon Duncan Hagar

Software Designers in Action: A Human-Centric Look at Design Work
André van der Hoek and Marian Petre

Fundamentals of Dependable Computing for Software Engineers
John Knight

Introduction to Combinatorial Testing
D. Richard Kuhn, Raghu N. Kacker, and Yu Lei

Building Enterprise Systems with ODP: An Introduction to Open Distributed Processing
Peter F. Linington, Zoran Milosevic, Akira Tanaka, and Antonio Vallecillo

Software Engineering: The Current Practice
Václav Rajlich

Software Development: An Open Source Approach
Allen Tucker, Ralph Morelli, and Chamindra de Silva

Software Metrics: A Rigorous and Practical Approach, Third Edition
Norman Fenton and James Bieman

CHAPMAN & HALL/CRC INNOVATIONS IN
SOFTWARE ENGINEERING AND SOFTWARE DEVELOPMENT

Software Metrics

A Rigorous and Practical Approach

THIRD EDITION

Norman Fenton

Queen Mary University of London, UK

James Bieman

Colorado State University, Fort Collins, USA

CRC Press
Taylor & Francis Group
Boca Raton London New York

CRC Press is an imprint of the
Taylor & Francis Group, an **informa** business

A CHAPMAN & HALL BOOK

CRC Press
Taylor & Francis Group
6000 Broken Sound Parkway NW, Suite 300
Boca Raton, FL 33487-2742

First issued in paperback 2020

© 2015 by Taylor & Francis Group, LLC
CRC Press is an imprint of Taylor & Francis Group, an Informa business

No claim to original U.S. Government works

ISBN-13: 978-1-4398-3822-8 (hbk)
ISBN-13: 978-0-367-65902-8 (pbk)

Library of Congress Cataloging-in-Publication Data

Fenton, Norman E., 1956-
  Software metrics : a rigorous and practical approach / Norman Fenton, James Bieman.
-- Third edition.
      pages cm -- (Chapman & Hall/CRC innovations in software engineering and
      software development)
  ISBN 978-1-4398-3822-8 (hardback)
  1. Software measurement. 2. Computer software--Evaluation. I. Bieman, James. II.
Title.

QA76.758.F46 2014
005.3028'7--dc23   2014028018

Visit the Taylor & Francis Web site at
http://www.taylorandfrancis.com

and the CRC Press Web site at
http://www.crcpress.com

Contents

Preface

SOFTWARE METRICS PLAY A KEY ROLE in good software engineering. Measurement is used to assess situations, track progress, evaluate effectiveness, and more. But the gap between how we *do* measure and how we *could* measure remains larger than it should be. A key reason for this gap between potential and practice was the lack of a coordinated, comprehensive framework for understanding and using measurement. The rigorous measurement framework introduced by the highly successful first edition of *Software Metrics: A Rigorous Approach* in 1991 and second edition of *Software Metrics: A Rigorous and Practical Approach* in 1997 has helped to advance the role of measurement by inspiring discussion of important issues, explaining essential concepts, and suggesting new approaches for tackling long-standing problems.

As one of the first texts on software metrics, the first edition broke new ground by introducing software engineers to measurement theory, graph-theoretic concepts, and new approaches to software reliability. The second edition added material on practical applications and expanded the framework to include notions of process visibility and goal-directed measurement. The new third edition reflects the great progress in the development and use of software metrics over the past decades. This progress includes the acceptance of quantitative analysis and empirical evaluation of software development methods in both research and practice. We have seen the emergence of numerous research journals and conferences that focus on quantitative and empirical methods applied to software engineering problems. The SEI Capability Maturity Model Integration (CMMI) for development, which relies on metrics, is now commonly used to evaluate the maturity of development organizations.

This third edition contains new material relevant to object-oriented design, design patterns, model-driven development, and agile development processes. Of particular note is the new chapter on causal models and Bayesian networks and their application to software engineering. The text also includes references to recent software metrics activities, including research results, industrial case studies, and standards. Along with the new material, the book contains numerous examples and exercises, and thus continues to provide an accessible and comprehensive introduction to software metrics.

This book is designed to suit several audiences. It is structured as the primary textbook for an academic or industrial course on software metrics and quality assurance. But it is also a useful supplement for any course in software engineering. Because of its breadth, the book is a major reference book for academics and practitioners, as it makes accessible important and interesting results that have appeared only in research-oriented publications. Researchers in software metrics will find special interest in the material reporting new results, and in the extensive bibliography of measurement-related information. Finally, the book offers help to software managers and developers who seek guidance on establishing or expanding a measurement program; they can focus on the practical guidelines for selecting metrics and planning their use.

The book is arranged in two parts. Part I offers the reader a basic understanding of why and how we measure. It examines and explains the fundamentals of measurement, experimentation, and data collection and analysis. Part II explores software engineering measurement in greater detail, with comprehensive information about a range of specific metrics and their uses, illustrated by many examples and case studies. The book also includes a bibliography and answers to selected exercises from the main chapters.

Acknowledgments

THIS BOOK HAS BEEN FASHIONED and influenced by many people over the last few years. We thank the reviewers of preliminary manuscripts, including Olga Ormandjieva and Salem N. Salloum. We thank our editors Alan Apt and Randi Cohen for their patience and help throughout this process.

We specially thank Shari Lawrence Pfleeger for her many contributions as co-author of the second edition. We are still indebted to Bev Littlewood, Barbara Kitchenham, and Peter Mellor for key contributions to the first edition, and for allowing us to embellish their ideas. Sadly, Peter passed away before this edition was completed.

Special thanks are also due to those who did so much to help with the original edition, notably Martin Bush, Ros Herman, Agnes Kaposi, David Mole, Meg Russell, and Robin Whitty. Other colleagues have given us helpful advice, including Nick Ashley, Richard Bache, Bill Bail, Albert Baker, Vic Basili, Sarah Brocklehurst, Tom DeMarco, Bob Grady, Dave Gustafson, Les Hatton, Gillian Hill, Chuck Howell, Bob Lockhart, Austin Melton, Steve Minnis, Margaret Myers, Martin Neil, Linda Ott, Stella Page, Gerald Robinson, Dieter Rombach, Chuck Pfleeger, Suzanne Robertson, James Robertson, Lorenzo Strigini, Alan Todd, Dave Wright, and Marv Zelkowitz. We also appreciate the efforts of Armstrong Takang, Faridah Yahya, Christof Ebert, Aimo Torn, Pim van den Broek, Rachel Harrison, Alexander Chatzigeorgiou, Cem Kaner, Robin Laney, P. van der Straaten, Metin Turan, and numerous others who have pointed out problems with previous editions that we have corrected here. We also thank the graduate teaching assistants, especially Aritra Bandyopadhyay and Chris Wilcox, who helped to develop several of the exercises included in this edition.

We continue to be indebted to the European Commission, whose ESPRIT-funded METKIT and PDCS projects partly supported the writing of the first edition; British Telecom Research Laboratories and NATO also supported some of the initial work. We are grateful to the UK Engineering and Physical Sciences Research Council and the UK Department of Trade and Industry, whose DESMET, SMARTIE, and DATUM projects supported some of the work.

Authors

Norman Fenton, PhD, is a professor of risk information management at Queen Mary London University and is also the chief executive officer of Agena, a company that specializes in risk management for critical systems. His experience in risk assessment covers a wide range of application domains such as critical software systems, legal reasoning (he has been an expert witness in major criminal and civil cases), medical trials, vehicle reliability, embedded software, transport systems, financial services, and football prediction. He has published 6 books and more than 140 refereed articles and has provided consulting to many major companies worldwide. His current projects are focused on using Bayesian methods of analysis to risk assessment. In addition to his research on risk assessment, he is renowned for his work in software engineering and software metrics, including the pioneering work in previous editions of this book. Further details can be found at http://www.eecs.qmul.ac.uk/~norman/.

James M. Bieman, PhD, is a professor of computer science at Colorado State University. He served as the founding director of the Software Assurance Laboratory, a Colorado State University Research Center. His research focus is on the evaluation of software designs and processes. His current research is directed toward ways to test nontestable software—software without a test oracle, which includes many scientific software systems. He is also studying techniques that support automated software repair. His long-term research involves studying the relationships between internal design attributes, such as coupling, cohesion, architectural contexts, and the use of design patterns, and external attributes such as maintainability, reliability, and testability. Professor Bieman served as the editor-in-chief of the *Software Quality Journal* published by Springer, from 2001 to 2009, and he serves on the editorial boards of the *Software Quality Journal* and the *Journal of Software and Systems Modeling*. Further details are available at http://www.cs.colostate.edu/~bieman/.

I

Fundamentals of Measurement and Experimentation

Measurement

What Is It and Why Do It?

S OFTWARE MEASUREMENT IS AN essential component of good software engineering. Many of the best software developers measure characteristics of their software to get some sense of whether the requirements are consistent and complete, whether the design is of high quality, and whether the code is ready to be released. Effective project managers measure attributes of processes and products to be able to tell when software will be ready for delivery and whether a budget will be exceeded. Organizations use process evaluation measurements to select software suppliers. Informed customers measure the aspects of the final product to determine if it meets the requirements and is of sufficient quality. Also, maintainers must be able to assess the current product to see what should be upgraded and improved.

This book addresses these concerns and more. The first seven chapters examine and explain the fundamentals of measurement and experimentation, providing you with a basic understanding of why we measure and how that measurement supports investigation of the use and effectiveness of software engineering tools and techniques. Chapters 8 through 11 explore software engineering measurement in further detail, with information about specific metrics and their uses. Collectively, the chapters offer broad coverage of software engineering measurement with enough depth so that you can apply appropriate metrics to your software processes, products, and resources. Even if you are a student, not yet experienced in working on projects with groups of people to solve interesting business or research problems, this book explains how measurement can become a natural and useful part of your regular development and maintenance activities.

This chapter begins with a discussion of measurement in our everyday lives. In the first section, we explain how measurement is a common and necessary practice for understanding, controlling, and improving our environment. In this section, the readers will see why measurement requires rigor and care. In the second section, we describe the role of measurement in software engineering. In particular, we look at how measurement needs are directly related to the goals we set and the questions we must answer when developing our software. Next, we compare software engineering measurement with measurement in other engineering disciplines, and propose specific objectives for software measurement. The last section provides a roadmap to the measurement topics discussed in the remainder of the book.

1.1 MEASUREMENT IN EVERYDAY LIFE

Measurement lies at the heart of many systems that govern our lives. Economic measurements determine price and pay increases. Measurements in radar systems enable us to detect aircraft when direct vision is obscured. Medical system measurements enable doctors to diagnose specific illnesses. Measurements in atmospheric systems are the basis for weather prediction. Without measurement, technology cannot function.

But measurement is not solely the domain of professional technologists. Each of us uses it in everyday life. Price acts as a measure of value of an item in a shop, and we calculate the total bill to make sure the shopkeeper gives us correct change. We use height and size measurements to ensure that our clothing will fit properly. When making a journey, we calculate distance, choose our route, measure our speed, and predict when we will arrive at our destination (and perhaps when we need to refuel). So, measurement helps us to understand our world, interact with our surroundings, and improve our lives.

1.1.1 What Is Measurement?

These examples present a picture of the variety in how we use measurement. But there is a common thread running through each of the described activities: in every case, some aspect of a thing is assigned a descriptor that allows us to compare it with others. In the shop, we can compare the price of one item with another. In the clothing store, we contrast sizes. And on a journey, we compare distance traveled to distance remaining. The rules for assignment and comparison are not explicit in the examples, but it is clear that we make our comparisons and calculations according to a well-defined set of rules. We can capture this notion by defining measurement formally in the following way:

> *Measurement* is the process by which numbers or symbols are assigned to attributes of entities in the real world in such a way so as to describe them according to clearly defined rules.

Thus, measurement captures information about *attributes* of *entities*. An *entity* is an object (such as a person or a room) or an event (such as a journey or the testing phase of a software project) in the real world. We want to describe the entity by identifying characteristics that are important to us in distinguishing one entity from another. An *attribute* is a feature or property of an entity. Typical attributes include the area or color (of a room), the cost (of a journey), or the elapsed time (of the testing phase). Often, we talk about entities and their attributes interchangeably, as in "It is cold today" when we really mean that the air temperature is cold today, or "she is taller than he" when we really mean "her height is greater than his height." Such loose terminology is acceptable for everyday speech, but it is unsuitable for scientific endeavors. Thus, it is wrong to say that we measure things or that we measure attributes; in fact, we measure attributes of things. It is ambiguous to say that we "measure a room," since we can measure its length, area, or temperature. It is likewise ambiguous to say that we "measure the temperature," since we measure the temperature of a specific geographical location under specific conditions. In other words, what is commonplace in common speech is unacceptable for engineers and scientists.

When we describe entities by using attributes, we often define the attributes using numbers or symbols. Thus, price is designated as a number of dollars or pounds sterling, while height is defined in terms of inches or centimeters. Similarly, clothing size may be "small," "medium," or "large," while fuel is "regular," "premium," or "super." These numbers and symbols are abstractions that we use to reflect our perceptions of the real world. For example, in defining the numbers and symbols, we try to preserve certain relationships that we see among the entities. Thus, someone who is 6 feet in height is taller than someone who is 5 feet in height. Likewise, a "medium" T-shirt is smaller than a "large" T-shirt. This number or symbol can be very useful and important. If we have never met Herman but are told that he is 7 feet tall, we can imagine his height in relation to ourselves without even having seen him. Moreover, because of his unusual height, we know that he will have to stoop when he enters the door of our office. Thus, we can make judgments about entities solely by knowing and analyzing their attributes.

Measurement is a process whose definition is far from clear-cut. Many different authoritative views lead to different interpretations about what constitutes measurement. To understand what measurement is, we must ask a host of questions that are difficult to answer. For example

- We have noted that color is an attribute of a room. In a room with blue walls, is "blue" a *measure* of the color of the room?

- The height of a person is a commonly understood attribute that can be measured. But what about other attributes of people, such as intelligence? Does an IQ test score adequately measure intelligence? Similarly, wine can be measured in terms of alcohol content ("proof"), but can wine quality be measured using the ratings of experts?

- The accuracy of a measure depends on the measuring instrument as well as on the definition of the measurement. For example, length can be measured accurately as long as the ruler is accurate and used properly. But some measures are not likely to be accurate, either because the measurement is imprecise or because it depends on the judgment of the person doing the measuring. For instance, the proposed measures of human intelligence or wine quality appear to have likely error margins. Is this a reason to reject them as bonafide measurements?

- Even when the measuring devices are reliable and used properly, there is margin for error in measuring the best understood physical attributes. For example, we can obtain vastly different measures for a person's height, depending on whether we make allowances for the shoes being worn or the standing posture. So how do we decide which error margins are acceptable and which are not?

- We can measure height in terms of meters, inches, or feet. These different *scales* measure the same attribute. But we can also measure height in terms of miles and kilometers—appropriate for measuring the height of a satellite above the Earth, but not for measuring the height of a person. When is a scale acceptable for the purpose to which it is put?

- Once we obtain measurements, we want to analyze them and draw conclusions about the entities from which they were derived. What kind of manipulations can we apply to the results of measurement? For example, why is it acceptable to say that Fred is twice as tall as Joe, but not acceptable to say that it is twice as hot today as it was yesterday? And why is it meaningful to calculate the mean of a set of heights (to say, e.g., that the average height of a London building is 200 m), but not the mean of the football jersey numbers of a team?

To answer these and many other questions, we examine the science of measurement in Chapter 2. This rigorous approach lays the groundwork for applying measurement concepts to software engineering problems. However, before we turn to measurement theory, we examine first the kinds of things that can be measured.

1.1.2 "What Is Not Measurable Make Measurable"

This phrase, attributable to Galileo Galilei (1564–1642), is part of the folklore of measurement scientists (Finkelstein 1982). It suggests that one of the aims of science is to find ways to measure attributes of interesting things. Implicit in Galileo's statement is the idea that measurement makes concepts more visible and therefore more understandable and controllable. Thus, as scientists, we should be creating ways to measure our world; where we can already measure, we should be making our measurements better.

In the physical sciences, medicine, economics, and even some social sciences, we can now measure attributes that were previously thought to be unmeasurable. Whether we like them or not, measures of attributes such as human intelligence, air quality, and economic inflation form the basis for important decisions that affect our everyday lives. Of course, some measurements are not as refined (in a sense to be made precise in Chapter 2) as we would like them to be; we use the physical sciences as our model for good measurement, continuing to improve measures when we can. Nevertheless, it is important to remember that the concepts of time, temperature, and speed, once unmeasurable by primitive peoples, are now not only commonplace but also easily measured by almost everyone; these measurements have become part of the fabric of our existence.

To improve the rigor of measurement in software engineering, we need not restrict the type or range of measurements we can make. Indeed, measuring the unmeasurable should improve our understanding of particular entities and attributes, making software engineering as powerful as other engineering disciplines. Even when it is not clear how we might measure an attribute, the act of proposing such measures will open a debate that leads to greater understanding. Although some software engineers may continue to claim that important software attributes like dependability, quality, usability, and maintainability are simply not quantifiable, we prefer to try to use measurement to advance our understanding of them.

We can learn strategies for measuring unmeasurable attributes from the business community. Businesses often need to measure intangible attributes that you might think are unmeasurable such as customer satisfaction, future revenues, value of intellectual property, a company's reputation, etc. A key concern of the business community, including the software development business community, is risk. Businesses want to reduce the risk of product failures, late release of a product, loss of key employees, bankruptcy, etc. Thus, Douglas Hubbard provides an alternative definition of measurement:

> Measurement: A quantitatively expressed reduction of uncertainty based on one or more observations.
>
> HUBBARD 2010, P. 23

Observations that can reduce uncertainty can quantitatively measure the risk of negative events or the likelihood of positive outcomes.

EXAMPLE 1.1

Douglas Hubbard lists the kinds of statements that a business executive might want to make involving the use of measurement to reduce uncertainty, for example (Hubbard 2010):

> There is an 85% chance we will win our patent dispute.
> We are 93% certain our public image will improve after the merger.
>
> HUBBARD 2010, P. 24

He also shows that you need very little data to reduce uncertainty by applying what he calls the *Rule of Five*:

> There is a 93.75% chance that the median of a population is between the smallest and largest values in any random sample of five from the population.
>
> HUBBARD 2010, P. 30

We can apply the *Rule of Five* to measurements relevant to software engineering. Assume that you want to learn whether your organization's developers are writing class method bodies that are short, because you have heard somewhere that shorter method bodies are better. Rather than examining all methods in your code base, you could randomly select five methods and count the lines of code in their bodies. Say that these methods have bodies with 10, 15, 25, 45, and 50 lines of code. Using the *Rule of Five*, you know that there is a 93.75% chance that the median size of all method bodies in your code base is between 10 and 50 lines of code. With a sample of only five methods, you can count lines of code manually—you may not need a tool. Thus, using this method, you can easily find the median of a population with a quantified level of uncertainty. Using a larger sample can further reduce the uncertainty.

Software development involves activities and events under uncertain conditions. Requirements and user communities change unpredictably. Developers can leave a project at unpredictable times. We will examine the use of metrics in decision-making under conditions of uncertainty in Chapter 7.

Strictly speaking, we should note that there are two kinds of quantification: measurement and calculation. *Measurement* is a direct quantification, as in measuring the height of a tree or the weight of a shipment of bricks. *Calculation* is indirect, where we take measurements and combine them into a quantified item that reflects some attribute whose value we are trying to understand. For example, when the city inspectors assign a valuation to a house (from which they then decide the amount of tax owed), they calculate it by using a formula that combines a variety of factors, including the number of rooms, the type of heating and cooling, the overall floor space, and the sale prices of similar houses in comparable locations. The valuation is quantification, not a measurement, and its expression as a number makes it more useful than qualitative assessment alone. As we shall see in Chapter 2, we use *direct* and *derived* to distinguish measurement from calculation.

Sport offers us many lessons in measuring abstract attributes like quality in an objective fashion. Here, the measures used have been accepted universally, even though there is often discussion about changing or improving the measures. In the following examples, we highlight measurement concepts, showing how they may be useful in software engineering:

EXAMPLE 1.2

In the decathlon athletics event, we measure the time to run various distances as well as the length covered in various jumping activities. These measures are subsequently combined into an *overall score*, computed using a complex weighting scheme that reflects the importance of each component measure. Over the years, the weights and scoring rules have changed as the relative importance of an event or measure changes. Nevertheless, the overall score is widely accepted as a description of the athlete's all-around ability. In fact, the winner of the Olympic decathlon is generally acknowledged to be the world's finest athlete.

EXAMPLE 1.3

In European soccer leagues, a points system is used to select the best all-around team over the course of a season. Until the early 1980s, two points were awarded for each win and one point was awarded for each draw. Thereafter, the points system was changed; a win yielded three points instead of two, while a draw still yielded one point. This change was made to reflect the consensus view that the *qualitative difference* between a win and a draw was greater than that between a draw and a defeat.

EXAMPLE 1.4

There are no universally recognized measures to identify the best individual soccer players (although number of goals scored is a fairly accurate measure of quality of a striker). Although many fans and players have argued that player quality is an unmeasurable attribute, there are organizations (such as optasports.com) that provide player ratings based on a wide range of measurable attributes such as tackles, saves, or interceptions made; frequency and distance of passes (of various types), dribbles, and headers. Soccer clubs, agents, betting, and media companies pay large sums to acquire these ratings. Often clubs and players' agents use the ratings as the basis for determining player value (both from a salary and transfer price perspective).

It is easy to see parallels in software engineering. In many instances, we want an overall score that combines several measures into a "big picture" of what is going on during development or maintenance. We want to be able to tell if a software product is good or bad, based on a set of measures, each of which captures a facet of "goodness." Similarly, we want to be able to measure an organization's ability to produce good software, or a model's ability to make good predictions about the software development process. The composite measures can be controversial, not only because of the individual measures comprising it, but also because of the weights assigned.

Likewise, controversy erupts when we try to capture qualitative information about some aspect of software engineering. Different experts have different opinions, and it is sometimes impossible to get consensus.

Finally, it is sometimes necessary to modify our environment or our practices in order to measure something new or in a new way. It may mean using a new tool (to count lines of code or evaluate code structure), adding a new step in a process (to report on effort), or using a new method (to make measurement simpler). In many cases, change is difficult for people to accept; there are management issues to be considered whenever a measurement program is implemented or changed.

1.2 MEASUREMENT IN SOFTWARE ENGINEERING

We have seen that measurement is essential to our daily lives, and measuring has become commonplace and well accepted. In this section, we examine the realm of software engineering to see why measurement is needed.

Software engineering describes the collection of techniques that apply an engineering approach to the construction and support of software products. Software engineering activities include managing, costing, planning, modeling, analyzing, specifying, designing, implementing, testing, and maintaining. By "engineering approach," we mean that each activity is understood and controlled, so that there are few surprises as the software is specified, designed, built, and maintained. Whereas computer science provides the theoretical foundations for building software, software engineering focuses on implementing the software in a controlled and scientific way.

The importance of software engineering cannot be understated, since software pervades our lives. From oven controls to automobiles, from banking transactions to air traffic control, and from sophisticated power plants to sophisticated weapons, our lives and the quality of life depend on software. For such a young profession, software engineering has usually done an admirable job of providing safe, useful, and reliable functionality. But there is room for a great deal of improvement. The literature is rife with examples of projects that have overrun their budgets and schedules.

Worse, there are too many stories about software that has put lives and businesses at risk.

Software engineers have addressed these problems by continually looking for new techniques and tools to improve process and product. Training supports these changes, so that software engineers are better prepared to apply the new approaches to development and maintenance. But methodological improvements alone do not make an engineering discipline.

1.2.1 Neglect of Measurement in Software Engineering

Engineering disciplines use methods that are underpinned by models and theories. For example, in designing electrical circuits, we appeal to theories like Ohm's law that describes the relationship between resistance, current, and voltage in the circuit. But the laws of electrical behavior have evolved by using the scientific method: stating a hypothesis, designing and running an experiment to test its validity, and analyzing the results. Underpinning the scientific process is measurement: measuring the variables to differentiate cases, measuring the changes in behavior, and measuring the causes and effects. Once the scientific method suggests the validity of a model or the truth of a theory, we continue to use measurement to apply the theory to practice. Thus, to build a circuit with a specific current and resistance, we know what voltage is required and we use instruments to measure whether we have such a voltage in a given battery.

It is difficult to imagine electrical, mechanical, and civil engineering without a central role for measurement. Indeed, science and engineering can be neither effective nor practical without measurement. But measurement has been considered a luxury in software engineering. For many development projects:

1. We fail to set measurable targets for our software products. For example, we promise that the product will be user-friendly, reliable, and maintainable without specifying clearly and objectively what these terms mean. As a result, from both is complete, we cannot tell if we have met our goals.

2. We fail to understand and quantify the component costs of software projects. For example, many projects cannot differentiate the cost of design from the cost of coding or testing. Since excessive cost is a frequent complaint from our customers, we cannot hope to control costs if we are not measuring the relative components of cost.

3. We do not quantify or predict the quality of the products we produce. Thus, we cannot tell a potential user how reliable a product will be in terms of likelihood of failure in a given period of use, or how much work will be needed to port the product to a different machine environment.

4. We allow anecdotal evidence to convince us to try yet another revolutionary new development technology, without doing a carefully controlled study to determine if the technology is efficient and effective. Promotional materials for software development tools and techniques typically include the following types of claims:

   a. "Our new technique guarantees 100% reliability."

   b. "Our tool improves productivity by 200%!!"

   c. "Build your code with half the staff in a quarter of the time."

   d. "Cuts test time by 2/3."

   These claims are generally not supported by scientific studies.

When measurements are made, they are often done infrequently, inconsistently, and incompletely. The incompleteness can be frustrating to those who want to make use of the results. For example, a developer may claim that 80% of all software costs involve maintenance, or that there are on average 55 faults in every 1000 lines of software code. But often we are not told how these results were obtained, how experiments were designed and executed, which entities were measured and how, and what were the realistic error margins. Without this additional information, we remain skeptical and unable to decide whether to apply the results to our own situations. In addition, we cannot do an objective study to repeat the measurements in our own environments. Thus, the lack of measurement in software engineering is compounded by the lack of a rigorous approach.

It is clear from other engineering disciplines that measurement can be effective, if not essential, in making characteristics and relationships more visible, in assessing the magnitude of problems, and in fashioning a solution to problems. As the pace of hardware innovation has increased, the software world has been tempted to relax or abandon its engineering underpinnings and hope for revolutionary gains. But now that software, playing a key role, involves enormous investment of energy and money, it is time for software engineering to embrace the engineering discipline that has been so successful in other areas.

1.2.2 Objectives for Software Measurement

Even when a project is not in trouble, measurement is not only useful but also necessary. After all, how can you tell if your project is healthy if you have no measures of its health? So, measurement is needed at least for assessing the status of your projects, products, processes, and resources. Since we do not always know what derails a project, it is essential that we measure and record characteristics of good projects as well as bad. We need to document trends, the magnitude of corrective action, and the resulting changes. In other words, we must control our projects, not just run them. In Chapter 3, you will see the key role that measurement plays in evaluating software development organizations and their software development processes.

There are compelling reasons to consider the measurement process scientifically, so that measurement will be a true engineering activity. Every measurement action must be motivated by a particular goal or need that is clearly defined and easily understandable. That is, it is not enough to assert that we must measure to gain control. The measurement objectives must be specific, tied to what the managers, developers, and users need to know. Thus, these objectives may differ according to the kind of personnel involved and at which level of software development and use they are generated. But it is the goals that tell us how the measurement information will be used once it is collected.

We now describe the kinds of information needed to understand and control a software development project, from both manager and developer perspectives.

1.2.2.1 Managers

- *What does each process cost?* We can measure the time and effort involved in the various processes that comprise software production. For example, we can identify the cost to elicit requirements, the cost to specify the system, the cost to design the system, and the cost to code and test the system. In this way, we gain understanding not only of the total project cost but also of the contribution of each activity to the whole.

- *How productive is the staff?* We can measure the time it takes for staff to specify the system, design it, code it, and test it. Then, using measures of the size of specifications, design, code, and test plans, for example, we can determine how productive the staff is at each activity. This information is useful when changes are proposed; the manager can use the productivity figures to estimate the cost and duration of the change.

- *How good is the code being developed?* By carefully recording faults, failures, and changes as they occur, we can measure software quality, enabling us to compare different products, predict the effects of change, assess the effects of new practices, and set targets for process and product improvement.

- *Will the user be satisfied with the product?* We can measure functionality by determining if all of the requirements requested have actually been implemented properly. And we can measure usability, reliability, response time, and other characteristics to suggest whether our customers will be happy with both functionality and performance.

- *How can we improve?* We can measure the time it takes to perform each major development activity, and calculate its effect on quality and productivity. Then we can weigh the costs and benefits of each practice to determine if the benefit is worth the cost. Alternatively, we can try several variations of a practice and measure the results to decide which is best; for example, we can compare two design methods to see which one yields the higher-quality code.

1.2.2.2 Developers

- *Are the requirements testable?* We can analyze each requirement to determine if its satisfaction is expressed in a measurable, objective way. For example, suppose a requirement states that a web-based system must be "fast"; the requirement can be replaced by one that states that the mean response time to a set of specific of inputs must be less than 2 s for specified browsers and number of concurrent users.

- *Have we found all the faults?* We can measure the number of faults in the specification, design, code, and test plans and trace them back to their root causes. Using models of expected detection rates, we can use this information to decide whether inspections and testing have been effective and whether a product can be released for the next phase of development.

- *Have we met our product or process goals?* We can measure characteristics of the products and processes that tell us whether we have met standards, satisfied a requirement, or met a process goal. For example, certification may require that fewer than 20 failures have been reported per beta-test site over a given period of time. Or a standard may mandate that all modules must pass code inspections. The testing process may require that unit testing must achieve 90% statement coverage.

- *What will happen in the future?* We can measure attributes of existing products and current processes to make predictions about future ones. For example, measures of size of specifications can be used to predict the size of the target system, predictions about future maintenance problems can be made from measures of structural properties of the design documents, and predictions about the reliability of software in operational use can be made by measuring reliability during testing.

1.2.3 Measurement for Understanding, Control, and Improvement

The information needs of managers and developers show that measurement is important for three basic activities. First, measurement can help us to *understand* what is happening during development and maintenance. We assess the current situation, establishing baselines that help us to set goals for future behavior. In this sense, the measurements make aspects of process and product more visible, giving us a better understanding of relationships among activities and the entities they affect.

Second, the measurement allows us to *control* what is happening in our projects. Using our baselines, goals, and understanding of relationships, we predict what is likely to happen and make changes to processes and products that help us to meet our goals. For example, we may monitor the complexity of code modules, giving thorough review only to those that exceed acceptable bounds.

Third, measurement encourages us to *improve* our processes and products. For instance, we may increase the number or type of design reviews we do, based on measures of specification quality and predictions of likely design quality.

No matter how measurements are used, it is important to manage the expectations of those who will make measurement-based decisions. Users of the data should always be aware of the limited accuracy of prediction

FIGURE 1.1   Software measurement—resource estimation.

and of the margin of error in the measurements. As with any other engineering discipline, there is room in software engineering for abuse and misuse of measurement. In particular, managers may pressure developers to produce precise measures with inadequate models, tools, and techniques (see Figure 1.1).

It is wrong to expect measurement to provide instant, easy solutions to your software engineering problems. Control and accurate prediction both require careful sets of measurements.

1.3 SCOPE OF SOFTWARE METRICS

Software metrics is a term that embraces many activities, all of which involve some degree of software measurement:

- Cost and effort estimation models and measures

- Data collection

- Quality models and measures

- Reliability models

- Security metrics

- Structural and complexity metrics

- Capability maturity assessment

- Management by metrics

- Evaluation of methods and tools

Each of these activities will be covered in some detail. Our theoretical foundations, to be described in Chapters 2 and 3, will enable us to consider the activities in a unified manner, rather than as diverse, unrelated topics.

The following brief introduction will give you a sense of the techniques currently in use for each facet of measurement. It provides signposts to where the material is covered in detail.

1.3.1 Cost and Effort Estimation

Managers provided the original motivation for deriving and using software measures. They wanted to be able to predict project costs during early phases in the software life cycle. As a result, numerous models for software cost and effort estimation have been proposed and used. Examples include Boehm's COCOMO II model (Boehm et al. 2000) and Albrecht's function point model (Albrecht 1979). These and other models often share a common approach: effort is expressed as a (predefined) function of one or more variables (such as size of the product, capability of the developers, and level of reuse). Size is usually defined as (predicted) lines of code or number of function points (which may be derived from the product specification). These cost and effort prediction models are discussed in Chapter 8.

1.3.2 Data Collection

The quality of any measurement program is clearly dependent on careful data collection. But collecting data is easier said than done, especially when data must be collected across a diverse set of projects. Thus, data collection is becoming a discipline in itself, where specialists work to ensure that measures are defined unambiguously, that collection is consistent and complete, and that data integrity is not at risk. But it is acknowledged that metrics data collection must be planned and executed in a careful and sensitive manner. We will see in Chapter 5 how useful data can be collected. Chapter 6 shows how to analyze and display collected data in order to draw valid conclusions and make decisions.

Data collection is also essential for scientific investigation of relationships and trends. We will see in Chapter 4 how good experiments, surveys, and case studies require carefully planned data collection, as well as thorough analysis and reporting of the results.

1.3.3 Quality Models and Measures

Since software quality involves many diverse factors, software engineers have developed models of the interaction between multiple quality factors. These models are usually constructed in a tree-like fashion, similar to Figure 1.2. The upper branches hold important high-level quality factors of software products, such as reliability and usability, that we would like to quantify. Each quality factor is composed of lower-level criteria, such as modularity and data commonality. The criteria are easier to understand and measure than the factors; thus, actual measures (metrics) are proposed for the criteria. The tree describes the pertinent relationships between factors and their dependent criteria, so we can measure the factors in terms of the dependent criteria measures. This notion of divide and conquer has been implemented as a standard approach to measuring software quality (IEEE 1061-2009). Quality models are described in Chapter 10.

1.3.4 Reliability Models

Most quality models include reliability as one of their component factors. But the need to predict and measure reliability itself has led to a separate specialization in reliability modeling and prediction. Chapter 11 describes

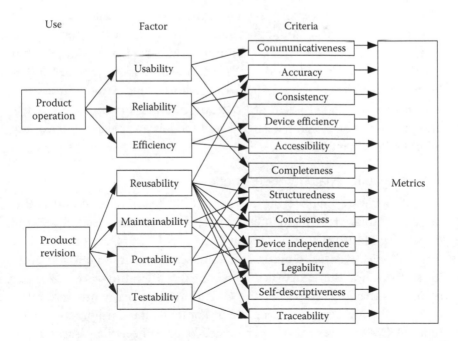

FIGURE 1.2 Software quality model.

software reliability models and measures starting with reliability theory. Chapter 11 shows how to apply reliability theory to analyze and predict software reliability.

1.3.5 Security Metrics

As computing has become part of almost every human activity, our concerns about the security of software systems have grown. We worry that attackers will steal or corrupt data files, passwords, and our accounts. Security depends on both the internal design of a system and the nature of the attacks that originate externally. In Chapter 10, we describe some standard ways to assess security risks in terms of impact, likelihood, threats, and vulnerabilities.

1.3.6 Structural and Complexity Metrics

Desirable quality attributes like reliability and maintainability cannot be measured until some operational version of the code is available. Yet, we wish to be able to predict which parts of the software system are likely to be less reliable, more difficult to test, or require more maintenance than others, even before the system is complete. As a result, we measure structural attributes of representations of the software that are available in advance of (or without the need for) execution; then, we try to establish empirically predictive theories to support quality assurance, quality control, and quality prediction. These representations include control flow graphs that usually model code and various unified modeling language (UML) diagrams that model software designs and requirements. Structural metrics can involve the arrangement of program modules, for example, the use and properties of design patterns. These models and related metrics are described in Chapter 9.

1.3.7 Capability Maturity Assessment

The US Software Engineering Institute (SEI) has developed and maintained a software development process evaluation technique, the Capability Maturity Model Integration (CMMI®) for Development. The original objective was to measure a contractor's ability to develop quality software for the US government, but CMMI ratings are now used by many other organizations. The CMMI assesses many different attributes of development, including the use of tools, standard practices, and more. A CMMI assessment is performed by an SEI certified assessor and involves determining the answers to hundreds of questions concerning software development practices in the organization being reviewed. The result of an assessment is a rating that is on a five-level scale, from *Level 1* (*Initial*—development is dependent on individuals) to *Level 5* (*Optimizing*—a development process that can be optimized based on quantitative process information).

The evaluation of process maturity has become much more common. Many organizations require contracted software development vendors to have CMMI certification at specified levels. In Chapter 3, we describe the CMMI evaluation mechanism and how process maturity can be useful in understanding what and when to measure, and how measurement can guide process improvements.

1.3.8 Management by Metrics

Measurement is an important part of software project management. Customers and developers alike rely on measurement-based charts and graphs to help them decide if a project is on track. Many companies and organizations define a standard set of measurements and reporting methods, so that projects can be compared and contrasted. This uniform collection and reporting is especially important when software plays a supporting role in an overall project. That is, when software is embedded in a product whose main focus is a business area other than software, the customer or ultimate user is not usually well versed in software terminology, so measurement can paint a picture of progress in general, understandable terms. For example, when a power plant designer asks a software developer to write control software, the power plant designer usually knows a lot about power generation and control, but very little about software development processes, programming languages, software testing, or computer hardware. Measurements must be presented in a way that tells both customer and developer how the project is doing. In Chapter 6, we will examine several measurement analysis and presentation techniques that are useful and understandable to all who have a stake in a project's success.

1.3.9 Evaluation of Methods and Tools

The literature is rife with descriptions of new methods and tools that may make your organization or project more productive and your products better and cheaper. But it is difficult to separate the claims from the reality. Many organizations perform experiments, run case studies, or administer surveys to help them decide whether a method or tool is likely to make a positive difference in their particular situations. These investigations cannot be done without careful, controlled measurement and analysis. As we will see in Chapter 4, an evaluation's success depends on good experimental design, proper identification of the factors likely to affect the outcome, and appropriate measurement of factor attributes.

1.4 SUMMARY

This introductory chapter has described how measurement pervades our everyday life. We have argued that measurement is essential for good engineering in other disciplines; it should likewise become an integral part of software engineering practice. In particular

- Owing to the lessons of other engineering disciplines, measurement now plays a significant role in software engineering.

- Software measurement is a diverse collection of topics that range from models for predicting software project costs at the specification stage to measures of program structure.

- General reasons for needing software engineering measurement are not enough. Engineers must have specific, clearly stated objectives for measurement.

- We must be bold in our attempts at measurement. Just because no one has measured some attribute of interest does not mean that it cannot be measured satisfactorily.

We have set the scene for a new perspective on software metrics. Our foundation, introduced in Chapter 2, supports a scientific and effective approach to constructing, calculating, and appropriately applying the metrics that we derive.

EXERCISES

1. Explain the role of measurement in determining the best players in your favorite sport.

2. Explain how measurement is commonly used to determine the best students in a university. Describe any problems with these measurement schemes.

3. How would you begin to measure the quality of a software product?

4. Consider some everyday measurements. What entities and attributes are being measured? What can you say about error margins in the measurements? Explain how the measuring process may affect the entity being measured.

5. Consider the user's viewpoint. What measurement objectives might a software user have?

6. A commonly used software quality measure in industry is the number of errors per thousand lines of product source code. Compare the usefulness of this measure for developers and users. What are the possible problems with relying on this measure as the sole expression of software quality?

7. Consider a web browsing client, such as Firefox, Chrome, Safari, or Explorer. Suggest some good ways to measure the quality of such a software system, *from an end-user's perspective.*

CHAPTER 2

The Basics of Measurement

IN CHAPTER 1, WE SAW how measurement pervades our world. We use measurement everyday to understand, control, and improve what we do and how we do it. In this chapter, we examine measurement in more depth, trying to apply general measurement lessons learned in daily activities to the activities we perform as part of software development.

Ordinarily, when we measure things, we do not think about the scientific principles we are applying. We measure attributes such as the length of physical objects, the timing of events, and the temperature of liquids or of the air. To do the measuring, we use both tools and principles that we now take for granted. However, these sophisticated measuring devices and techniques have been developed over time, based on the growth of understanding of the attributes we are measuring. For example, using the length of a column of mercury to capture information about temperature is a technique that was not at all obvious to the first person who wanted to know how much hotter it is in summer than in winter. As we understood more about temperature, materials, and the relationships between them, we developed a framework for describing temperature as well as tools for measuring it.

Unfortunately, we have no comparably deep understanding of software attributes. Nor do we have the associated sophisticated measurement tools. Questions that are relatively easy to answer for non-software entities are difficult for software. For example, consider the following questions:

1. How much must we know about an attribute before it is reasonable to consider measuring it? For instance, do we know enough about "complexity" of programs to be able to measure it?

2. How do we know if we have really measured the attribute we wanted to measure? For instance, does a count of the number of "bugs" found in a system during integration testing measure the quality of the system? If not, what does the count tell us?

3. Using measurement, what meaningful statements can we make about an attribute and the entities that possess it? For instance, is it meaningful to talk about doubling a design's quality? If not, how do we compare two different designs?

4. What meaningful operations can we perform on measures? For instance, is it sensible to compute average productivity for a group of developers, or the average quality of a set of modules?

To answer these questions, we must establish the basics of a theory of measurement. We begin by examining formal measurement theory, developed as a classical discipline from the physical sciences. We see how the concepts of measurement theory apply to software, and we explore several examples to determine when measurements are meaningful and useful in decision-making. This theory tells us not only when and how to measure, but also how to analyze and depict data, and how to tie the results back to our original questions about software quality and productivity.

2.1 THE REPRESENTATIONAL THEORY OF MEASUREMENT

In any measurement activity, there are rules to be followed. The rules help us to be consistent in our measurement, as well as providing a basis for interpretation of data. Measurement theory tells us the rules, laying the groundwork for developing and reasoning about all kinds of measurement. This rule-based approach is common in many sciences. For example, recall that mathematicians learned about the world by defining axioms for a geometry. Then, by combining axioms and using their results to support or refute their observations, they expanded their understanding and the set of rules that govern the behavior of objects. In the same way, we can use rules about measurement to codify our initial understanding, and then expand our horizons as we analyze our software.

However, just as there are several kinds of geometry (e.g., Euclidean and non-Euclidean) with each depending on the set of rules chosen, there are also several theories of measurement. In this book, we present an overview of the *representational* theory of measurement.

2.1.1 Empirical Relations

The *representational theory of measurement* seeks to formalize our intuition about the way the world works. That is, the data we obtain as measures should represent attributes of the entities we observe, and manipulation of the data should preserve relationships that we observe among the entities. Thus, our intuition is the starting point for all measurement.

Consider the way we perceive the real world. We tend to understand things by comparing them, not by assigning numbers to them. For example, Figure 2.1 illustrates how we learn about height. We observe

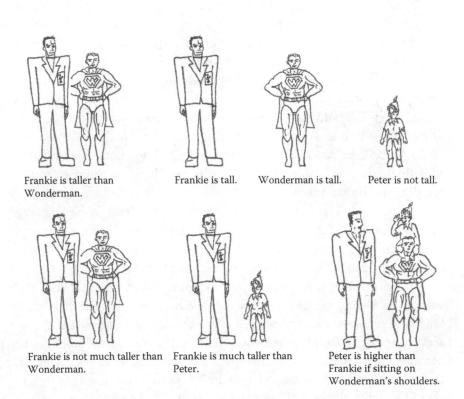

Frankie is taller than Wonderman.

Frankie is tall.   Wonderman is tall.   Peter is not tall.

Frankie is not much taller than Wonderman.

Frankie is much taller than Peter.

Peter is higher than Frankie if sitting on Wonderman's shoulders.

FIGURE 2.1 Some empirical relations for the attribute *height*.

that certain people are *taller than* others without actually measuring them. It is easy to see that Frankie is taller than Wonderman who in turn is taller than Peter; anyone looking at this figure would agree with this statement. However, our observation reflects a set of rules that we are imposing on the set of people. We form pairs of people and define a binary relation on them. In other words, *taller than* is a binary relation defined on the set of pairs of people. Given any two people, x and y, we can observe that

- x is taller than y, or
- y is taller than x

Therefore, we say that *taller than* is an empirical relation for height.

When the two people being compared are very close in height, we may find a difference of opinion; you may think that Jack is taller than Jill, while we are convinced that Jill is taller than Jack. Our empirical relations permit this difference by requiring only a consensus of opinion about relationships in the real world. A (binary) empirical relation is one for which there is a reasonable consensus about which pairs are in the relation.

We can define more than one empirical relation on the same set. For example, Figure 2.1 also shows the relation *much taller than*. Most of us would agree that both Frankie and Wonderman are much taller than Peter (although there is less of a consensus about this relation than *taller than*).

Empirical relations need not be binary. That is, we can define a relation on a single element of a set, or on collections of elements. Many empirical relations are unary, meaning that they are defined on individual entities. The relation *is tall* is an example of a unary relation in Figure 2.1; we can say Frankie is tall but Peter is not tall. Similarly, we can define a ternary relationship by comparing groups of three; Figure 2.1 shows how Peter sitting on Wonderman's shoulders is higher than Frankie.

We can think of these relations as mappings from the empirical, real world to a formal mathematical world. We have entities and their attributes in the real world, and we define a mathematical mapping that preserves the relationships we observe. Thus, height (that is, tallness) can be considered as a mapping from the set of people to the set of real numbers. If we can agree that Jack is *taller than* Jill, then any measure of height should assign a higher number to Jack than to Jill. As we shall see later in

TABLE 2.1   Sampling 100 Users to Express Preferences among
Products A, B, C, and D

	More Functionality				More User-Friendly			
	A	B	C	D	A	B	C	D
A	—	80	10	80	—	45	50	44
B	20	—	5	50	55	—	52	50
C	90	95	—	96	50	48	—	51
D	20	50	4	—	54	50	49	—

this chapter, this preservation of intuition and observation is the notion behind the representation condition of measurement.

EXAMPLE 2.1

Suppose we are evaluating the four best-selling contact management programs: A, B, C, and D. We ask 100 independent computer users to rank these programs according to their functionality, and the results are shown on the left-hand portion of Table 2.1. Each cell of the table represents the percentage of respondents who preferred the row's program to the column's program; for instance, 80% rated program A as having greater functionality than B. We can use this survey to define an empirical relation *greater functionality than* for contact management programs; we say that program x has greater functionality than program y if the survey result for cell (x,y) exceeds 60%. Thus, the relation consists of the pairs (C,A), (C,B), (C,D), (A,B), and (A,D). This set of pairs tells us more than just five comparisons; for example, since C has greater functionality than A, and A in turn has greater functionality than B and D, then C has greater functionality than B and D. Note that neither pair (B,D) nor (D,B) is in the empirical relation; there is no clear consensus about which of B and D has greater functionality.

Suppose we administer a similar survey for the attribute *user-friendliness*, with the results shown on the right-hand side of Table 2.1. In this case, there is no real consensus at all. At best, we can deduce that *greater user-friendliness* is an empty empirical relation. This statement is different from saying that all the programs are equally user-friendly, since we did not specifically ask the respondents about indifference or equality. Thus, we deduce that our understanding of user-friendliness is so immature that there are no useful empirical relations.

Example 2.1 shows how we can start with simple user surveys to gain a preliminary understanding of relationships. However, as our understanding grows, we can define more sophisticated measures.

TABLE 2.2   Historical Advances in Temperature Measurement

2000 BC	Rankings, *hotter than*
1600 AD	First thermometer measuring *hotter than*
1720 AD	Fahrenheit scale
1742 AD	Celsius scale
1854 AD	Absolute zero, Kelvin scale

EXAMPLE 2.2

Table 2.2 shows that people had an initial understanding of temperature thousands of years ago. This intuition was characterized by the notion of *hotter than*. Thus, for example, by putting your hand into two different containers of liquid, you could feel if one were hotter than the other. No measurement is necessary for this determination of temperature difference. However, people needed to make finer discriminations in temperature. In 1600, the first device was constructed to capture this comparative relationship; the thermometer could consistently assign a higher number to liquids that were *hotter than* others.

Example 2.2 illustrates an important characteristic of measurement. We can begin to understand the world by using relatively unsophisticated relationships that require no measuring tools. Once we develop an initial understanding and have accumulated some data, we may need to measure in more sophisticated ways and with special tools. Analyzing the results often leads to the clarification and re-evaluation of the attribute and yet more sophisticated empirical relations. In turn, we have improved accuracy and increased understanding.

Formally, we define *measurement* as the mapping from the empirical world to the formal, relational world. Consequently, a *measure* is the number or symbol assigned to an entity by this mapping in order to characterize an attribute.

Sometimes, the empirical relations for an attribute are not yet agreed, especially when they reflect personal preference. We see this lack of consensus when we look at the ratings of wine or the preference for a design technique, for example. Here, the raters have some notion of the attribute they want to measure, but there is not always a common understanding. We may find that what is tasteless or difficult for one rater is delicious or easy for another rater. In these cases, we can still perform a subjective

Likert Scale
Give the respondent a statement with which to agree or disagree. Example:
This software program is reliable.

Strongly Agree	Agree	Neither agree nor disagree	Disagree	Strongly Disagree

Forced Ranking
Give n alternatives, ordered from 1 (best) to n (worst). Example:
Rank the following five software modules in order of maintenance difficulty, with 1 = least complex, 5 = most complex:

—	Module A
—	Module B
—	Module C
—	Module D
—	Module E

Verbal Frequency Scale
Example: How often does this program fail?

Always　　　Often　　　Sometimes　　　Seldom　　　Never

Ordinal Scale
List several ordered alternatives and have respondents select one. For example:
How often does the software fail?

1. Hourly
2. Daily
3. Weekly
4. Monthly
5. Several times a year
6. Once or twice a year
7. Never

Comparative Scale

Very superior			About the same				Very inferior
1	2	3	4	5	6	7	8

Numerical Scale

Unimportant							Important
1	2	3	4	5	6	7	8

FIGURE 2.2　Subjective rating schemes.

assessment, but the result is not necessarily a measure, in the sense of measurement theory. For example, Figure 2.2 shows several rating formats, some of which you may have encountered in taking examinations or opinion polls. These questionnaires capture useful data. They enable us to establish the basis for empirical relations, characterizing properties so that formal measurement may be possible in the future.

2.1.2 The Rules of the Mapping

We have seen how a measure is used to *characterize* an attribute. We begin in the real world, studying an entity and trying to understand more about it. Thus, the real world is the *domain* of the mapping, and the mathematical world is the *range*. When we map the attribute to a mathematical system, we have many choices for the mapping and the range. We can use real numbers, integers, or even a set of non-numeric symbols.

EXAMPLE 2.3

To measure a person's height, it is not enough to simply specify a number. If we measure height in inches, then we are defining a mapping from the set of people into inches; if we measure height in centimeters, then we have a different mapping. Moreover, even when the domain and range are the same, the mapping definition may be different. That is, there may be many different mappings (and hence different ways of measuring) depending on the conventions we adopt. For example, we may or may not allow shoes to be worn, or we may measure people standing or sitting.

Thus, a measure must specify the domain and range as well as the rule for performing the mapping.

EXAMPLE 2.4

In some everyday situations, a measure is associated with a number, the assumptions about the mapping are well known, and our terminology is imprecise. For example, we say "Felix's age is 11," or "Felix is 11." In expressing ourselves in this way, we really mean that we are measuring age by mapping each person into years in such a way that we count only whole years since birth. But there are many different rules that we can use. For example, the Chinese measure age by counting from the time of conception;

their assumptions are therefore different, and the resulting number is different. For this reason, we must make the mapping rules explicit.

We encounter some of the same problems in measuring software. For example, many organizations measure the size of their source code in terms of the number of lines of code (LOC) in a program. But the definition of a line of code must be made clear. The US Software Engineering Institute developed a checklist to assist developers in deciding exactly what is included in a line of code (Park 1992). This standard is still used by major organizations and metrics tool providers (see, e.g., the *unified code counter* tool produced by University of Southern California in collaboration with the Aerospace Corporation (Pfeiffer, 2012). Figure 2.3 illustrates part of the checklist, showing how different choices result in different counting rules. Thus, the checklist allows you to tailor your definition of lines-of-code to your needs. We will examine the issues addressed by this checklist in more depth in Chapter 8.

Many systems consist of programs in a variety of languages. For example, the GNU/Linux distribution includes code written in at least 19 different languages (Wheeler 2002). In order to deal with code written in such a variety of languages, David Wheeler's code analysis tool uses a simple scheme for counting LOC: "a physical source line of code is a line ending in a newline or end-of-file marker, and which contains at least one non-whitespace non-comment character."

2.1.3 The Representation Condition of Measurement

We saw that, by definition, each relation in the empirical relational system corresponds via the measurement to an element in a number system. We want the properties of the measures in the number system to be the same as the corresponding elements in the real world, so that by studying the numbers, we learn about the real world. Thus, we want the mapping to preserve the relation. This rule is called the representation condition, and it is illustrated in Figure 2.4.

The *representation condition* asserts that a measurement mapping M must map entities into numbers and empirical relations into numerical relations in such a way that the empirical relations preserve and are preserved by the numerical relations. In Figure 2.4, we see that the empirical relation *taller than* is mapped to the numerical relation >. In particular, we can say that

A is *taller than* B if and only if $M(A) > M(B)$

Statement type		
	Include	*Exclude*
Executable		
Nonexecutable		
Declarations		
Compiler directives		
Comments		
On their own lines		
On lines with source code		
Banners and nonblank spacers		
Blank (empty) comments		
Blank lines		
How produced		
	Include	*Exclude*
Programmed		
Generated with source code generators		
Converted with automatic translators		
Copied or reused without change		
Modified		
Removed		
Origin		
	Include	*Exclude*
New work: no prior existence		
Prior work: taken or adapted from		
A previous version, build, or release		
Commercial, off-the-shelf software, other than libraries		
Government furnished software, other than reuse libraries		
Another product		
A vendor-supplied language support library (unmodified)		
A vendor-supplied operating system or utility (unmodified)		
A local or modified language support library or operating system		
Other commercial library		
A reuse library (software designed for reuse)		
Other software component or library		

FIGURE 2.3   Portion of US Software Engineering Institute checklist for lines-of-code count.

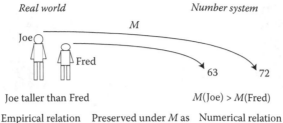

FIGURE 2.4   Representation condition.

This statement means that:

- Whenever Joe is taller than Fred, then M(Joe) must be a bigger number than M(Fred).

- We can map Jill to a higher number than Jack only if Jill is taller than Jack.

EXAMPLE 2.5

In Section 2.1.1, we noted that there can be many relations on a given set, and we mentioned several for the attribute *height*. The representation condition has implications for each of these relations. Consider these examples:

For the (binary) empirical relation *taller than*, we can have the numerical relation

$$x > y$$

Then, the representation condition requires that for any measure M,

A taller than B if and only if $M(A) > M(B)$

For the (unary) empirical relation *is-tall*, we might have the numerical relation

$$x > 70$$

The representation condition requires that for any measure M,

A is-tall if and only if $M(A) > 70$

For the (binary) empirical relation *much taller than,* we might have the numerical relation

$$x > y + 15$$

The representation condition requires that for any measure M,

A *much taller than B* if and only if $M(A) > M(B) + 15$

For the (ternary) empirical relation x *higher than y if sitting on z's shoulders,* we could have the numerical relation

$$0.7x + 0.8z > y$$

The representation condition requires that for any measure M,

A *higher than B if sitting on C's shoulders* if and only if $0.7M(A) + 0.8M(C) > M(B)$

Consider the actual assignment of numbers M given in Figure 2.5. Wonderman is mapped to the real number 72 (i.e., $M(\text{Wonderman}) = 72$), Frankie to 84 ($M(\text{Frankie}) = 84$), and Peter to 42 ($M(\text{Peter}) = 42$). With this particular mapping M, the four numerical relations hold whenever the four empirical relations hold. For example,

- Frankie is taller than Wonderman, and $M(\text{Frankie}) > M(\text{Wonderman})$.
- Wonderman is tall, and $M(\text{Wonderman}) = 72 > 70$.
- Frankie is much taller than Peter, and $M(\text{Frankie}) = 84 > 57 = M(\text{Peter}) + 15$. Similarly Wonderman is much taller than Peter and $M(\text{Wonderman}) = 72 > 57 = M(\text{Peter}) + 15$.
- Peter is higher than Frankie when sitting on Wonderman's shoulders, and $0.7M(\text{Peter}) + 0.8M(\text{Wonderman}) = 87 > 84 = M(\text{Frankie})$

Since all the relations are preserved in this way by the mapping, we can define the mapping as a *measure* for the attribute. Thus, if we think of the

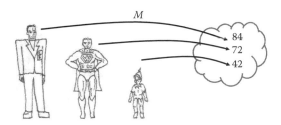

FIGURE 2.5   A measurement mapping.

measure as a measure of height, we can say that Frankie's height is 84, Peter's is 42, and Wonderman's is 72.

Not every assignment satisfies the representation condition. For instance, we could define the mapping in the following way:

$$M(\text{Wonderman}) = 72$$

$$M(\text{Frankie}) = 84$$

$$M(\text{Peter}) = 60$$

Then three of the above relations are satisfied, but *much taller than* is not. This is because *Wonderman is much taller than Peter* is not true under this mapping.

The mapping that we call a measure is sometimes called a *representation* or *homomorphism*, because the measure represents the attribute in the numerical world. Figure 2.6 summarizes the steps in the measurement process.

There are several conclusions we can draw from this discussion. First, we have seen that there may be many different measures for a given attribute. In fact, we use the notion of representation to define validity: any measure that satisfies the representation condition is a *valid* measure. Second, the richer the empirical relation system, the fewer the valid measures. We consider a relational system to be rich if it has a large number of

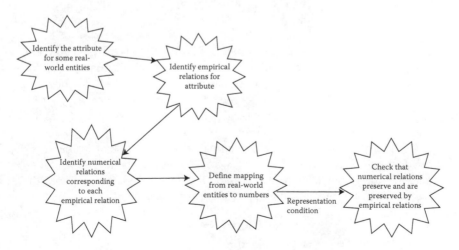

FIGURE 2.6   Key stages of formal measurement.

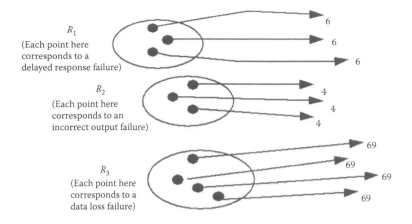

FIGURE 2.7   Measurement mapping.

relations that can be defined. But as we increase the number of empirical relations, we increase the number of conditions that a measurement mapping must satisfy in the representation condition.

EXAMPLE 2.6

Suppose we are studying the entity *software failures*, and we look at the attribute *criticality*. Our initial description distinguishes among only three types of failures:

- Delayed-response
- Incorrect output
- Data-loss

where every failure lies in exactly one failure class (based on which outcome happens first). This categorization yields an empirical relation system that consists of just three unary relations: R_1 for delayed response, R_2 for incorrect output, and R_3 for data loss. We assume every failure is in either R_1, R_2, or R_3. At this point, we cannot judge the relative criticality of these failure types; we know only that the types are different.

To find a representation for this empirical relation system in the set of real numbers, we need to choose only any three distinct numbers, and then map members from different classes into different numbers. For example, the mapping M, illustrated in Figure 2.7, assigns the mapping as:

$$M(\text{each delayed response}) = 6$$

$$M(\text{each incorrect output}) = 4$$

$$M(\text{each data loss}) = 69$$

This assignment is a representation, because we have numerical relations corresponding to R_1, R_2, and R_3. That is, the numerical relation corresponding to R_1 is the relation is 6; likewise, the numerical relation corresponding to R_2 is the relation is 4, and the numerical relation corresponding to R_3 is the relation is 69.

Suppose next we have formed a deeper understanding of failure criticality in a particular environment. We want to add to the above relation system a new (binary) relation, *is more critical than*. We now know that each data-loss failure is more critical than each incorrect output failure and delayed response failure; each incorrect output failure is more critical than each delayed response failure. Thus, *x more critical than y* contains all those pairs (x,y) of failures for which either

x is in R_3 and y is in R_2 or R_1, or

x is in R_2 and y is in R_1

To find a representation in the real numbers for this enriched empirical relation system, we now have to be much more careful with our assignment of numbers. First of all, we need a numerical relation to correspond to *more critical than*, and it is reasonable to use the binary relation >. However, it is not enough to simply map different failure types to different numbers. To preserve the new relation, we must ensure that data-loss failures are mapped to a higher number than incorrect output failures, which in turn are mapped to a higher number than delayed-response failures. One acceptable representation is the mapping:

M(each delayed response) = 3

M(each incorrect output) = 4

M(each data-loss) = 69

Note that the mapping defined initially in this example would *not* be a representation, because > does not preserve *is more critical than;* incorrect output failures were mapped to a lower number than delayed response failures.

There is nothing wrong with using the same representation in different ways, or using several representations for the same attribute. Table 2.3 illustrates a number of examples of specific measures used in software engineering. In it, we see that examples 1 and 2 in Table 2.3 give different measures of program length, while examples 9 and 10 give different measures of program reliability. Similarly, the same measure (although

TABLE 2.3   Examples of Specific Measures Used in Software Engineering

	Entity	Attribute	Measure
1	Completed project	Duration	Months from start to finish
2	Completed project	Duration	Days from start to finish
3	Program code	Length	Number of lines of code (LOC)
4	Program code	Length	Number of executable statements
5	Integration testing process	Duration	Hours from start to finish
6	Integration testing process	Rate at which faults are found	Number of faults found per KLOC (thousand LOC)
7	Test set	Efficiency	Number of faults found per number of test cases
8	Test set	Effectiveness	Number of faults found per KLOC (thousand LOC)
9	Program code	Reliability	Mean time to failure (MTTF) in CPU hours
10	Program code	Reliability	Rate of occurrence of failures (ROCOF) in CPU hours

of course not the same measurement mapping), *faults found per thousand lines of code (KLOC)*, is used in examples 6, 7, and 8.

How good a measure is faults per KLOC? The answer depends entirely on the entity–attribute pair connected by the mapping. Intuitively, most of us would accept that faults per KLOC is a good measure of the *rate at which faults are found* for the *testing process* (example 6). However, it is not such a good measure of *efficiency* of the *tester* (example 7), because intuitively we feel that we should also take into account the difficulty of understanding and testing the program under scrutiny. This measure may be reasonable when comparing two testers of the same program, though. Faults per KLOC is not likely to be a good measure of *quality* of the *program code*; if integration testing revealed program X to have twice as many faults per KLOC than program Y, we would probably not conclude that the quality of program Y was twice that of program X.

2.2 MEASUREMENT AND MODELS

In Chapter 1, we have discussed several types of *models*: cost estimation models, quality models, capability maturity models, and more. In general, a *model* is an abstraction of reality, allowing us to strip away detail and view an entity or concept from a particular perspective. For example, cost models permit us to examine only those project aspects that contribute to the project's final cost. Models come in many different forms: as equations, mappings, or diagrams, for instance. These show us how the component parts relate to one another, so that we can examine and understand these relationships and make judgments about them.

In this chapter, we have seen that the representation condition requires every measure to be associated with a model of how the measure maps the entities and attributes in the real world to the elements of a numerical system. These models are essential in understanding not only how the measure is derived, but also how to interpret the behavior of the numerical elements when we return to the real world. But we also need models even before we begin the measurement process.

Let us consider more carefully the role of models in measurement definition. Previous examples have made clear that if we are measuring height of people, then we must understand and declare our assumptions to ensure unambiguous measurement. For example, in measuring height, we would have to specify whether or not we allow shoes to be worn, whether or not we include hair height, and whether or not we specify a certain posture. In this sense, we are actually defining a *model* of a person, rather than the person itself, as the entity being measured. Thus, the model of the mapping should also be supplemented with a model of the mapping's domain—that is, with a model of how the entity relates to its attributes.

EXAMPLE 2.7

To measure the length of programs using LOC, we need a model of a program. The model would specify how a program differs from a subroutine, whether or not to treat separate statements on the same line as distinct LOC, whether or not to count comment lines, whether or not to count data declarations, etc. The model would also tell us what to do when we have programs written in different languages. It might also distinguish delivered operational programs from those under development, and it would tell us how to handle situations where different versions run on different platforms.

Process measures are often more difficult to define than product and resource measures, in large part because the process activities are less understood.

EXAMPLE 2.8

Suppose we want to measure the attributes of the testing process. Depending on our goals, we might measure the time or effort spent on this process, or the number of faults found during the process. To do this, we need a careful definition of what is meant by the testing process; at the very least, we must be able to identify unambiguously when the process starts and ends. A model of the testing process can show us which activities are included, when they start and stop, and what inputs and outputs are involved.

2.2.1 Defining Attributes

When measuring, there is always a danger that we focus too much on the formal, mathematical system, and not enough on the empirical one. We rush to create mappings and then manipulate numbers, without given careful thought to the relationships among entities and their attributes in the real world. Figure 2.8 presents a whimsical view of what can happen when we rush to manipulate numbers without considering their real meaning.

The dog in Figure 2.8 is clearly an exceptionally intelligent dog, but its intelligence is not reflected by the result of an IQ test. It is clearly wrong to *define* the intelligence of dogs in this way. Many people have argued that defining the intelligence of people by using IQ tests is just as problematic. What is needed is a comprehensive set of characteristics of intelligence, appropriate to the entity (so that dog intelligence will have a different set of characteristics from people intelligence) and associated by a model.

FIGURE 2.8   Using a suspect definition.

The model will show us how the characteristics relate. Then, we can try to define a measure for each characteristic, and use the representation condition to help us understand the relationships as well as overall intelligence.

EXAMPLE 2.9

In software development, our intuition tells us that the complexity of a program can affect the time it takes to code it, test it, and fix it; indeed, we suspect that complexity can help us to understand when a module is prone to contain faults. But there are few researchers who have built models of exactly what it means for a module to be complex. Instead, we often assume that we know what complexity is, and we measure complexity without first defining it in the real world. For example, many software developers still define program complexity as the cyclomatic number proposed by McCabe and illustrated in Figure 2.9 (McCabe 1976). This number, based on a graph-theoretic concept, counts the number of linearly independent paths through a program. We will discuss this measure (and its use in testing) in more detail in Chapter 9.

McCabe felt that the number of such paths was a key indicator not just of testability but also of complexity. Hence, he originally called this number, v, the cyclomatic complexity of a program. On the basis of empirical research, McCabe claimed that modules with high values of v were those most likely to be fault-prone and unmaintainable. He proposed a threshold value of 10 for each module; that is, any module with v greater than 10 should be redesigned to reduce v. However, the cyclomatic number presents only a partial view of complexity. It can be shown mathematically that the cyclomatic number is equal to one more than the number of decisions in a program, and there are many programs that have a large number of decisions but are easy to understand, code, and maintain. Thus, relying only on the cyclomatic number to measure actual program complexity can be misleading. A more complete model of program complexity is needed.

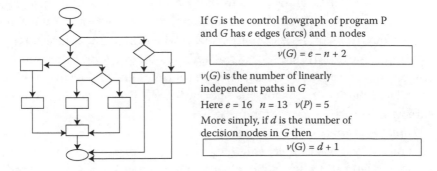

If G is the control flowgraph of program P and G has e edges (arcs) and n nodes

$$v(G) = e - n + 2$$

$v(G)$ is the number of linearly independent paths in G

Here $e = 16$ $n = 13$ $v(P) = 5$

More simply, if d is the number of decision nodes in G then

$$v(G) = d + 1$$

FIGURE 2.9   Computing McCabe's cyclomatic number.

Directed graphs are probably the most commonly used abstraction for modeling software designs and implementations. To develop a measure of software design attributes, we should establish relations relevant to design attributes in terms of graph models of designs. Then we can use the relations to derive and validate a measure of the attributes.

EXAMPLE 2.10

To evaluate existing software design measures, Briand and his colleagues developed relations on directed graph models of software designs, described as properties, relevant to a set of attributes including module size, module coupling, and system complexity (Briand et al. 1996).

One property of any module size measure is *module additivity*—the size of a system is the sum of the sizes of its disjoint modules. One property of any module coupling measure is that if you merge modules *m1* and *m2* to create module *M*, then *Coupling(M) ≤ (Coupling(m1) + Coupling(m2))*. The coupling of *M* may be less than *Coupling(m1) + Coupling(m2)* because *m1* and *m2* may have common intermodule relationships. Complexity properties are defined in terms of systems of modules, where complexity is defined in terms of the number of relationships between elements in a system. One complexity property is that the complexity of a system consisting of disjoint modules is the sum of the complexity of its modules.

We can use the set of properties for a software attribute as an empirical relation system to evaluate whether measures that are purported to be size, coupling, or complexity measures are really consistent with the properties. That is, we can determine if the measure satisfies the representation condition of measurement.

2.2.2 Direct and Derived Measurement

Once we have a model of the entities and attributes involved, we can define the measure in terms of them. Many of the examples we have used employ direct mappings from attribute to number, and we use the number to answer questions or assess situations. But when there are complex relationships among attributes, or when an attribute must be measured by combining several of its aspects, then we need a model of how to combine the related measures. It is for this reason that we distinguish direct measurement from derived measurement.

Direct measurement of an attribute of an entity involves no other attribute or entity. For example, *length* of a physical object can be measured without reference to any other object or attribute. On the other hand, measures of the *density* of a physical object can be derived in terms of *mass* and *volume*; we then use a model to show us that the relationship between the three is

$$Density = Mass/Volume$$

Similarly, the speed of a moving object is most accurately measured using direct measures of distance and time. Thus, direct measurement forms the building blocks for our assessment, but many interesting attributes can be measured only by derived measurement.

The following direct measures are commonly used in software engineering:

- *Size* of source code (measured by LOC)

- *Schedule* of the testing process (measured by elapsed time in hours)

- *Number of defects discovered* (measured by counting defects)

- *Time* a programmer spends on a project (measured by months worked)

Table 2.4 provides examples of some derived measures that are commonly used in software engineering. The most common of all, and the most controversial, is the measure for programmer productivity, as it emphasizes size of output without taking into consideration the code's functionality or complexity. The defect detection efficiency measure is computed with respect to a specific testing or review phase; the total number of defects refers to the total number discovered during the entire product life cycle. Japanese software developers routinely compute the system spoilage measure; it indicates how much effort is wasted in fixing faults, rather than in building new code.

TABLE 2.4 Examples of Common Derived Measures Used in Software Engineering

Programmer productivity	LOC produced/person-months of effort
Module defect density	Number of defects/module size
Defect detection efficiency	Number of defects detected/total number of defects
Requirements stability	Number of initial requirements/total number of requirements
Test coverage	Number of test requirements covered/total number of test requirements
System spoilage	Effort spent fixing faults/total project effort

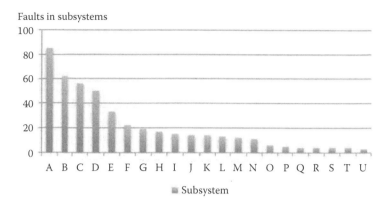

FIGURE 2.10 Using direct measurement to assess a product (from a major system made up of several subsystems).

Derived measurement is often useful in making visible the interactions between direct measurements. That is, it is sometimes easier to see what is happening on a project by using combinations of measures. To see why, consider the graph in Figure 2.10.

The graph shows the (anonymized) number of faults in each subsystem of a large, important software system in the United Kingdom. From the graph, it appears as if there are five subsystems that contain the most problems for the developers maintaining this system. However, Figure 2.11 depicts the same data with one big difference: instead of using the direct measurement of faults, it shows fault density (i.e., the derived measure defined as faults per KLOC). From the derived measurement, it is very clear that one subsystem is responsible for the majority of the problems.

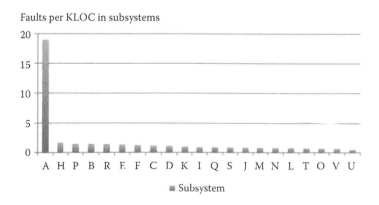

FIGURE 2.11 Using derived measurement to assess a product.

In fact, subsystem A is only 4000 LOC out of two million, but A is a big headache for the maintainers. Here, the derived measurement helps the project team to focus their maintenance efforts more effectively.

The representational theory of measurement, as described in this chapter, is initially concerned with direct measurement of attributes. Where no previous measurement has been performed, direct measurement constitutes the natural process of trying to understand entities and the attributes they possess. However, simple models of direct measurement do not preclude the possibility of more accurate subsequent measurement that will be achieved indirectly. For example, temperature can be measured as the length of a column of mercury under given pressure conditions. This measure is derived because we are examining the column, rather than the entity whose temperature we want to know.

2.2.3 Measurement for Prediction

When we talk about measuring something, we usually mean that we wish to assess some entity that already exists. This measurement for *assessment* is very helpful in understanding what exists now or what has happened in the past. However, in many circumstances, we would like to *predict* an attribute of some entity that does not yet exist. For example, suppose we are building a software system that must be highly reliable, such as the control software for an aircraft, power plant, or x-ray machine. The software construction may take some time, and we want to provide early assurance that the system will meet reliability targets. However, reliability is defined in terms of operational performance, something we clearly cannot measure before the product is finished. To provide reliability indicators *before* the system is complete, we can build a model of the factors that affect reliability, and then predict the likely reliability based on our understanding of the system while it is still under development.

Similarly, we often need to predict how much a development project will cost, or how much time and effort will be needed, so that we can allocate the appropriate resources to the project. Simply waiting for the project to end, and then measuring cost and schedule attributes are clearly not acceptable.

The distinction between measurement for assessment and prediction is not always clear-cut. For example, suppose we use a globe to determine the distance between London and Washington, DC. This derived measurement helps us to assess how far apart the cities are. However, the same activity is also involved when we want to predict the distance we will travel on a future journey. Notice that the action we take in assessing distance involves the globe as a model of the real world, plus prediction procedures that describe how to use the model.

In general, measurement for prediction always requires some kind of *mathematical model* that relates the attributes to be predicted to some other attributes that we can measure now. The model need not be complex to be useful.

EXAMPLE 2.11

Suppose we want to predict the number of pages, m, that will print out as a source code program, so that we can order sufficient paper or estimate the time it will take to do the printing. We could use the very simple model

$$m = x/a$$

where x is a variable representing a measure of source-code program length in LOC, and a is a constant representing the average number of lines per page.

Project managers universally need effort prediction.

EXAMPLE 2.12

A common generic model for predicting the effort required in software projects has the form

$$E = aS^b$$

where E is effort in person-months, S is the size (in LOC) of the system to be constructed, and a and b are constants. We will examine many of these models in Chapter 11.

Sometimes, the same model is used both for assessment and prediction, as we saw with the example of the globe, above. The extent to which it applies to each situation depends on how much is known about the parameters of the model. In Example 2.11, suppose a is known to be 55 in a specific environment. If a program exists with a known x, then the derived measure of hard copy pages computed by the given formula is not really a prediction problem (except in a very weak sense), particularly if the hard copy already exists. However, if we have only a program specification, and we wish to know roughly how many hard copy pages the final implementation will involve, then we *would* be using the model to solve a prediction problem. In this case, we need some kind of procedure for determining the unknown value of x based on our knowledge of the program specification. The same is true in Example 2.12, where invariably we need some means of determining the parameters a, b, and S based on our knowledge of the project to be developed.

These examples illustrate that the model alone is not enough to perform the required prediction. In addition, we need some means of determining the model parameters, plus a procedure to interpret the results. Therefore, we must think in terms of a *prediction system*, rather than of the model itself. A *prediction system* consists of a mathematical model together with a set of prediction procedures for determining unknown parameters and interpreting results (Littlewood 1988).

EXAMPLE 2.13

Suppose we want to predict the cost of an automobile journey from London to Bristol. The entity we want to predict is the journey and the attribute is its cost. We begin by obtaining measures (in the assessment sense) of:

- a: the distance between London and Bristol
- b: the cost per gallon of fuel
- c: the average distance we can travel on a gallon of fuel in our car

Next, we can predict the journey's cost using the formula

$$cost = ab/c$$

In fact, we are using a *prediction system* that involves:

1. A *model*: that is, the formula $cost = ab/c$.
2. A *set of procedures for determining the model parameters*: that is, how we determine the values of a, b, and c. For example, we may consult with the local automobile association, or simply ask a friend.
3. *Procedures for interpreting the results*: for example, we may use Bayesian probability to determine likely margins of error.

Using the same model will generally yield different results, if we use different prediction procedures. For instance, in Example 2.13, the model parameters supplied by a friend may be very different from those supplied by the automobile association. This notion of changing results is especially important when predicting software reliability.

EXAMPLE 2.14

A well-known reliability model is based on an exponential distribution for the time to the *i*th failure of the product. This distribution is described by the formula

$$F(t) = 1 - e^{-(N-i+1)at}$$

Here, N represents the number of faults initially residing in the program, while a represents the overall rate of occurrence of failures. There are many ways that the model parameters N and a can be estimated, including sophisticated techniques such as maximum likelihood estimation. The details of these prediction systems will be discussed in Chapter 11.

Accurate predictive measurement is always based on measurement in the assessment sense, so the need for assessment is especially critical in software engineering. Everyone wants to be able to predict key determinants of success, such as the effort needed to build a new system, or the reliability of the system in operation. However, there are no magic models. The models are dependent on high-quality measurements of past projects (as well as the current project during development and testing) if they are to support accurate predictions. Since software development is more a creative process than a manufacturing one, there is a high degree of risk when we undertake to build a new system, especially if it is very different from systems we have developed in the past. Thus, software engineering involves risk, and there are some clear parallels with gambling.

> Testing your methods on a sample of past data gets to the heart of the scientific approach to gambling. Unfortunately this implies some preliminary spadework, and most people skimp on that bit, preferring to rely on blind faith instead. (Drapkin and Forsyth 1987)

We can replace "gambling" with *software prediction*, and then heed the warning. In addition, we must recognize that the quality of our predictions is based on several other assumptions, including the notion that the future will be like the past, and that we understand how data are distributed. For instance, many reliability models specify a particular distribution, such as Gaussian or Poisson. If our new data does not behave like the distribution in the model, our prediction is not likely to be accurate.

2.3 MEASUREMENT SCALES AND SCALE TYPES

We have seen how direct measurement of an attribute assigns a representation or mapping M from an observed (empirical) relation system to some numerical relation system. The purpose of performing the mapping is to be able to manipulate data in the numerical system and use the results to draw conclusions about the attribute in the empirical system. We do this sort of analysis all the time. For example, we use a thermometer to measure air temperature, and then we conclude that it is hotter today than yesterday; the numbers tell us about the characteristic of the air.

But not all measurement mappings are the same. And the differences among the mappings can restrict the kind of analysis we can do. To understand these differences, we introduce the notion of a *measurement scale*, and then we use the scale to help us understand which analyses are appropriate.

We refer to our measurement mapping, M, together with the empirical and numerical relation systems, as a *measurement scale*. Where the relation systems (i.e., the domain and range) are obvious from the context, we sometimes refer to M alone as the scale. There are three important questions concerning representations and scales:

1. How do we determine when one numerical relation system is preferable to another?

2. How do we know if a particular empirical relation system has a representation in a given numerical relation system?

3. What do we do when we have several different possible representations (and hence many scales) in the same numerical relation system?

Our answer to the first question is pragmatic. Recall that the formal relational system to which the scale maps need not be numeric; it can be symbolic. However, symbol manipulation may be far more unwieldy than numerical manipulation. Thus, we try to use the real numbers wherever possible, since analyzing real numbers permits us to use techniques with which we are familiar.

The second question is known as the *representation problem*, and its answer is sought not just by software engineers but also by all scientists who are concerned with measurement. The representation problem is one of the basic problems of measurement theory; it has been solved for various types of relation systems characterized by certain types of axioms. Rather than addressing it in this book, we refer the readers to the classical literature on measurement theory.

Our primary concern in this chapter is with the third question. Called the *uniqueness problem*, this question addresses our ability to determine which representation is the most suitable for measuring an attribute of interest.

In general, there are many different representations for a given empirical relation system. We have seen that the more relations there are, the fewer are the representations. This notion of shrinking representations can be best understood by a formal characterization of scale types. In this section, we classify measurement scales as one of five major types:

1. Nominal

2. Ordinal

3. Interval

4. Ratio

5. Absolute

There are other scales that can be defined (such as a logarithmic scale), but we focus only on these five, as they illustrate the range of possibilities and the issues that must be considered when measurement is done.

One relational system is said to be *richer* than another if all relations in the second are contained in the first. Using this notion, the scale types listed above are shown in increasing level of richness. That is, the richer the empirical relation system, the more restrictive the set of representations, and so the more sophisticated the scale of measurement.

The idea behind the formal definition of scale types is quite simple. If we have a satisfactory measure for an attribute with respect to an empirical relation system (i.e., it captures the empirical relations in which we are interested), we want to know what other measures exist that are also acceptable. For example, we may measure the length of physical objects by using a mapping from length to inches. But there are equally acceptable measures in feet, meters, furlongs, miles, and more. In this example, all of the accept-able measures are very closely related, in that we can map one into another by multiplying by a suitable positive constant (such as converting inches into feet by multiplying by 1/12). A mapping from one acceptable measure to another is called an *admissible transformation*. When measuring length, the class of admissible transformations is very restrictive, in the sense that all admissible transformations are of the form

$$M' = aM$$

where M is the original measure, M' is the new one, and a is a constant.

In particular, transformations of the form

$$M' = b + aM \quad (b \neq 0)$$

or

$$M' = aM^b \quad (b \neq 1)$$

are not acceptable. Thus, the set of admissible transformations for length is smaller than the set of all possible transformations. We say that the more restrictive the class of admissible transformations, the more *sophisticated* the measurement scale.

2.3.1 Nominal Scale Type

Suppose we define classes or categories, and then place each entity in a particular class or category, based on the value of the attribute. This cat-egorization is the basis for the most primitive form of measurement, the *nominal scale*. Thus, the nominal scale has two major characteristics:

1. The empirical relation system consists only of different classes; there is no notion of ordering among the classes.

2. Any distinct numbering or symbolic representation of the classes is an acceptable measure, but there is no notion of magnitude associ-ated with the numbers or symbols.

In other words, nominal scale measurement places elements in a classifi-cation scheme. The classes are not ordered; even if the classes are numbered from 1 to n for identification, there is no implied ordering of the classes.

EXAMPLE 2.15

Suppose that we are investigating the set of all known software faults in our code, and we are trying to capture the *location* of the faults. Then we seek a measurement scale with faults as entities and location as the attribute. We can use a common but primitive mapping to identify the fault location: we denote a fault as *specification, design*, or *code*, according to where the fault was first introduced. Notice that this classification imposes no judgment about which class of faults is more severe or important than another. However, we have a clear distinction among the classes, and every fault belongs to exactly one class. This is a very simple empirical relation system. Any mapping, M, that assigns the three different classes to three different numbers satisfies the representation condition and is therefore an acceptable measure. For example, the mappings M_1 and M_2 defined by

$$M_1(x) = \begin{cases} 1, & \text{if } x \text{ is specification fault} \\ 2, & \text{if } x \text{ is design fault} \\ 3, & \text{if } x \text{ is code fault} \end{cases}$$

$$M_2(x) = \begin{cases} 101, & \text{if } x \text{ is specification fault} \\ 2.73, & \text{if } x \text{ is design fault} \\ 69, & \text{if } x \text{ is code fault} \end{cases}$$

are acceptable. In fact, any two mappings, M and M', will always be related in a special way: M' can be obtained from M by a one-to-one mapping. The mappings need not involve numbers; distinct symbols will suffice. Thus, the class of admissible transformations for a nominal scale measure is the set of all one-to-one mappings.

2.3.2 Ordinal Scale Type

We can often augment the nominal scale with information about an ordering of the classes or categories creating an *ordinal scale*. The ordering leads to analysis not possible with nominal measures. The ordinal scale has the following characteristics:

- The empirical relation system consists of classes that are ordered with respect to the attribute.

- Any mapping that preserves the ordering (i.e., any monotonic function) is acceptable.

- The numbers represent ranking only, so addition, subtraction, and other arithmetic operations have no meaning.

However, classes can be combined, as long as the combination makes sense with respect to the ordering.

EXAMPLE 2.16

Suppose our set of entities is a set of software modules, and the attribute we wish to capture quantitatively is *complexity*. Initially, we may define five distinct classes of module complexity: *trivial, simple, moderate, complex,* and *incomprehensible*. There is an implicit order relation of *less complex than* on these classes; that is, all trivial modules are less complex than simple modules, which are less complex than moderate modules, etc. In this case, since the measurement mapping must preserve this ordering, we cannot be as free in our choice of mapping as we could with a nominal measure. Any mapping, M, must map each distinct class to a different number, as with nominal measures. But we must also ensure that the more complex classes are mapped to bigger numbers. Therefore, M must be a monotonically increasing function. For example, each of the mappings M_1, M_2, and M_3 is a valid measure, since each satisfies the representation condition.

$$M_1(x) = \begin{cases} 1 \text{ if } x \text{ is trivial} \\ 2 \text{ if } x \text{ is simple} \\ 3 \text{ if } x \text{ is moderate} \\ 4 \text{ if } x \text{ is complex} \\ 5 \text{ if } x \text{ is incomprehensible} \end{cases} \qquad M_2(x) = \begin{cases} 1 \text{ if } x \text{ is trivial} \\ 2 \text{ if } x \text{ is simple} \\ 3 \text{ if } x \text{ is moderate} \\ 4 \text{ if } x \text{ is complex} \\ 10 \text{ if } x \text{ is incomprehensible} \end{cases}$$

$$M_3(x) = \begin{cases} 0.1 \text{ if } x \text{ is trivial} \\ 1001 \text{ if } x \text{ is simple} \\ 1002 \text{ if } x \text{ is moderate} \\ 4570 \text{ if } x \text{ is complex} \\ 4573 \text{ if } x \text{ is incomprehensible} \end{cases}$$

However, neither M_4 nor M_5 is valid:

$$M_4(x) = \begin{cases} 1 \text{ if } x \text{ is trivial} \\ 1 \text{ if } x \text{ is simple} \\ 3 \text{ if } x \text{ is moderate} \\ 4 \text{ if } x \text{ is complex} \\ 5 \text{ if } x \text{ is incomprehensible} \end{cases} \qquad M_5(x) = \begin{cases} 1 \text{ if } x \text{ is trivial} \\ 3 \text{ if } x \text{ is simple} \\ 2 \text{ if } x \text{ is moderate} \\ 4 \text{ if } x \text{ is complex} \\ 10 \text{ if } x \text{ is incomprehensible} \end{cases}$$

Since the mapping for an ordinal scale preserves the ordering of the classes, the set of ordered classes $<C_1, C_2, \ldots, C_n>$ is mapped to an increasing series of numbers $< a_1, a_2, \ldots, a_n>$ where a_i is greater than a_j when i is greater than j. Any acceptable mapping can be transformed to any other as long as the series of a_i is mapped to another increasing series. Thus, in the ordinal scale, any two measures can be related by a monotonic mapping, so the class of admissible transformations is the set of all monotonic mappings.

2.3.3 Interval Scale Type

We have seen how the ordinal scale carries more information about the entities than does the nominal scale, since ordinal scales preserve ordering. The interval scale carries more information still, making it more powerful than nominal or ordinal. This scale captures information about the size of the intervals that separate the classes, so that we can in some sense understand the size of the jump from one class to another. Thus, an interval scale can be characterized in the following way:

- An interval scale preserves order, as with an ordinal scale.

- An interval scale preserves differences but not ratios. That is, we know the difference between any two of the ordered classes in the range of the mapping, but computing the ratio of two classes in the range does not make sense.

- Addition and subtraction are acceptable on the interval scale, but not multiplication and division.

To understand the difference between ordinal and interval measures, consider first an example from everyday life.

EXAMPLE 2.17

We can measure air temperature on a Fahrenheit or Celsius scale. Thus, we may say that it is usually 20° Celsius on a summer's day in London, while it may be 30° Celsius on the same day in Washington, DC. The interval from one degree to another is the same, and we consider each degree to be a class related to heat. That is, moving from 20° to 21° in London increases the heat in the same way that moving from 30° to 31° does in Washington. However, we cannot say that it is two-third as hot in London as Washington; neither can we say that it is 50% hotter in Washington than in London. Similarly, we cannot say that a 90° Fahrenheit day in Washington is twice as hot as a 45° Fahrenheit day in London.

There are fewer examples of interval scales in software engineering than of nominal or ordinal.

EXAMPLE 2.18

Recall the five categories of complexity described in Example 2.16. Suppose that the difference in complexity between a trivial and simple system is the same as that between a simple and moderate system. Then any interval measure of complexity must preserve these differences. Where this equal step applies to each class, we have an attribute measurable on an interval scale. The following measures have this property and satisfy the representation condition:

$$M_1(x) = \begin{cases} 1 \text{ if } x \text{ is trivial} \\ 2 \text{ if } x \text{ is simple} \\ 3 \text{ if } x \text{ is moderate} \\ 4 \text{ if } x \text{ is complex} \\ 5 \text{ if } x \text{ is incomprehensible} \end{cases} \qquad M_2(x) = \begin{cases} 0 \text{ if } x \text{ is trivial} \\ 2 \text{ if } x \text{ is simple} \\ 4 \text{ if } x \text{ is moderate} \\ 6 \text{ if } x \text{ is complex} \\ 8 \text{ if } x \text{ is incomprehensible} \end{cases}$$

$$M_3(x) = \begin{cases} 3.1 \text{ if } x \text{ is trivial} \\ 5.1 \text{ if } x \text{ is simple} \\ 7.1 \text{ if } x \text{ is moderate} \\ 9.1 \text{ if } x \text{ is complex} \\ 11.1 \text{ if } x \text{ is incomprehensible} \end{cases}$$

Suppose an attribute is measurable on an interval scale, and M and M' are mappings that satisfy the representation condition. Then we can always find numbers a and b such that

$$M = aM' + b$$

We call this type of transformation an *affine transformation*. Thus, the class of admissible transformations of an interval scale is the set of affine transformations. In Example 2.17, we can transform Celsius to Fahrenheit by using the transformation

$$F = 9/5C + 32$$

Likewise, in Example 2.18, we can transform M_1 to M_3 by using the formula

$$M_3 = 2M_1 + 1.1$$

EXAMPLE 2.19

The timing of an event's occurrence is a classic use of interval scale measurement. We can measure the timing in units of years, days, hours, or some other standard measure, where each time is noted relative to a given fixed event. We use this convention everyday by measuring the year with respect to an event (i.e., by saying "2014 AD"), or by measuring the hour from midnight. Software development projects can be measured in the same way, by referring to the project's start day. We say that we are on day 87 of the project, when we mean that we are measuring 87 days from the first day of the project. Thus, using these conventions, it is meaningless to say "Project X started twice as early as project Y" but meaningful to say "the time between project X's beginning and now is twice the time between project Y's beginning and now."

On a given project, suppose the project manager is measuring time in months from the day work started: April 1, 2013. But the contract manager is measuring time in years from the day that the funds were received from the customer: January 1, 2014. If M is the project manager's scale and M' the contract manager's scale, we can transform the contract manager's time into the project manager's by using the following admissible transformation:

$$M = 12M' + 9$$

2.3.4 Ratio Scale Type

Although the interval scale gives us more information and allows more analysis than either nominal or ordinal, we sometimes need to be able to do even more. For example, we would like to be able to say that one liquid is twice as hot as another, or that one project took twice as long as another. This need for ratios gives rise to the ratio scale, the most useful scale of measurement, and one that is common in the physical sciences. A *ratio scale* has the following characteristics:

- It is a measurement mapping that preserves ordering, the size of intervals between entities, and ratios between entities.

- There is a zero element, representing total lack of the attribute.

- The measurement mapping must start at zero and increase at equal intervals, known as units.

- All arithmetic can be meaningfully applied to the classes in the range of the mapping.

The key feature that distinguishes ratio from nominal, ordinal, and interval scales is the existence of empirical relations to capture ratios.

EXAMPLE 2.20

The length of physical objects is measurable on a ratio scale, enabling us to make statements about how one entity is twice as long as another. The zero element is theoretical, in the sense that we can think of an object as having no length at all; thus, the zero-length object exists as a limit of things that get smaller and smaller. We can measure length in inches, feet, centimeters, meters, and more, where each different measure preserves the relations about length that we observe in the real world. To convert from one length measure into another, we can use a transformation of the form $M = aM'$, where a is a constant. Thus, to convert feet into inches, we use the transformation $I = 12F$.

In general, any acceptable transformation for a ratio scale is a mapping of the form

$$M = aM'$$

where a is a positive scalar. This type of transformation is called a *ratio transformation*.

EXAMPLE 2.21

The length of software code is also measurable on a ratio scale. As with other physical objects, we have empirical relations like *twice as long*. The notion of a zero-length object exists—an empty piece of code. We can measure program length in a variety of ways, including LOC, thousands of LOC, the number of characters contained in the program, the number of executable statements, and more. Suppose M is the measure of program length in LOC, while M' captures length as number of characters. Then we can transform one to the other by computing $M' = aM$, where a is the average number of characters per line of code.

2.3.5 Absolute Scale Type

As the scales of measurement carry more information, the defining classes of admissible transformations have become increasingly restrictive. The absolute scale is the most restrictive of all. For any two measures, M and M', there is only one admissible transformation: the identity transformation. That is, there is only one way in which the measurement can be made, so M and M' must be equal. The *absolute scale* has the following properties:

- The measurement for an absolute scale is made simply by counting the number of elements in the entity set.

- The attribute always takes the form "number of occurrences of x in the entity."

- There is only one possible measurement mapping, namely the actual count, and there is only one way to count elements.

- All arithmetic analysis of the resulting count is meaningful.

There are many examples of absolute scale in software engineering. For instance, the number of failures observed during integration testing can be measured only in one way: by counting the number of failures observed. Hence, a count of the number of failures is an absolute scale measure for the number of failures observed during integration testing. Likewise, the number of people working on a software project can be measured only in one way: by counting the number of people.

Since there is only one possible measure of an absolute attribute, the set of acceptable transformations for the absolute scale is simply the identity transformation. The uniqueness of the measure is an important difference between the ratio scale and absolute scale.

EXAMPLE 2.22

We saw in Example 2.21 that the number of LOC is a ratio scale measure of length for source code programs. A common mistake is to assume that LOC is an *absolute* scale measure of length, because it is obtained by counting. However, it is the *attribute* (as characterized by empirical relations) that determines the scale type. As we have seen, the length of programs cannot be absolute, because there are many different ways to measure it (such as LOC, thousands of LOC, number of characters, and number of bytes). It is incorrect to say that LOC is an absolute scale measure of program length. However, LOC is an absolute scale measure of the attribute "number of lines of code" of a program. For the same reason, "number of years" is a ratio scale measure of a person's age; it cannot be an absolute scale measure of age, because we can also measure age in months, hours, minutes, or seconds.

TABLE 2.5   Scales of Measurement

Scale Type	Admissible Transformations (How Measures M and M' must be Related)	Examples
Nominal	1–1 mapping from M to M'	Labeling, classifying entities
Ordinal	Monotonic increasing function from M to M', that is, $M(x)$ $M(y)$ implies $M'(x)$ $M'(y)$	Preference, hardness, air quality, intelligence tests (raw scores)
Interval	$M' = aM + b \ (a > 0)$	Relative time, temperature (Fahrenheit, Celsius), intelligence tests (standardized scores)
Ratio	$M' = aM \ (a > 0)$	Time interval, length, temperature (Kelvin)
Absolute	$M' = M$	Counting entities

Table 2.5 summarizes the key elements distinguishing the measurement scale types discussed in this chapter. This table is similar to those found in other texts on measurement. However, most texts do not point out the possible risk of mis-interpretation of the examples column. Since scale types are defined with respect to the set of admissible transformations, we should never give examples of attributes without specifying the empirical relation system that characterizes an attribute. We have seen that as we enrich the relation system for an attribute by preserving more information with the measurement mapping, so we may arrive at a more restrictive (and hence different) scale type. Thus, when Table 2.5 says that the attributes length, time interval, and (absolute) temperature are on the ratio scale, what it really means is that we have developed sufficiently refined empirical relation systems to allow ratio scale measures for these attributes.

2.4 MEANINGFULNESS IN MEASUREMENT

There is more than just academic interest in scale types. Understanding scale types enables one to determine when statements about measurement make sense. For instance, we have seen how it is inappropriate to compute ratios with nominal, ordinal, and interval scales. In general, measures often map attributes to real numbers, and it is tempting to manipulate the real numbers in familiar ways: adding, averaging, taking logarithms, and performing sophisticated statistical analysis. But we must remember that the analysis is constrained by the scale type. We can perform only those calculations that are permissible for the given scale, reflecting the type of attribute and mapping that generated the data. In other words, knowledge of scale type tells us about limitations on the kind of mathematical manipulations that can be performed. Thus, the key question we should ask after having made our measurements is: can we deduce meaningful statements about the entities being measured?

This question is harder to answer than it first appears. To see why, consider the following statements:

1. The number of errors discovered during the integration testing of program X was at least 100.

2. The cost of fixing each error in program X is at least 100.

3. A semantic error takes twice as long to fix a syntactic error.

4. A semantic error is twice as complex as a syntactic error.

Intuitively, Statement 1 seems to make sense, but Statement 2 does not; the number of errors may be specified without reference to a particular scale, but the cost of fixing an error cannot be. Statement 3 seems to make sense (even if we think it cannot possibly be true) because the ratio of time taken is the same, regardless of the scale of measurement used (i.e., if a semantic error takes twice as many minutes to repair as a syntactic error, it also takes twice as many hours, seconds, or years to repair). Statement 4 does not appear to be meaningful, and we require clarification. If "complexity" means time to understand, then the statement makes sense. But other definitions of complexity may not admit measurement on a ratio scale; in those instances, Statement 4 is meaningless.

Our intuitive notion of a statement's meaningfulness involving measurement is quite distinct from the notion of the statement's *truth*. For example, the statement

The President of the United States is 125 years old

is a meaningful statement about the age measure, even though it is clearly false. We can define *meaningfulness* in a formal way.

We say that a statement involving measurement is *meaningful* if its truth value is invariant of transformations of allowable scales.

EXAMPLE 2.23

We can examine the transformations to decide on meaningfulness. Consider these statements:

Fred is twice as tall as Jane

This statement implies that the measures are at least on the ratio scale, because it uses scalar multiplication as an admissible transformation. The statement is meaningful because no matter which measure of height we use (inches, feet, centimeters, etc.), the truth or falsity of the statement remains consistent. In other words, if the statement is true and if M and M' are different measures of height, then both the statements

$$M(\text{Fred}) = 2M(\text{Jane})$$

and

$$M'(\text{Fred}) = 2M'(\text{Jane})$$

are true. This consistency of truth is due to the relationship $M = aM'$ for some positive number a.

The temperature in Tokyo today is twice that in London

This statement also implies ratio scale but is not meaningful, because we measure (air) temperature only on two scales, Fahrenheit and Celsius. Suppose that the temperature in Tokyo is 40°C and in London 20°C. Then on the Celsius scale, the statement is true. However, on the Fahrenheit scale, Tokyo is 104°F while London is 68°F.

The difference in temperature between Tokyo and London today is twice what it was yesterday

This statement implies that the distance between two measures is meaningful, a condition that is part of the interval scale. The statement is meaningful, because Fahrenheit and Celsius are related by the affine transformation $F = 9/5C + 32$, ensuring that ratios of differences (as opposed to just ratios) are preserved. For example, suppose yesterday's temperatures on the Celsius scale were 35°C in Tokyo and 25°C (a difference of 10) in London, while today it is 40°C in Tokyo and 20°C in London (a difference of 20). If we transform these temperatures to the Fahrenheit scale, we find that yesterday's temperatures were 95°F in Tokyo and 77°F London (a difference of 18); today's are 104°F in Tokyo and 68°F in London (a difference of 36). Thus, the truth value of the statement is preserved with the transformation.

Failure x is twice as critical as failure y

This statement is not meaningful, since we have only an ordinal scale for failure criticality. To see why, suppose we have four classes of failures, class$_i$, for i from 1 to 4. We can define two mappings, M and M', to be valid ordinal measures as follows:

Failure Class	Mapping M	Mapping M'
Class$_1$	1	3
Class$_2$	3	4
Class$_3$	6	5
Class$_4$	7	10

Suppose y is in class$_2$ and x in class$_3$. Notice that $M(x) = 6$ and $M(y) = 3$ while $M'(x) = 5$ and $M'(y) = 4$. In this case, the statement is true under M but false under M'.

Meaningfulness is often clear when we are dealing with measures with which we are familiar. But sometimes we define new measures, and it is not as easy to tell if the statements about them are meaningful.

EXAMPLE 2.24

Suppose we define a crude notion of *speed* of software programs, and we rank three programs A, B, and C with respect to a single empirical binary relation *faster than*. Suppose further that the empirical relation is such that A is faster than B, which is faster than C. This notion of program speed is measurable on an ordinal scale, and any mapping M in which $M(A) > M(B) > M(C)$ is an acceptable measure. Now consider the statement "Program A is faster than both Programs B and C" where we mean that A is faster than B and A is faster than C. We can show that this statement is meaningful in the following way. Let M and M' be any two acceptable measures. Then we know that, for any pair of programs x and y, $M(x) > M(y)$ if and only if $M'(x) > M'(y)$. Using the scale M, the statement under scrutiny corresponds to

$$M(A) > M(B) \text{ and } M(A) > M(C)$$

which is true. But then

$$M'(A) > M'(B) \text{ and } M'(A) > M'(C)$$

is also true because of the relationship between M and M'.

By similar argument, we can show that the statement "Program *B* is faster than both Programs *A* and *C*" is meaningful even though it is false.

However, consider the statement "Program *A* is more than twice as fast as Program *C*." This statement is not meaningful. To see why, define acceptable measures *M* and *M'* as follows:

$$M(A) = 3; \; M(B) = 2; \; M(C) = 1$$

$$M(A) = 3; \; M(B) = 2.5; \; M(C) = 2$$

Using scale *M* the statement is true, since $3 = M(A) > 2M(C) = 2$. However, using *M'* the statement is false. Although the statement seems meaningful given our understanding of speed, the sophistication of the notion *twice as fast* was not captured in our over-simplistic empirical relation system, and hence was not preserved by all measurement mappings.

The terminology often used in software engineering can be imprecise and misleading. Many software practitioners and researchers mistakenly think that to be meaningful, a measure must be useful, practical, worthwhile, or easy to collect. These characteristics are not part of meaningfulness. Indeed, such issues are difficult to address for any measure, whether it occurs in software or in some other scientific discipline. For example, carbon-dating techniques for measuring the age of fossils may not be practical or easy to do, but the measures are certainly valid and meaningful! Thus, meaningfulness should be viewed as only one attribute of a measure.

2.4.1 Statistical Operations on Measures

The scale type of a measure affects the types of operations and statistical analyses that can be sensibly applied to the data. Many statistical analyses use arithmetic operators:

$$+, -, \div, \times$$

The analysis need not be sophisticated. At the very least, we would like to know something about how the whole data set is distributed. We use two basic measures to capture this information: measures of central tendency and measures of dispersion. A *measure of central tendency*, usually called an average, tells us something about where the "middle" of the set is

We have measured an attribute for 13 entities, and the resulting data points in ranked order are:

$$2, 2, 4, 5, 5, 8, 8, 10, 11, 11, 11, 15, 16$$

The *mean* of this set of data (i.e., the sum divided by the number of items) is *8.3*.

The *median* (i.e., the value of the middle-ranked item) is *8*.

The *mode* (i.e., the value of the most commonly occurring item) is *11*.

FIGURE 2.12   Different ways to compute the average of a set of numbers.

likely to be, while a *measure of dispersion* tells us how far the data points stray from the middle.

Figure 2.12 shows the computation of measures of central tendency for a given set of data. Measures of dispersion include the maximum and minimum values, as well as the variance and standard deviation; these measures give us some indication of how the data are clustered around a measure of central tendency.

But even these simple analytical techniques cannot be used universally. In particular, nominal and ordinal measures do not permit computation of mean, variance, and standard deviation. That is, the notion of mean is not meaningful for nominal and ordinal measures.

EXAMPLE 2.25

Suppose the data points $\{x_1, \ldots, x_n\}$ represent a measure of understandability for each module in system X, while $\{y_1, \ldots, y_m\}$ represent the understandability values for each module in system Y. We would like to know which of the two systems has the higher average understandability. The statement "The average of the x_is is greater than the average of the y_js" must be meaningful; that is, the statement's truth value should be invariant with respect to the particular measure used.

Suppose we assess every module's understandability according to the following classification: trivial, simple, moderate, complex, and incomprehensible. In this way, our notion of understandability is representable on an ordinal scale. From this, we can define two valid measures of understandability, M and M', as in Table 2.6.

Suppose that X consists of exactly five modules, and the understandability of each is rated as:

TABLE 2.6   Measures of Understandability

	Trivial	Simple	Moderate	Complex	Incomprehensible
M	1	2	3	4	5
M'	1	2	3	4	10

TABLE 2.7   Different Measure M''

	Trivial	Simple	Moderate	Complex	Incomprehensible
M''	0.5	3.8	69	104	500

x_1 trivial
x_2 simple
x_3 simple
x_4 moderate
x_5 incomprehensible

while Y's seven modules have understandability

y_1 simple
y_2 moderate
y_3 moderate
y_4 moderate
y_5 complex
y_6 complex
y_7 complex

Using M, the mean of the X values is 2.6, while the mean of the Y values is 3.1; thus, the "average" of the Y values is greater than the average of the X values. However, using M', the mean of the X values is 3.6, while the mean of the Y values is 3.1. Since the definition of meaningfulness requires the relation to be preserved, then *mean* is not a meaningful measure of central tendency for ordinal scale data.

On the other hand, the *median* (i.e., the middle-ranked item) is a meaningful measure of central tendency. Using both M and M', the median of the Y values (in both cases 3) is greater than the median of the X values (in both cases 2). Similarly, if we define M'' as a radically different measure according to Table 2.7, the median of the Y values, 69, is still greater than the median of the X values, 3.8.

Example 2.25 confirms that the mean cannot be used as a measure of central tendency for ordinal scale data. However, the mean is acceptable for interval and ratio scale data. To see why, let $\{X_1, \ldots, X_n\}$ and $\{Y_1, \ldots, Y_n\}$

be two sets of entities for which some attribute can be measured on a ratio scale. We must show that the statement "The mean of the X_is is greater than the mean of Y_js" is meaningful. To do so, let M and M' be two measures for the attribute in question. Then we want to show that the means preserve the relation. In mathematical terms, we must demonstrate that

$$\frac{1}{n}\sum_{i=1}^{n} M(x_i) > \frac{1}{m}\sum_{j=1}^{m} M(y_j) \quad \text{if and only if } \frac{1}{n}\sum_{i=1}^{n} M'(x_i) > \frac{1}{m}\sum_{j=1}^{m} M'(y_j)$$

The ratio scale gives us the extra information we need to show that the assertion is valid. Thanks to the relationship between acceptable transformations for a ratio scale, we know that $M = aM'$ for some $a > 0$. When we substitute aM' for M in the above equation, we get a statement that is clearly valid.

The same investigation can be done for any statistical technique, using scale and transformation properties to verify that a certain analysis is valid for a given scale type. Table 2.8 presents a summary of the meaningful statistics for different scale types. The entries are inclusive reading downwards. That is, every meaningful statistic of a nominal scale type is also meaningful for an ordinal scale type; every meaningful statistic of an ordinal scale type is also meaningful for an interval scale type, etc. We will return to the appropriateness of analysis when we discuss experimental design and analysis in Chapter 4, and again when we investigate the analysis of software measurement data in Chapter 6.

2.4.2 Objective and Subjective Measures

When measuring attributes of entities, we strive to keep our measurements objective. By doing so, we make sure that different people produce the same measurement results, regardless of whether they are measuring product, process, or resource. This consistency of measurement is very important. Subjective measures depend on the environment in which they are made. The measures can vary with the person measuring, and they reflect the judgment of the measurer. What one judge considers bad, another may consider good, and it may be difficult to reach consensus on attributes such as process, product, or resource quality.

Nevertheless, it is important to recognize that subjective measurements can be useful, as long as we understand their imprecision. For example,

TABLE 2.8   Summary of Measurement Scales and Statistics Relevant to Each

Scale Type	Defining Relations	Examples of Appropriate Statistics	Appropriate Statistical Tests
Nominal	Equivalence	Mode	Nonparametric
		Frequency	
Ordinal	Equivalence	Median	Nonparametric
	Greater than	Percentile	
		Spearman r_s	
		Kendall τ	
		Kendall W	
Interval	Equivalence	Mean	Non-parametric
	Greater than	Standard deviation	
	Known ratio of any intervals	Pearson product-moment correlation	
		Multiple product-moment correlation	
Ratio	Equivalence	Geometric mean	Nonparametric and parametric
	Greater than	Coefficient of variation	
	Known ratio of any intervals		
	Known ratio of any two scale values		

Source: Siegel S. and Castellan N.J. Jr., *Nonparametrics Statistics for the Behavioral Sciences*, 2nd Edition. McGraw-Hill, New York, 1988.

suppose we want to measure the quality of requirements before we turn the specification over to the test team, who will then define test plans from them. Any of the techniques shown in Figure 2.2 would be acceptable. For example, we may ask the test team to read and rate each requirement on a scale from 1 to 5, where '1' means "I understand this requirement completely and can write a complete test script to determine if this requirement is met," to '5:' "I do not understand this requirement and cannot begin to write a test script." Suppose the results of this assessment look like the chart in Table 2.9.

Even though the measurement is subjective, the measures show us that we may have problems with our interface requirements; perhaps the interface requirements should be reviewed and rewritten before proceeding to test plan generation or even to design. It is the general picture that is important, rather than the exactness of the individual measure, so the subjectivity, although a drawback, does not prevent us from gathering useful information about the entity. We will see other examples throughout

TABLE 2.9   Results of Requirements Assessment

Requirement Type	1 (good)	2	3	4	5 (bad)
Performance requirements	12	7	2	1	0
Database requirements	16	12	2	0	0
Interface requirements	3	4	6	7	1
Other requirements	14	10	1	0	0

this book where measurement is far from ideal but still paints a useful picture of what is going on in the project.

2.4.3 Measurement in Extended Number Systems

In many situations we cannot measure an attribute directly. Instead, we must derive a measure in terms of the more easily understood *sub-attributes*. For example, suppose that we wish to assess the *quality* of the different types of transport available for traveling from our home to another city. We may not know how to measure quality directly, but we know that quality involves at least two significant sub-attributes, journey time and cost per mile. Hence, we accumulate data in Table 2.10 to describe these attributes.

Intuitively, given two transport types, A and B, we would rank A superior to B (i.e., A is of higher quality than B) if

journey time (A) < journey time (B) **AND** cost per mile

(A) < cost per mile (B)

Using this rule with the data collected for each journey type, we can depict the relationships among the candidates as shown in Figure 2.13. In the figure, an arrow from transport type B to transport type A indicates the superiority of A to B. Thus, Car is superior to both Train and Plane because, in each case, the journey time is shorter *and* the cost per mile is less. Figure 2.13 also shows us that Car, Train, and Plane are all superior to Coach.

Notice that in this relation Train and Plane are incomparable; that is, neither is superior to the other. Train is slower but cheaper than Plane.

TABLE 2.10   Transportation Attributes

Option	Journey Time (h)	Cost Per Mile ($)
Car	3	1.5
Train	5	2.0
Plane	3.5	3.5
Executive coach	7	4.0

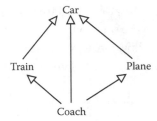

FIGURE 2.13 Quality relationships based on rule and collected data.

It would be inappropriate to force an ordering because of the different underlying attributes. We could impose an ordering only if we had additional information about the relative priorities of cost and timing. If cost is more important to us, then Train is preferable; if speed is more important, we would prefer Plane.

Now suppose we wish to use the representation condition to define a measure to characterize the notion of journey quality given by the above relation system. It is easy to prove that there is *no* possible measure that is a single-valued real number. Suppose that such a measure exists, then Plane would be mapped to some real number m(Plane), while Train would be mapped to some real number m(Train). Then, exactly one of the following must be true:

1. m(Plane) < m(Train)

2. m(Plane) > m(Train)

3. m(Plane) = m(Train)

If the first statement were true, then the representation condition implies that Plane must be superior to Train. This is false, because Train is cheaper. Similarly, the second statement is false because Train is slower than Plane. But the third statement is also false, since it implies an equality relation that does not exist.

The reason we cannot find a measure satisfying the representation condition is because we are looking at too narrow a number system. When we have genuinely incomparable entities, we have a *partial order*, as opposed to what is called a *strict weak order*, so we cannot measure in the set of real numbers. (A *strict weak order* has two properties: it is asymmetric and negatively transitive. By *asymmetric*, we mean that if the pair (x,y) is in the relation, then (y,x) is not in the relation. A relation is *negatively transitive* if, whenever (x,y) is in the relation, then for every z, either (x,z)

or (z,y) is in the relation.) What we need instead is a mapping into *pairs* of real numbers, that is, into the set . In the transport example, we can define a representation in the following way. First, we define a measure m that takes a transport type into a pair of elements:

$$m(\text{Transport}) = (\text{Journey time, Cost per mile})$$

Then, we define the actual pairs:

$$m(\text{Car}) = (3, 1.5)$$

$$m(\text{Train}) = (5, 2)$$

$$m(\text{Plane}) = (3.5, 3.5)$$

$$m(\text{Coach}) = (7, 4)$$

The numerical binary relation over that corresponds to the empirical superiority relation is defined as:

$$(x,y) \text{ superior to } (x',y') \quad \text{if } x < x' \quad \text{and} \quad y < y'$$

The numerical relation preserves the empirical relation. That too is only a partial order in because it contains incomparable pairs. For example, the pair (5,2) is not superior to (3.5,3.5); nor is (3.5,3.5) superior to (5,2).

EXAMPLE 2.26

Suppose we wish to assess the *quality* of four different C compilers. We determine that our notion of quality is defined in terms of two sub-attributes: *speed* (average KLOC compiled per second) and *resource* (minimum Kbytes of RAM required). We collect data about each compiler, summarized in Table 2.11.

Using the same sort of analysis as above, we can show that it is not possible to find a measure of this attribute in the real numbers that satisfies the representation condition.

These examples are especially relevant to software engineering. The International Standards Organization has published a standard, (ISO/IEC

TABLE 2.11   Comparing Four Compilers

	Speed	Resource
A	45	200
B	30	400
C	20	300
D	10	600

25010:2011), for measuring software quality that explicitly defines *software quality* as the combination of eight distinct sub-attributes. We will discuss the details of this standard in Chapter 10. However, it is important to note here that the standard reflects a widely held view that no single real-valued number can characterize such a broad attribute as quality. Instead, we look at *n*-tuples that characterize a set of *n* sub-attributes. The same observation can be made for *complexity* of programs.

EXAMPLE 2.27

Many attempts have been made to define a single, real-valued metric to characterize program complexity. For instance, in Example 2.9, we were introduced to one of the most well known of these metrics, the cyclomatic number. This number, originally defined by mathematicians on graphs, is the basis of an intuitive notion of program complexity. The number corresponds to an intuitive relation, *more complex than*, that allows us to compare program flowgraphs and then make judgments about the programs from which they came. That is, the cyclomatic number is a mapping from the flowgraphs into real numbers, intended to preserve the complexity relation. As we have seen in examining journey quality, if the relation *more complex than* is not a strict weak order, then cyclomatic number cannot be an ordinal scale measure of complexity. (Indeed, a theorem of measurement theory asserts that a strict weak order is a necessary *and* sufficient condition for an ordinal scale representation in $\mathcal{R}$.) We contend that no general notion of complexity can give rise to such an order. To see why, consider the graphs depicted in Figure 2.14. Flowgraph y represents a conditional choice structure, x represents a sequence of two such structures, and z represents a looping construct. Intuitively, it seems reasonable that graph x is more complex than graph y. If *more complex than* produced a strict weak order, we should be able to show that this relation is negatively transitive. That is, we should be able to show that for any z, either x is related to z or z is related to y. But neither of the following statements is obviously true:

x is more complex than z

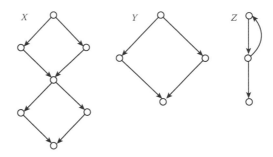

FIGURE 2.14   Three program flowgraphs.

and

z is more complex than y

Some programmers would argue that x is more complex than z; for instance, while others would say that z is more complex than x; we cannot reach consensus. In other words, some of the graphs are not comparable, so the relation is not a strict weak order and the cyclomatic number cannot be on an ordinal scale. Notice that the cyclomatic number for x is 3, for y is 2, and for z is 2, forcing us to conclude that x should be more complex than z. Thus, the cyclomatic number, clearly useful in counting the number of linearly independent paths in the program flowgraph, should not be used as a comprehensive measure of complexity.

In spite of theoretical problems, there are many situations when we must combine sub-attributes to impose a strict ranking, and hence an ordinal scale. That is, we need to define a single real-valued number as a measure. For example, if we are buying a coat, we may take into account the price, quality of material, fit, and color. But in the end, we are forced to determine preference. Consciously or subconsciously, we must define some combination of component measures to arrive at a preference ranking.

EXAMPLE 2.28

We want to buy a contact management program. It is likely that the decision will be based on a collection of attributes, such as price, reliability, and usability. If the program is for a single user's home PC, we may give price a heavier weighting than reliability when ranking the programs. However, if we are buying it for network use in a major organisation, it is likely that reliability would get a larger weighting than price.

Other, similar problems can arise. We may need to determine which program is safest, based on a set of criteria. Or we may wish to choose from among a variety of design techniques, based on survey data that captures developer preferences, design quality assessments, and cost of training and tools. Each of these instances presents a problem in making a decision with multiple criteria. There is an extensive literature on multi-criteria decision theory, and the measurement theory that relates to it. We discuss this type of analysis in Chapter 6, when we address data analysis techniques. However, here we must look at how to combine measures in a way that remains true to the spirit of measurement theory.

2.4.4 Derived Measurement and Meaningfulness

When we measure a complex attribute in terms of simpler sub-attributes, we are measuring indirectly. In doing so, we must adhere to the same basic rules of measurement theory that apply to direct measures. We must pay particular attention to issues of scale types and meaningfulness.

Scale types for derived measures are similar to those for direct ones. Our concerns include the uniqueness of the representation, as well as the admissible transformations for each scale type. We call an admissible transformation a rescaling, and we define rescaling in the following way. Suppose that we measure each of n sub-attributes with measure M_i. Let M be a derived measure involving components $M_1, M_2, \ldots, M_n$. That is, $M = f(M_1, M_2, \ldots, M_n)$ for some function f. We say that M' is a *rescaling* of M if there are rescalings $M_1', M_2', \ldots, M_n'$ of $M_1, M_2, \ldots, M_n$, respectively, such that $M' = f(M_1', M_2', \ldots, M_n')$.

Strictly speaking, this defines rescaling in the *wide sense*. Rescaling in the *narrow sense* requires us to verify that $M' = f(M_1, M_2, \ldots, M_n)$.

EXAMPLE 2.29

Density d is a derived measure of mass m and volume V. The specific relationship is expressed as

$$d = m/V$$

Every rescaling of d is of the form $d' = \alpha d$ (for $\alpha > 0$). To see why, we must demonstrate two things: that a function of this form is a rescaling, and that every rescaling has this form. For the first part, we have to find rescalings m' and V' of m and V, respectively, such that $\alpha d = m'/V'$. Both m and V are ratio scale measures, so αm and V are acceptable rescalings of m and V, respectively. Since

$$\alpha d = \alpha \left(\frac{m}{V} \right) = \frac{\alpha m}{V}$$

therefore we have a rescaling.

To show that every rescaling is of the appropriate form, notice that since m and V are ratio scale measures, every rescaling of m must be of the form $\alpha_1 m$ for some α_1 and every rescaling of V must be of the form $\alpha_2 V$ for some α_2. Therefore, every rescaling of d has the form

$$\frac{\alpha_1 m}{\alpha_2 V} = \frac{\alpha_1}{\alpha_2} \left(\frac{m}{V} \right) = \frac{\alpha_1}{\alpha_2} d = \alpha d \quad \left(\text{where } \alpha = \frac{\alpha_1}{\alpha_2} \right)$$

Now, we can define scale types for derived scales in exactly the same way as for direct scales. Example 2.29 shows us that the scale for density d is ratio, because all the admissible transformations have the form $d \leftarrow \alpha d$. In the same way, we can show that the scale type for a derived measure M will generally be no stronger than the weakest of the scale types of the M_is. Thus, if the M_is contain a mixture of ratio, interval, and nominal scale types, then the scale type for M will at best be nominal, since it is weakest.

EXAMPLE 2.30

A derived measure of testing efficiency T is D/E, where D is the number of defects discovered and E is effort in person-months. Here D is an absolute scale measure, while E is on the ratio scale. Since absolute is stronger than ratio scale, it follows that T is a ratio scale measure. Consequently, the acceptable rescalings of T arise from rescalings of E into other measures of effort (person-days, person-years, etc.)

Many of the measures we have used in our examples are assessment measures. But derived measures proliferate as prediction measures, too.

EXAMPLE 2.31

In Example 2.12, we saw that many software resource prediction models predict effort E (in person-months) by using an equation of the form

$$E = aS^b$$

where S is a measure of software size, and a and b are constants. Some researchers have doubted the meaningfulness of these derived effort measures. For example, DeMillo and Lipton looked at the Walston and Felix model. Walston and Felix assert that effort can be predicted by the equation

$$E = 5.2S^{0.91}$$

where S is measured in lines of code (see Perlis et al. 1981). DeMillo and Lipton contend that the prediction equation is an example of a meaningless measure. They assert that "both E and S are expressed as a ratio scale... but the measurement is not invariant under the transformation SS and so is meaningless" (DeMillo and Lipton 1981). In fact, this argument is relevant only when we consider scales defined in the *narrow* sense. In the more usual wide sense, it is easy to show that the equation is meaningful and that the scale type for effort is ratio. However, demonstrating scale type and meaningfulness is very different from asserting that the relationship is valid.

Many times, models of effort involve several levels of derived measurement. That is, a derived measure is defined to be a combination of other measures, both direct and derived.

EXAMPLE 2.32

Halstead developed a theory of software physics (discussed in Chapter 8) that defines attributes as combinations of counts of operators and operands. His equation for software effort, E, is

$$E = V/L$$

where V, the program volume, is on a ratio scale, but L, the estimated program level, appears to be only on an ordinal scale. Thus, E cannot be a ratio scale. However, Halstead claims that E represents the number of mental discriminations necessary to implement the program, which is *necessarily* a ratio scale measure of effort. Therefore, Halstead's effort equation is not meaningful.

The unit in which the measure is expressed can affect the scale of the measure.

EXAMPLE 2.33

Consider another effort measure

$$E = 2.7v + 121w + 26x + 12y + 22z - 497$$

cited by DeMillo and Lipton (1981). E is supposed to represent person-months, v is the number of program instructions, and w is a subjective complexity rating. The value of x is the number of internal documents generated on the project, while y is the number of external documents. Finally, z is the size of the program in words. DeMillo and Lipton correctly point out that, as in Example 2.32, effort should be on a ratio scale, but it cannot be ratio in this equation because w, an ordinal measure, restricts E to being ordinal. Thus, the equation is meaningless. However, E could still be an ordinal scale measure of effort if we drop the pre-condition that E expresses effort in person-months.

In this chapter, we have laid a foundation of principles on which to base valid measurement. The next chapter builds on this foundation by introducing a framework for how to choose measures, based on needs and process.

2.5 SUMMARY

Measurement requires us to identify intuitively understood attributes possessed by clearly defined entities. Then, we assign numbers or symbols to the entities in a way that captures our intuitive understanding about the attribute. Thus, intuitive understanding of that attribute must precede direct measurement of a particular attribute. This intuitive understanding leads to the identification of relations between entities. For example, the attribute height for the entity person gives rise to relations like *is tall*, *taller than*, and *much taller than*.

To measure the attribute, we define corresponding relations in some number system; then measurement assigns numbers to the entities in such a way that these relations are preserved. This relationship between the domain and range relationships is called the *representation condition*.

In general, there may be many ways of assigning numbers that satisfy the representation condition. The nature of different assignments determines the scale type for the attribute. There are five well-known scale types: nominal, ordinal, interval, ratio, and absolute. The scale type for a measure determines what kind of statements we can meaningfully make using the measure. In particular, the scale type tells us what kind of operations we can perform. For example, we can compute means for ratio scale measures, but not for ordinal measures; we can compute medians for ordinal scale measures but not for nominal scale measures.

Many attributes of interest in software engineering are not directly measurable. This situation forces us to use vectors of measures, with rules for combining the vector elements into a larger, derived measure. We define scale types for these in a similar way to direct measures, and hence can determine when statements and operations are meaningful.

In the next chapter, we build on this foundation to examine a framework for measurement that helps us to select appropriate measures to meet our needs.

EXERCISES

1. At the beginning of this chapter, we posed four questions:

   a. How much we must know about an attribute before it is reasonable to consider measuring it? For instance, do we know enough about "complexity" of programs to be able to measure it?

   b. How do we know if we have really measured the attribute we wanted to measure? For instance, does a count of the number of "bugs" found in a system during integration testing measure the quality of the system? If not, what does the count tell us?

   c. Using measurement, what meaningful statements can we make about an attribute and the entities that possess it? For instance, is it meaningful to talk about doubling a design's quality? If not, how do we compare two different designs?

   d. What meaningful operations can we perform on measures? For instance, is it sensible to compute average productivity for a group of developers, or the average quality of a set of modules?

   Based on what you have learned in this chapter, answer these questions.

2. a. List, in increasing order of sophistication, the five most important measurement scale types.

   b. Suppose that the attribute "complexity" of software modules is ranked as a whole number between 1 and 5, where 1 means "trivial," 2 "simple," 3 "moderate," 4 "complex," and 5 "incomprehensible." What is the scale type for this definition of complexity? How do you know? With this measure, how could you meaningfully measure the average of a set of modules?

3. We commonly use ordinal measurement scales. For example, we can use an ordinal scale to rank the understandability of programs as *trivial, simple, moderate, complex,* or *incomprehensible.* For each of the two other common measurement scale types, give an example of a useful software measure of that type. State exactly which software entity is being measured and which attribute. State whether the entity is a product, process, or resource.

4. Define measurement, and briefly summarize the representation condition for measurement.

5. For the empirical and numerical relation system of Example 2.5, determine which of the following numerical assignments satisfy the representation condition:

   a. $M(\text{Wonderman}) = 100$; $M(\text{Frankie}) = 90$; $M(\text{Peter}) = 60$

   b. $M(\text{Wonderman}) = 100$; $M(\text{Frankie}) = 120$; $M(\text{Peter}) = 60$

   c. $M(\text{Wonderman}) = 100$; $M(\text{Frankie}) = 120$; $M(\text{Peter}) = 50$

   d. $M(\text{Wonderman}) = 68$; $M(\text{Frankie}) = 75$; $M(\text{Peter}) = 40$

6. For the relation systems in Example 2.6, determine which of the following mappings are representations. Explain your answers in terms of the representation condition.

   a. $M(\text{each delayed response}) = 6$; $M(\text{each incorrect output}) = 6$; $M(\text{each data-loss}) = 69$

   b. $M(\text{each delayed response}) = 1$; $M(\text{each incorrect output}) = 2$; $M(\text{each data-loss}) = 3$

   c. $M(\text{each delayed response}) = 6$; $M(\text{each incorrect output}) = 3$; $M(\text{each data-loss}) = 2$

   d. $M(\text{each delayed response}) = 0$; $M(\text{each incorrect output}) = 1$; $M(\text{each data-loss}) = 0.5$

7. Suppose that we could classify every software failure as either a) syntactic, b) semantic, or c) system crash. Suppose additionally that

we agree that every system crash failure is more critical than every semantic failure, which in turn is more critical than every syntactic failure. Use this information to define two different measures of the attribute of *criticality* of software failures. How are these measures related? What is the scale of each?

8. Explain why you would not conclude that the quality of program X was twice as great as program Y if integration testing revealed program X to have twice as many faults per KLOC than program Y.

9. Explain why it is wrong to assert that lines of code is a bad software measure.

10. Explain why neither M_4 nor M_5 is a valid mapping in Example 2.16.

11. In Example 2.18, determine the affine transformations from:

   a. M_1 to M_2

   b. M_2 to M_1

   c. M_2 to M_3

   d. M_3 to M_2

   e. M_1 to M_3

12. Explain why duration of processes is measurable on a ratio scale. Give some example measures and the ratio transformations that relate them.

13. Determine which of the following statements are meaningful:

   a. The length of Program A is 50.

   b. The length of Program A is 50 executable statements.

   c. Program A took 3 months to write.

   d. Program A is twice as long as Program B.

   e. Program A is 50 lines longer than Program B.

   f. The cost of maintaining program A is twice that of maintaining Program B.

   g. Program B is twice as maintainable as Program A.

   h. Program A is more complex than Program B.

14. Formally, what do we mean when we say that a statement about measurement is meaningful? Discuss the meaningfulness of the following statements:

   "The average size of a Windows application program is about four times that of a similar Linux program."

   "Of the two Ada program analysis tools recommended in the Ada coding standard, tool A achieved a higher average usability rating than tool B." For this example, program usability was rated on a four-point scale:

   4: can be used by a non-programmer

   3: requires some knowledge of Ada

   2: usable only by someone with at least 5 years Ada programming experience

   1: totally unusable.

15. Show that the mean can be used as a measure of central tendency for interval scale data.

16. Show that, for nominal scale measures, the median is not a meaningful notion of average, but the mode (i.e., the most commonly occurring class of item) is meaningful.

17. Suppose that "complexity" of individual software modules is ranked (according to some specific criteria) as one of the following:

   {trivial, simple, moderate, complex, very complex, incomprehensible}

   Let M be any measure (in the representation sense) for this notion of complexity, and let S be a set of modules for each of which M has been computed.

   a. You want to indicate the average complexity of the modules in S. How would you do this in a meaningful way? (Briefly explain your choice.)

   b. Explain why it is not meaningful to compute the mean of the Ms. (You should construct a statement involving means that you can prove is not meaningful.)

c. Give two examples of criteria that might be used to enable an assessor objectively to determine which of the complexity values a given module should be.

State carefully any assumptions you are making.

18. Example 2.26 defines a quality attribute for compilers. Draw a diagram to illustrate the empirical relation of quality. Explain why it is not possible to find a measure for this attribute in the set of real numbers that satisfies the representation condition. Define a measurement mapping into an alternative number system that does satisfy the representation condition.

19. A commonly used derived measure of programmer productivity P is $P = L/E$, where L is the number of lines of code produced and E is effort in person-months. Show that every rescaling of P is of the form $P' = P$ (for >0).

20. Show that the Walston–Felix effort equation in Example 2.32 defines a ratio scale measure.

21. Construct a representation for the relations *greater functionality* and *greater user-friendliness* characterized by Table 2.1.

22. Consider the attribute, "number of bugs found," for software testing processes. Define an absolute scale measure for this attribute. Why is "number of bugs found" not an absolute scale measure of the attribute of *program correctness*?

23. Consider the following software *application domain (AD)* measure and some data concerning software applications:

Application Domain (AD) Measure

AD	Domain Description
1	WWW browsers
2	WWW servers
3	Compilers
4	Embedded systems
5	Operating systems
6	Word processors

Application Domain (AD) Data

System	AD
A	5
B	1
C	1
D	1
E	2
F	6
G	5

a. What is the measurement scale of the application domain measure?

b. How would you measure the central tendency of application domain? Calculate this central tendency value for the applications A–G listed above.

c. Using the measurement theory definition of "meaningful", give an example statement concerning the applications listed above that is not meaningful.

24. Consider the attribute *program adaptability*, which is defined as "the difficulty of adding new features to a program."

a. Define a measure for some aspect of program adaptability, and give an example of an empirical relation.

b. Give the measurement scale for the proposed measure and justify your choice.

c. Explain why the measure satisfies the representation condition.

d. Describe one advantage and one disadvantage of the measure that you defined.

REFERENCES

Littlewood B., Forecasting software reliability, In: *Software Reliability, Modelling and Identification* (Ed: Bittanti S.), Lecture Notes in Computer Science 341, Springer-Verlag Berlin Heidelberg, pp. 141–209, 1988.

McCabe T, A software complexity measure, *IEEE Transactions on Software Engineering*, SE 2(4), pp. 308–320, 1976.

Perlis A.J., Sayward F.G., and Shaw M. (Eds.), *Software Metrics: An Analysis and Evaluation*, MIT Press, Cambridge, Massachusetts, 1981.

Pfeiffer R.E. *Unified Code Counter (UCC) Software Design*. AEROSPACE REPORT NUMBER: TOR-2012(3906)-72, UNIFIED CODE COUNTER (UCC) SOFT-WARE DESIGN (10-JUL-2012), 2012. http://www.everyspec.com/USAF/TORs/TOR2012-3906-72_10JUL2012_47673/. 7 August 2014.

FURTHER READING

There is no elementary textbook as such on measurement theory. The most readable book on the representational theory of measurement is:

Roberts F.S., *Measurement Theory with Applications to Decision Making, Utility, and the Social Sciences*, Addison-Wesley, Reading, Massachusetts, 1979.

A more formal mathematical treatment of the representational theory of measurement (only for the mathematically gifted) is

Krantz D.H., Luce R.D., Suppes P., and Tversky A., *Foundations of Measurement*, Volume 1, Academic Press, New York, 1971.

A very good introduction to measurement using non-scientific examples, including attributes like public image, religiosity, and aspects of political ideology, may be found in the following:

Finkelstein L., What is not measurable, make measurable, *Measurement and Control*, 15, pp. 25–32, 1982.
Hubbard D.W., *How to Measure Anything: Finding the Value of Intangibles in Business*, 2nd Edition, John Wiley and Sons, Hoboken, New Jersey, 2010.

Detailed discussions of alternative definitions of meaningfulness in measurement may be found in the following:

Falmagne, J.-C. and Narens, L., Scales and meaningfulness of quantitative laws, *Synthese*, 55, pp. 287–325, 1983.
Roberts, F. S., Applications of the theory of meaningfulness to psychology, *Journal of Mathematical Psychology*, 29, pp. 311–332, 1985.

The origin of the definition of the hierarchy of measurement scale types (nominal, ordinal, interval, and ratio) is the classic paper:

Stevens S.S., On the theory of scale types and measurement, *Science*, 103, pp. 677–680, 1946.

A criticism of this basic approach appears in:

Velleman P.F. and Wilkinson L., Nominal, ordinal, interval and ratio typologies are misleading, *The American Statistician*, 47(1), pp. 65–72, February 1993.

Other relevant references are:

Abran A., *Software Metrics and Software Metrology*, John Wiley and Sons Inc., Hoboken, New Jersey, 2010.

Belton V., A comparison of the analytic hierarchy process and a simple multi-attribute utility function, *European Journal of Operational Research*, 26, pp. 7–21, 1986.

Briand L.C., Morasca S., and Basili V.R., Property-based software engineering measurement, *IEEE Transactions on Software Engineering*, 22(1), pp. 68–86, 1996.

Campbell N.R., *Physics: The Elements*, Cambridge University Press, Cambridge, MA, 1920. Reprinted as *Foundations of Science: The Philosophy of Theory and Experiment*, Dover, New York, 1957.

Ellis B., *Basic Concepts of Measurement*, Cambridge University Press, Oxford, England, 1966.

Finkelstein L., A review of the fundamental concepts of measurement, *Measurement*, 2(1), pp. 25–34, 1984.

Finkelstein L., Representation by symbol systems as an extension of the concept of measurement, *Kybernetes*, Volume 4, pp. 215–223, 1975.

Kyburg H.E., *Theory and Measurement*, Cambridge University Press, Cambridge, England, 1984.

Sydenham P.H. (Ed.), *Handbook of Measurement Science*, Volume 1, John Wiley, New York, 1982.

Vincke P., *Multicriteria Decision Aids*, John Wiley, New York, 1992.

Whitmire S., *Object-Oriented Design Measurement,* John Wiley and Sons, 1997.

Zuse H., *Software Complexity: Measures and Methods*, De Gruyter, Berlin, 1991.

CHAPTER 3

A Goal-Based Framework for Software Measurement

CHAPTER 1 DESCRIBES MEASUREMENT's essential role in good software engineering practice, and Chapter 2 shows how to apply a general theory of measurement to software. In this chapter, we present a conceptual framework for diverse software measurement activities that you can apply to an organization's software development practices. These practices may include development and maintenance activities, plus experiments and case studies that evaluate new techniques and tools.

The framework presented here is based on two principle activities: classifying the entities to be examined and using measurement goals to identify relevant metrics. We show how such a goal-based framework supports evaluations of software products, processes, and software development organizations. We also look at measurement validation: both the process of insuring that we are measuring what we say we are, so that we satisfy the representation condition introduced in Chapter 2, as well as demonstrating the utility of a measure.

3.1 CLASSIFYING SOFTWARE MEASURES

As we have seen in Chapter 2, the first obligation of any software measurement activity is identifying the entities and attributes that we want to measure. Software entities can be classified as follows:

- *Processes*: Software-related activities.

- *Products*: Artifacts, deliverables, or documents that result from a process activity.

- *Resources*: Entities required by a process activity.

A process is usually associated with some time scale. Process activities have duration—they occur over time, and they may be ordered or related in some way that depends on time, so that one activity must be completed before another can begin. The timing can be explicit, as when design must be complete by October 31, or implicit, as when a flow diagram shows that design must be completed before coding can begin.

Resources and products are associated with a process. Each process activity uses resources and products, and produces products. Thus, the product of one activity may feed another activity. For example, a design document can be the product of the design activity, which is then used as an input to the coding activity.

Many of the examples that we use in this book relate to the development process or maintenance process. But the concepts that we introduce apply to any process: the reuse process, the configuration management process, the testing process, etc. In other words, measurement activities may focus on any process and need not be concerned with the comprehensive development process. As we show in Section 3.2, our choice of measurements depends on our measurement goals.

Within each class of entity, we distinguish between internal and external attributes of a product, process, or resource:

- *Internal attributes*: Attributes that can be measured purely in terms of the product, process, or resource itself. An internal attribute can be measured by examining the product, process, or resource on its own, without considering its behavior.

- *External attributes*: Attributes that can be measured only with respect to how the product, process, or resource relates to its environment. Here, the behavior of the process, product, or resource is important, rather than the entity itself.

To understand the difference between internal and external attributes, consider a set of software modules. Without actually executing the code, we can determine several important internal attributes: its size (perhaps in terms of lines of code, LOC or number of operands), its complexity (perhaps in terms of the number of decision points in the code), and the dependencies among modules. We may even find faults in the code as we read it: misplaced commas, improper use of a command, or failure to consider a particular case. However, there are other attributes of the code that can be measured only when the code is executed: the number of failures experienced by the user, the difficulty that the user has in navigating among the screens provided, or the length of time it takes to search the database and retrieve requested information, for instance. It is easy to see that these attributes depend on the behavior of the code, making them external attributes rather than internal. Table 3.1 provides additional examples of types of entities and attributes.

Managers often want to be able to measure and predict external attributes. For example, the cost-effectiveness of an activity (such as design inspections) or the productivity of the staff can be very useful in ensuring that the quality stays high while the price stays low. Users are also interested in external attributes, since the behavior of the system affects them directly; the system's reliability, usability, and portability affect maintenance and purchase decisions. However, external attributes are usually more difficult to measure than internal ones, and they are measured quite late in the development process. For example, reliability can be measured only after development is complete and the system is ready for use.

Moreover, it can be difficult to define the attributes in measurable ways that satisfy all stakeholders. For example, we all want to build and purchase systems of high quality, but we do not always agree on what we mean by quality, and it is often difficult to measure quality in a comprehensive way. Thus, we tend to define these high-level attributes in terms of other, more-concrete attributes that are well defined and measurable. The McCall model introduced in Chapter 1 is a good example of this phenomenon, where software quality is defined as a composite of a large number of narrower, more easily measurable terms.

In many cases, developers and users focus their efforts on only one facet of a broad attribute. For example, some measure quality, an external product attribute, as the number of faults found during formal testing, an internal process attribute. Using internal attributes to make judgments about external attributes can lead to invalid conclusions. But there is a clear need to be able to use internal attribute measurements decision-making about external attributes. One of the goals of software measurement research is

TABLE 3.1 Components of Software Measurement

Entities	Attributes	
Products	**Internal**	**External**
Requirements Use case models and scenarios	Size, reuse, modularity, redundancy, functionality, syntactic correctness, etc.	Comprehensibility, maintainability, etc.
Designs, Design models	Size, reuse, modularity, coupling, cohesiveness, functionality, etc.	Quality, complexity, maintainability, etc.
Code	Size, reuse, modularity, coupling, functionality, algorithmic complexity, control-flow structuredness, etc.	Reliability, usability, maintainability, etc.
Test requirements	Size, etc.	Effectiveness, etc.
Test data	Size, coverage,	Fault-finding ability
Test harness	Languages supported, features	Ease of use
...	...	...
Processes		
Constructing requirements	Time, effort, number of requirements changes, etc.	Quality, cost, stability, etc.
Detailed design	Time, effort, number of specification faults found, etc.	Cost, cost-effectiveness, etc.
Testing	Time, effort, number of coding faults found, etc.	Cost, cost-effectiveness, stability, etc.
...	...	...
Resources		
Personnel	Age, price, etc.	Productivity, experience, intelligence, etc.
Teams	Size, communication level, structuredness, etc.	Productivity, quality, etc.
Software	Price, size, etc.	Usability, reliability, etc.
Hardware	Price, speed, memory size, etc.	Reliability, etc.
Offices	Size, temperature, light, etc.	Comfort, quality, etc.
...	...	...

to identify the relationships between internal and external attributes, as well as to find new and useful methods for directly measuring the attributes of interest.

We also need to be able to track changes in product, process, and resource attributes as systems evolve over time. For example, as a system evolves from version 2 to version 3, the design, code, regression tests, test requirements, processes, and personnel will change. Changes in key software measures can potentially identify problems such as design decay, improvements or reduction in reliability, and personnel changes. Measurements can also be used to determine the success or failure of the response to identified problems.

3.1.1 Processes

We often have questions about our software development activities and processes that measurement can help us to answer. We want to know how long it takes for a process to complete, how much it will cost, whether it is effective or efficient, and how it compares with other processes that we could have selected. However, only a limited number of internal process attributes can be measured directly. These measures include:

- The duration of the process or one of its activities.

- The effort associated with the process or one of its activities.

- The number of incidents of a specified type arising during the process or one of its activities.

For example, we may be reviewing our requirements to ensure their quality before turning them over to the designers. To measure the effectiveness of the review process, we can measure the number of requirement errors found during the review as well as the number of errors found during later activities. Likewise, to determine how well we are doing integration testing, we can measure the number of faults found during integration testing, as well as those found later. And the number of personnel working on the project during a specified period can give us insight into the resources needed for the development process.

Many of these measures can be used in combination with other measures to gain a better understanding of what is happening on a project:

EXAMPLE 3.1

During formal testing, we can use the indirect measure

$$\frac{\text{Cost}}{\text{Number of errors}}$$

as a measure of the average cost of each error found during the process.

EXAMPLE 3.2

AT&T developers wanted to know the effectiveness of their software inspections. In particular, managers needed to evaluate the cost of the inspections against the benefits received. To do this, they measured the average amount of effort expended per thousand lines of code reviewed. As we will see later in this chapter, this information, combined with measures of the number of faults discovered during the inspections, allowed the managers to perform a cost–benefit analysis.

In some cases, we may want to measure properties of a process that consists of a number of distinct sub-processes:

EXAMPLE 3.3

The testing process may be composed of unit testing, integration testing, system testing, and acceptance testing. Each component process can be measured to determine how effectively it contributes to overall testing. We can track the number of errors identified in each subprocess, along with the duration and cost of identifying each error, to see if each subprocess is cost-effective.

Cost is not the only process attribute that we can examine. Controllability, observability, and stability are also important in managing a large project. These attributes are clearly external ones. For example, stability of the design process can depend on the particular period of time, as well as on which designers are involved. Attributes such as these may not yet be sufficiently well understood to enable numeric measurements according to the principles described in Chapter 2, so they are often indicated by subjective ratings on an ordinal scale. However, the subjective rankings based on informal observations can form the basis for the empirical relations required for subsequent objective measurement.

We often use objective measures of internal attributes as surrogate measures of external attributes. For example, surrogate measures of the effectiveness of code maintenance can be defined as a composite measure in terms of the number of faults discovered and the number of faults corrected. In the AT&T study of Example 3.2, inspection effectiveness was measured as average faults detected per thousand lines of code (KLOC) inspected.

In each case, we examine the process of interest and decide what kind of information would help us to understand, control, or improve the process. We will investigate process attributes in Chapter 13.

3.1.2 Products

Products are not restricted to the items that management is committed to deliver as the final software product. Any artifact or document produced during the software life cycle can be measured and assessed. For example, developers often model the requirements and design using various diagrams defined in the Unified Modeling Language, and they build prototypes. The purpose of models and prototypes is to help developers to understand the requirements or evaluate possible designs; these models and prototypes may be measured in some way. Likewise, test harnesses that are constructed to assist in system testing may be measured; system size measurements should include this software if they are to be used to determine team productivity. And documents, such as the user guide or the customer specification document, can be measured for size, quality, and more.

3.1.2.1 External Product Attributes

There are many examples of external product attributes. Since an external product attribute depends on both product behavior and environment, each attribute measure should take these constraints into account. For example, if we are interested in measuring the reliability of code, we must consider the machine and system configuration on which the program is run as well as the mode of operational usage. That is, someone who uses a word-processing package only to type letters may find its reliability to be different from someone who uses the same package to merge tables and link to spreadsheets. Similarly, the understandability of a document depends on the experience and credentials of the person reading it; a nuclear engineer reading the specification for power plant software is likely to rate its understandability higher than a mathematician reading the same document. Or the maintainability of a system may depend on the skills of the maintainers and the tools available to them.

Usability, integrity, efficiency, testability, reusability, portability, and interoperability are other external attributes that we can measure. These attributes describe not only the code but also the other documents that support the development effort. Indeed, the maintainability, reusability, and even testability of specifications and designs are as important as the code itself.

3.1.2.2 Internal Product Attributes

Internal product attributes are sometimes easy to measure. We can determine the size of a product by measuring the number of pages it fills or the number of words it contains, for example. Since the products are concrete, we have a better understanding of attributes like size, effort, and cost. Other internal product attributes are more difficult to measure, as opinions differ as to what they mean and how to measure them. For example, there are many aspects of code complexity, and no consensus about what best measures it. We will explore this issue in Chapter 9.

Because products are relatively easy to examine in an automated fashion, there is a set of commonly measured internal attributes. For instance, you can assess specifications in terms of their length, functionality, modularity, reuse, redundancy, and syntactic correctness. Formal designs and code can be measured in the same way; we can also measure attributes such as structuredness (of control and data flow, for example) as well as module coupling and cohesiveness.

Users sometimes dismiss many of these internal attributes as unimportant, since a user is interested primarily in the ultimate functionality, quality, and utility of the software. However, the internal attributes can be very helpful in suggesting what we are likely to find as the external attributes.

EXAMPLE 3.4

Consider a software purchase to be similar to buying an automobile. If we want to evaluate a used car, we can perform dynamic testing by actually driving the car in various conditions in order to assess external attributes such as performance and reliability. But usually we cannot make a complete dynamic assessment before we make a purchase decision, because not every type of driving condition is available (such as snowy or slick roads). Instead, we supplement our limited view of the car's performance with measures of static properties (i.e., internal attributes) such as water level, oil type and level, brake fluid type and level, tire tread and wear pattern, brake wear, shock absorber response, fan belt flexibility and wear, etc. These internal attributes provide insight into the likely external attributes; for example, uneven tire wear may indicate that the tires have been under-inflated, and the owner may have abused the car by not performing necessary maintenance. Indeed, when a car is serviced, the mechanic measures only internal attributes and makes adjustments accordingly; the car is rarely driven for any length of time to verify the conditions indicated by the internal attributes. In the same way, measures of internal software product attributes can tell us what the software's performance and reliability may be. Changes to the inspection process, for instance, may be based on measures of faults found, even though the ultimate goal of reliability is based on failures, not faults.

Just as processes may be composed of subprocesses, there are products that are collections of subproducts.

TABLE 3.2   Definition of Example Design Measurements

Entity	Entity Type	Attribute	Proposed Measure	Type
Design-level class diagram D_i	Product	Size	Number_of_classes$_i$	Direct
System design $\{D_1, \ldots, D_n\}$	Product	Average class diagram size	$1/n \sum$ Number_of_classes$_i$	Indirect

---

EXAMPLE 3.5

A software system design may consist of a number of Unified Modeling Language design-level class diagrams. Average class diagram size, an attribute of the system design, can be derived by calculating the size of each diagram. We can measure size by number of classes in each diagram, and then calculate the comprehensive measure according to Table 3.2.

3.1.2.3 The Importance of Internal Attributes

Many software engineering methods proposed and developed in the last 40 years provide rules, tools, and heuristics for producing software products. Almost invariably, these methods give structure to the products and the common wisdom is that this structure makes software products easier to understand, analyze, test, and modify. The structure involves two aspects of development:

1. The development process, as certain products need to be produced at certain stages, and

2. The products themselves, as the products must conform to certain structural principles.

In particular, levels of internal attributes such as modularity, coupling, or cohesion usually characterize product structure.

EXAMPLE 3.6

One of the most widely respected books in software engineering is Brooks' *Mythical Man-Month* (Brooks 1975, 1995). There, Brooks describes the virtues of top-down design:

A good top-down design avoids bugs in several ways. First, the clarity of structure and representation makes the precise statement of requirements and functions of the modules easier. Second, the partitioning and independence of modules avoids system bugs. Third, the suppression of detail makes flaws in the structure more apparent. Fourth, the design can be tested at each of its refinement steps, so testing can start earlier and focus on the proper level of detail at each step.

Similarly, the notion of high module cohesion and low module coupling is the rationale for most design methods from structured design (Yourdon and Constantine 1979) to object-oriented design (Larman 2004). Designs that possess these attributes are assumed to lead to more reliable and maintainable code.

Brooks, Yourdon and Constantine, and Larman assume what most other software engineers assume: good internal structure leads to good external quality. Although intuitively appealing, the connection between internal attribute values and the resulting external attribute values has rarely been established, in part because (as we shall see in Chapter 4) it is difficult to perform controlled experiments and confirm relationships between attributes. At the same time, validating measures of attributes such as cohesion and coupling is difficult but necessary. Valid software measurement can help us to understand and confirm relationships empirically; a key to the success of this analysis is the ability to provide accurate and meaningful measures of internal product attributes.

3.1.2.4 Internal Attributes and Quality Control and Assurance

A major reason that developers want to use internal attributes to predict external ones is the need to monitor and control the products during development. For example, we want to be able to identify modules at the design stage whose profile, in terms of measures of internal attributes, shows that they are likely to be error-prone or difficult to maintain or test later on. To do this, we need to know the relationship between internal design attributes and failures. Figure 3.1 is an irreverent view of this process.

3.1.2.5 Validating Composite Measures

Software engineers frequently use the term "quality" to describe an internal attribute of design or code. However, "quality" is multi-dimensional; it does not reflect a single aspect of a particular product.

EXAMPLE 3.7

Consider the gross national product (GNP), a measure of all the goods and services produced by a country in a given time period. Economists look

FIGURE 3.1   Using internal measures for quality control and assurance.

at the trend in GNP over time, hoping that it will rise—that the country is becoming more productive. The GNP is a weighted combination of the values of key goods and services produced; the weights reflect the priorities and opinions of the economists defining the measure. The value of individual goods and services can be measured directly, but the GNP itself is an indirect measure.

In the same way that economists want to control the economy and make a country more productive, we want to measure and control the quality of our products, and we usually do so by measuring and controlling a number of internal (structural) product attributes. Without articulating the specific attributes that contribute to the general notion of quality, many people assume that a single number, much like GNP, can capture the various cognitive and structural notions of complexity, maintainability, and usability. This number should therefore be a powerful indicator of all the attributes that one normally associates with high-quality systems, such as high reliability and high maintainability.

However, this approach ignores the question of whether the component internal attributes are a complete and accurate depiction of the comprehensive one, and whether the weighting is appropriate. Just as economists question whether GNP is valid (does it capture values like happiness, beauty, and environmental quality?), we must question the validity of measures that paint only a partial picture of the attribute of interest. In Chapters 8 through 11, we shall give scrutiny to conventional approaches to measuring size, quality, complexity, and more.

3.1.3 Resources

The resources that we are likely to measure include any input for software production. Thus, personnel (individual or teams), materials (including office supplies), tools (both software and hardware), and methods are candidates for measurement. We measure resources to determine their magnitude (how many staff are working on this project?), their cost (how much are we paying for testing tools?), and their quality (how experienced are our designers?). These measures help us to understand and control the process by telling us how the process is using and changing inputs to outputs. For example, if we are producing poor-quality software, resource measurements may show us that the software quality is the result of using too few people or people with the wrong skills.

Cost is often measured across all types of resources, so that managers can see how the cost of the inputs affects the cost of the outputs. For instance, the division chief may want to know if a large investment in software modeling or testing tools is yielding benefits in terms of more productive staff or better quality products. Cost is often defined in terms of its components, so that managers can see which aspects of cost are having the biggest effect.

Productivity is always important, and managers are keen not only to measure it but also to understand how to improve it. Although a measure of staff, productivity is an external resource attribute, as it depends on the underlying development process. That is, a productive worker using one process may become less productive if the process changes. Productivity is usually measured as some form of the following equation:

$$\frac{\text{Amount of output}}{\text{Effort input}}$$

Notice that this resource measure combines a process measure (input) with a product measure (output). The general notion is an economic one, where businesses or markets are judged by comparing what goes in with what comes out. For software development, the measure of output is usually computed as the amount of code or functionality produced as the final product, while the input measure is the number of person-months used to specify, design, code, and test the software. However, the economic analogy is incomplete for software, as the amount of software output is not related to the input in the same way as that in manufacturing. That is, most manufacturing processes involve replication, so that one car is much like another coming off the assembly line. But software development is a creation or design process, not a replication process, and the relationship between inputs and outputs is defined differently.

There are many staff attributes that we can measure, whose values may have an influence on the process or product. For example, the education, experience, age, or intelligence of a developer may affect the quality of the design or code. Similarly, the size, structure, and communication patterns of the development team are important.

We can also classify and analyze tools and methods. Languages are strongly typed or not, object-oriented or not, etc. Techniques can be rated as manual or automated, and tools can require special training or experience. These attributes of resources help us to understand how to use tools and methods in more effective ways.

3.1.4 Change and Evolution

Software that is used is continually revised. New features are added and errors are fixed. To understand and manage evolving software, we need to measure attributes of products, processes, and resources at various points in time. Processes change over time. For example, improved inspection processes may be added to improve reliability. Clearly, software products change as new features are added—modules are added and expanded. Resources also change as developers move from project to project and adopt new tools. Measurement can help us to understand and manage evolving systems by allowing us to track the changes in attributes over time. We can track the growth of modules, as well as changes in our processes and resources. We can also track changes in external attributes to identify, for example, improvements (hopefully) in product reliability and availability. Thus, we are interested in measurement over time, and the changes and trends of measured attributes. Measuring trends is how we can demonstrate product and process improvement.

No developer has the time to measure, analyze, and track everything. It is important to focus measurement activities on those areas needing the most visibility, understanding, and improvement. In Section 3.2, we present a technique for determining which attributes to measure first.

3.2 DETERMINING WHAT TO MEASURE

Measurement is useful only if it helps you to understand an underlying process or one of its resultant products. Determining the appropriate attributes to measure depends on your objectives—you select specific measurements based on what information you need to meet your goals. The Goal-Question-Metric approach (GQM) for metrics selection, first suggested by Basili and his colleagues, is an effective approach (Basili and Weiss 1984; Basili and Rombach 1988; Basili et al. 2009). To use GQM, you first identify the overall goals of your organization. Depending on the situation, the organization can be an entire corporation, an individual department or lab, or a single project or group. Then, you generate questions whose answers you must know in order to determine whether your goals are being met. Finally, you analyze each question to identify measurements you need in order to answer each question.

One common goal is to evaluate the maturity of a software organization and its process in order to improve it. We shall see how the Goal-Question-Metric paradigm is applied to do such an evaluation.

3.2.1 Goal-Question-Metric Paradigm

Many metrics programs begin by measuring what is convenient or easy to measure, rather than by measuring what is needed. Such programs often fail because the resulting data are not useful to the developers and maintainers of the software. A measurement program can be more successful if it is designed with the goals of the project in mind. The GQM approach provides a framework involving three steps:

1. List the major goals of the development or maintenance project.

2. Derive from each goal the questions that must be answered to determine whether the goals are being met.

3. Decide what must be measured in order to be able to answer the questions adequately.

By deriving the measurements in this way, it becomes clear how to use the resulting data. As an example, Figure 3.2 illustrates how several metrics might be generated from a single goal.

Suppose your overall goal is to evaluate the effectiveness of using a coding standard, as shown in Figure 3.2. That is, you want to know whether

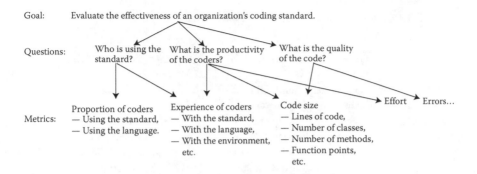

Goal: Evaluate the effectiveness of an organization's coding standard.

Questions: Who is using the standard? What is the productivity of the coders? What is the quality of the code?

Metrics:

Proportion of coders
— Using the standard,
— Using the language.

Experience of coders
— With the standard,
— With the language,
— With the environment, etc.

Code size
— Lines of code,
— Number of classes,
— Number of methods,
— Function points, etc.

Effort   Errors...

FIGURE 3.2   Example of deriving metrics from goals and questions.

code produced by following the standard is superior in some way to code produced without it. To decide whether the standard is effective, you must ask several key questions. First, it is important to know who is using the standard, so that you can compare the productivity of the coders who use the standard with the productivity of those who do not. Likewise, you probably want to compare the quality of the code produced with the standard with the quality of non-standard code.

Once these questions are identified, you must analyze each question to determine what must be measured in order to answer the question. For example, to understand who is using the standard, it is necessary to know what proportion of coders is using the standard. However, it is also important to have an experience profile of the coders, explaining how long they have worked with the standard, the environment, the language, and other factors that will help to evaluate the effectiveness of the standard. The productivity question requires a definition of productivity, which is usually some measure of effort divided by some measure of product size. As shown in Figure 3.2, the metric can be in terms of LOCs, function points, or any other metric that will be useful to you. Similarly, quality may be measured in terms of the number of errors found in the code, plus any other quality measures that you would like to use.

In this way, you generate only those measures that are related to the goal. Notice that, in many cases, several measurements may be needed to answer a single question. Likewise, a single measurement may apply to more than one question. The goal provides the purpose for collecting the data, and the questions tell you and your project how to use the data.

TABLE 3.3   Examples of AT&T Goals, Questions, and Metrics

Goal	Questions	Metrics
Plan	How much does the inspection process cost?	Average effort per KLOC Percentage of reinspections
	How much calendar time does the inspection process take?	Average effort per KLOC Total KLOC inspected
Monitor and control	What is the quality of the inspected software?	Average faults detected per KLOC Average inspection rate Average preparation rate
	To what degree did the staff conform to the procedures?	Average inspection rate Average preparation rate Average lines of code inspected Percentage of reinspections
	What is the status of the inspection process?	Total KLOC inspected
Improve	How effective is the inspection process?	Defect removal efficiency Average faults detected per KLOC Average inspection rate Average preparation rate Average lines of code inspected
	What is the productivity of the inspection process?	Average effort per fault detected Average inspection rate Average preparation rate Average lines of code inspected

EXAMPLE 3.8

AT&T used GQM to help determine which metrics were appropriate for assessing their inspection process (Barnard and Price 1994). Their goals, with the questions and metrics derived, are shown in Table 3.3.

What is not evident from the GQM tree or table is the model needed to combine the measurements in a sensible way so that the questions can be answered. For example, the tree in Figure 3.2 suggests measuring coder productivity; this attribute may be measured in terms of effort per LOC, but that relationship is not explicit in the tree. Thus, one or more models that express the relationships among the metrics must supplement the GQM approach.

EXAMPLE 3.9

Once AT&T researchers and developers generated the list of metrics in Example 3.8, they specified metrics equations and the data items that describe what the metrics really mean. For instance, the average preparation rate is a function of the total number of lines of code inspected, the preparation time for each inspection, and the number of inspectors. A model of the metric expresses the average preparation rate as an equation. First, the preparation time for each inspection is divided by the number of inspectors; then, the sum over all inspections is calculated and used to normalize the total number of lines of code.

Even when the metrics are expressed as an equation or relationship, the definition is not always clear and unambiguous. The tree does not tell you how to measure a line of code or a function point, only that some measure of code size is needed to answer a question about productivity. Additional work is needed to define each metric. In cases where no objective measure is available, subjective measures must be identified.

In general, typical goals are expressed in terms of productivity, quality, risk, and customer satisfaction and the like, coupled with verbs expressing the need to assess, evaluate, improve, or understand. It is important that the goals and questions be understood in terms of their audience: a productivity goal for a project manager may be different from that for a department manager or corporate director. To aid in generating the goals, questions, and metrics, Basili and Rombach provided a series of templates:

Templates for goal definition:

- *Purpose:* To (characterize, evaluate, predict, motivate, etc.) the (process, product, model, metric, etc.) in order to (understand, assess, manage, engineer, learn, improve, etc.) it.

  Example: To evaluate the maintenance process in order to improve it.

- *Perspective:* Examine the (cost, effectiveness, correctness, defects, changes, product measures, etc.) from the viewpoint of the (developer, manager, customer, etc.)

  Example: Examine the *cost* from the viewpoint of the *manager.*

- *Environment:* The environment consists of the following: process factors, people factors, problem factors, methods, tools, constraints, etc.

  Example: The maintenance staff consists of poorly motivated programmers who have limited access to tools.

A wide range of goals has been derived along with questions and associated metrics to develop measurement programs in industry.

EXAMPLE 3.10

Fuggetta et al. report on the use of GQM on a software process at the Digital Software Engineering Center in Gallarate, Italy (Fuggetta et al. 1998). The group identified five goals including the following goal, which has a format consistent with our templates:

> Analyze the design and qualification phases of the development process for the purpose of evaluating failure detection effectiveness from the viewpoint of the management and the development team.

The measurement group derived 35 questions that could be answered using data from approximately 50 metrics.

Guidelines are available to help in defining both product-related questions and process-related questions (Basili and Rombach 1988; van Solingen and Berghout 1999). Steps involve defining the process or product, defining the relevant attributes, and obtaining feedback related to the attributes. What constitutes a goal or question may be vague, and several levels of refinement may be required for certain goals and questions; that is, a goal may first have to be related to a series of subgoals before questions can be derived.

We can relate the GQM templates to the attribute framework introduced in Section 3.1. A goal or question can be associated with at least one pair of entities and attributes. Thus, a goal is stated, leading to a question that should be answered so that we can tell whether we have met our goal; the answer to the question requires that we measure some attribute of an entity (and possibly several attributes of an entity, or several attributes of several entities). The use of the measure is determined by the goals and questions, so that assessment, prediction, and motivation are tightly linked to the data analysis and reporting.

Thus, GQM complements the entity-attribute measurement framework. The results of a GQM analysis are a collection of measurements related by goal tree and overall model. However, GQM does not address issues of measurement scale, objectivity, or feasibility. So the GQM measures should be used with care, remembering the overall goal of providing useful data that can help to improve our processes, products, and resources.

3.2.2 Measurement for Process Improvement

One common goal in the software industry is process improvement. Software processes can range from chaotic and ad hoc, to well defined and well managed. A more mature process is more likely to develop software that is reliable, adaptable, delivered on time, and within budget. Measurement quantifies the relationships between the processes, products, resources, methods, and technologies of software development. Thus, measurement plays a key role in evaluating and improving software processes. We will look at a popular process evaluation technique, the Software Engineering Institute's (SEI's) Capability Maturity Model Integration (CMMI®) for Development, from the perspective of GQM (CMMI Product Team 2010).

The CMMI for Development provides an ordinal ranking of development organizations from *initial* (the least predictable and controllable, and least understood) to *optimizing* (the most predictable and controllable), which is described by the CMMI Product Team as follows:

1. *Initial*: Level 1 processes are ad hoc and "success depends on the competence and heroics of the people in the organization."

2. *Managed*: Level 2 processes are planned; "the projects employ skilled people ... have adequate resources ... involve relevant stakeholders; are monitored, controlled, and reviewed ...."

3. *Defined*: Level 3 "processes are well characterized and understood, and are described in standards procedures, tools, and methods."

4. *Quantitatively managed*: A Level 4 "organization and projects establish quantitative objectives for quality and process performance and use them as criteria in managing projects."

5. *Optimizing*: A Level 5 "organization continually improves its processes based on a quantitative understanding of its business objectives and performance needs" (CMMI Product Team 2010).

The SEI CMMI distinguishes one level from another in terms of key process activities going on at each level. Although it appears that measurement is only important at Level 4, actually measurement is used in evaluation at each level. Specific goals, questions, and metrics are developed to assess whether an organization has reached a particular level. Generally, to reach a particular level, an organization must have measurement values that give the desired answer to each question at that level.

For example, to reach CMMI-Development version 1.3 Level 2 Managed, a process must satisfy 15 goals in 7 process areas:

1. *Configuration management goals*: Establish baselines, track and control changes, establish integrity.

2. *Measurement and analysis goals*: Align measurement and analysis activities, provide measurement results.

3. *Project monitoring and control goals*: Monitor project against plan, manage corrective actions to closure.

4. *Project planning goals*: Establish estimates, develop a project plan, and obtain commitment to the plan.

5. *Process and quality assurance goals*: Objectively evaluate processes and work products, provide objective insight.

6. *Requirements management goals*: Manage requirements.

7. *Supplier agreement management goals*: Establish supplier agreements, satisfy supplier agreements.

Answers to questions related to the goals in each process area determine whether a goal is achieved. These questions tend to be measured by yes or no answers—either a process performs an activity at a required level or it does not. For example, answers to the following questions determine whether configuration management process area goals are met:
Does the development process

1. Identify configuration items?

2. Establish a configuration management system?

3. Create or release baselines?

4. Track change requests?

5. Control changes to configuration items?

6. Establish configuration management records?

7. Perform configuration audits?

Trained evaluators determine whether each question is answered yes or no based on interviews with process participants and evaluations of process documents. Evaluators determine whether the process performs each activity as required for Level 2 certification. To achieve a Level 2 rating, the answer must be yes to questions concerning 65 practices related to 17 goals.

To achieve Level 3 Defined, an organization must satisfy all Level 2 goals, plus 27 additional goals in 11 process areas. The answers to questions concerning 88 different practices determine whether the goals are met. Levels 4 and 5 add additional process areas, goals, and questions about practices.

The CMMI and other models, such as ISO-9000, share a common goal and approach, namely that they use process visibility as a key discriminator among a set of "maturity levels." That is, the more visibility into the overall development process, the higher the maturity and the better managers and developers can understand and control their development and maintenance activities. At the lowest levels of maturity, the process is not well understood at all; as maturity increases, the process is better understood and better defined. At each maturity level, measurement and visibility are closely related: a developer can measure only what is visible in the process, and measurement helps to enable and increase visibility. Thus, the five-level maturity scale such as the one employed by the CMMI is a convenient context for determining what to measure first and how to plan an increasingly comprehensive measurement program.

Successful metrics programs start small and grow according to the goals and needs of a particular project (Rifkin and Cox 1991). To be successful, a metrics program should be planned in context with an organization's "processes, structures, climate, and power" (Frederiksen and Mathiassen 2005). A metrics program should begin by addressing the critical problems or goals of the project, viewed in terms of what is meaningful or realistic at the project's maturity level. The process maturity framework then acts as a guideline for how to expand and build a metrics program that not only takes advantage of visibility and maturity but also enhances process improvement activities.

Next, we explain how to use the process maturity framework, coupled with understanding of goals, to build a comprehensive measurement program.

3.2.3 Combining GQM with Process Maturity

Suppose you are using the Goal-Question-Metric paradigm to decide what your project should measure. You may have identified at least one of the following high-level goals:

- Improving productivity

- Improving quality

- Reducing risk

Within each category, you can represent the goal's satisfaction as a set of subgoals, each of which can be examined for its implications for resources, products, and process. For example, the goal of improving productivity can be interpreted as several subgoals affecting resources:

- Assuring adequate staff skills

- Assuring adequate managerial skills

- Assuring adequate host software engineering technology

Similarly, improving productivity with products can mean

- Identifying problems early in the life cycle

- Using appropriate technology

- Reusing previously built products

Next, for each subgoal, you generate questions that reflect the areas of deepest concern. For example, if improving productivity is your primary goal, and then assuring adequate staff skills is an important subgoal, you create a list of questions that you may be interested in having answered, such as:

1. Does project staffing have the right assortment of skills?

2. Do the people on the project have adequate experience?

Similarly, if you have chosen improving quality with a subgoal of improving the quality of the requirements, then the related questions might include:

1. Is the set of requirements clear and understandable?

2. Is the set of requirements testable?

3. Is the set of requirements reusable?

4. Is the set of requirements maintainable?

5. Is the set of requirements correct?

However, before you identify particular measurements to help you answer these questions, it is useful to determine your process maturity level. Since process maturity suggests that you measure only what is visible, the incorporation of process maturity with GQM paints a more comprehensive picture of what measures will be most useful to you.

For example, suppose you want to answer the question: Is the set of requirements maintainable?

If you have specified a process at Level 1, then the project is likely to have ill-defined requirements. Measuring requirements characteristics is difficult at this level, so you may choose to count the number of requirements and changes to those requirements to establish a baseline. If your process is at Level 2, the requirements are well defined and you can collect additional information: the type of each requirement (database requirements, interface requirements, performance requirements, etc.) and the number of changes to each type. At Level 3, your visibility into the process is improved, and intermediate activities are defined, with entry and exit criteria for each activity. For this level, you can collect a richer type of measurement: measuring the traceability of each requirement to the corresponding design, code, and test components, for example, and noting the effects of each change on the related components. Thus, the goal and question analysis is the same, but the metric recommendations vary with maturity. The more mature your process, the richer your measurements. In other words, the more mature your process, the more mature your measurements.

Moreover, maturity and measurement work hand-in-hand. As the measurements provide additional visibility, aiding you in solving project problems, you can use this approach again to expand the measurement set and address other pressing problems. Thus, your measurement program can start small and grow carefully, driven by process and project needs.

Using goals to suggest a metrics program has been successful in many organizations and is well documented in the literature. (For examples, see Grady and Caswell (1987), Rifkin (1991), Pfleeger (1993), Mendonça and Basili (2000), and van Solingen and Berghout (1999).) The GQM paradigm, in concert with process maturity, has been used as the basis for several tools that assist managers in designing measurement programs (Pulford et al. 1995).

GQM has helped us to understand why we measure an attribute, and process maturity suggests whether we are capable of measuring it in a meaningful way. Together, they provide a context for measurement. Without such a context, measurements can be used improperly or inappropriately, giving us a false sense of comfort with our processes and products. In Section 3.3, we look at the need for care in capturing and evaluating measures.

3.3 APPLYING THE FRAMEWORK

In Chapter 1, we introduced several diverse topics involving software measurement. In particular, we saw how the topics relate to essential software engineering practices. When divorced from any conceptual high-level view of software measurement and its objectives, many of the activities may have seemed unrelated. In this section, we revisit the topics to see how each fits into our unifying framework and to describe some of the issues that will be addressed in Chapters 8 through 11. That is, we look at which processes, products, and resources are relevant in each case, which attributes (and whether these are internal or external) we are measuring, and whether the focus is on assessment or prediction. The discussion here is not intended to be a detailed explanation of the various topics. Instead, we will indicate where the details will be provided later in the book.

3.3.1 Cost and Effort Estimation

Cost and effort estimation focuses on predicting the attributes of cost or effort for the development process. Here, the process includes activities from detailed specification through implementation. Most of the estimation techniques present a model; we examine a popular approach to see how cost and effort estimation are usually done.

EXAMPLE 3.11

Boehm's original, basic COCOMO model asserts that the effort required for the process of developing a software system (measured by E person-months) and size (measured by S thousands of delivered source statements) are related by

$$E = aS^b$$

where a and b are parameters determined by the type of software system to be developed (Boehm 1981).

The model is intended for use during requirements capture, when estimates of effort are needed. To use the technique, we must determine parameters a and b. Boehm provides three choices, which are based on the type of the software system. We must also determine the size S of the eventual system; since the system has yet to be built, we must predict S in order to predict E.

Thus, it is more correct to view COCOMO as a prediction system rather than as a model. The model is expressed as the equation in Example 3.11. But the prediction system includes the inference procedures for calculating the model's parameters. The calculations involve a combination of calibration based on past history, assessment based on expert judgment, and the subjective rating of attributes. The underlying theory provides various means for interpreting the results based on the choice of parameters. It is ambiguous to talk of *the* COCOMO cost estimation model, since the same data can produce results that vary according to the particular prediction procedures used. These observations apply to most cost and effort estimation models.

Cost models often reflect a variety of attributes, representing all of the entity types. For example, for more advanced versions of COCOMO as well as other cost estimation approaches, the inference procedures involve numerous internal and external attributes of the requirements specification (a product), together with process and resource attributes subjectively provided by users.

Both basic COCOMO and Albrecht's function point cost estimation model (described in more detail in Chapter 8) assert that effort is an indirect measure of a single product attribute: size. In Albrecht's model, the number of function points, an attribute of the specifications, measures size. Here "size" really means a more specific attribute, namely functionality. In the latest version of COCOMO, called COCOMO II, size is defined in different ways, depending on when during the development process the estimate is being made (Boehm et al. 2000). Early on, size is described in terms of object points, which can be derived from a prototype or initial specification. As more is known about the system, function points provide the size measure. When yet more information is available, size is defined as the number of thousands of delivered source statements, an attribute of the final implemented system. Thus, to use the model for effort prediction, you must predict this product attribute at the specification phase; that is, using the model involves predicting an attribute of a future product in order to predict effort. This approach can be rather unsatisfactory, since predicting size may be just as hard as predicting effort. In addition, the number of source statements captures only one specific view of size. In Chapter 8, we propose several ways to address these problems.

3.3.2 Productivity Measures and Models

To analyze and model productivity, we must measure a resource attribute, namely the number of personnel (either as teams or individuals) active during particular processes. The most commonly used model for productivity measurement expresses productivity as the ratio "process output influenced by the personnel" divided by "personnel effort or cost during the process." In this equation, productivity is viewed as a resource attribute, captured as a derived measure of a product attribute measure and a process attribute measure.

Inspired by Japanese notions of *spoilage*, where engineers measure how much effort is spent fixing things that could have been put right before delivery, productivity of software engineers can also incorporate the idea of cost of fault prevention compared with the cost of fault detection and correction.

EXAMPLE 3.12

Watts Humphrey has defined a Personal Software Process (PSP) that encourages software engineers to evaluate their individual effectiveness at producing quality code (Humphrey 1995, 2005). As part of the PSP, Humphrey suggests that we capture time and effort information about appraisal, failure, and prevention. The appraisal cost is measured as the percentage of development time spent in design and test reviews. The failure cost is the percentage of time spent during compilation and testing, while the prevention cost is that time spent preventing defects before they occur (in such activities as prototyping and formal specification). The ratio of appraisal cost to failure cost then tells us the relative effort spent in early defect removal. Such notions help us to understand how to improve our productivity.

3.3.3 Data Collection

Even the simplest models depend on accurate measures. Often, product measures may be extracted with a minimum of human intervention, since the products can be digitized and analyzed automatically. But automation is not usually possible for process and resource measures. Much of the work involved in data collection is therefore concerned with how to set in place rigorous procedures for gathering accurate and consistent measures of process and resource attributes.

Chillarege and his colleagues at IBM have suggested a scheme for collecting defect information that adds the rigor usually missing in such data capture. In their scheme, the defects are classified *orthogonally*, meaning that any defect falls in exactly one category (Chillarege et al. 1992; Huber 2000). Such consistency of capture and separation of classes of defects by causes allow us to analyze not only the product attributes but also process effectiveness.

3.3.4 Quality Models and Measures

Quality models usually involve product measurement, as their ultimate goal is predicting product quality. We saw in Chapter 1 that both cost and productivity are dependent on the quality of products output during various processes. The productivity model in Figure 1.5 explicitly considers the quality of output products; it uses an internal process attribute, defects discovered, to measure quality. The quality factors used in most quality models usually correspond to external product attributes. The criteria into which the factors are broken generally correspond to internal product or process attributes. The metrics that measure the criteria then correspond to proposed measures of the internal attributes. In each case, the terms used are general and are usually not well defined and precise.

Gilb has suggested a different approach to quality. He breaks high-level quality attributes into lower-level attributes; his work is described in Chapter 10 (Gilb 1988).

3.3.5 Reliability Models

Reliability is a high-level, external product attribute that appears in all quality models. The accepted view of reliability is described as the likelihood of successful operation, so reliability is a relevant attribute only for executable code. Many people view reliability as the single most-important quality attribute to consider, so research on measuring and predicting reliability is sometimes considered a separate sub-discipline of software measurement.

In Chapter 10, we consider proposals to measure reliability using internal process measures, such as number of faults found during formal testing or mean time to failure during testing. We also describe in Chapter 10 the pitfalls of such an approach. In Chapter 11, we describe how reliability theory is applied to software development.

As we have seen, many developers observe failures (and subsequent fixes) during operation or testing, and then use this information to determine the probability of the next failure (which is a random variable). Measures of reliability are defined in terms of the subsequent distributions (and are thus defined in terms of process attributes). For example, we can consider the distribution's mean or median, or the rate of occurrence of failures. But this approach presents a prediction problem: We are trying to say something about future reliability on the basis of past observations.

EXAMPLE 3.13

The Jelinski–Moranda model for software reliability assumes an exponential probability distribution for the time of the ith failure. The mean of this distribution (and hence the mean time to ith failure, or $MTTF_i$) is given by

$$MTTF_i = \frac{\alpha}{N - i + 1}$$

where N is the number of faults assumed to be initially contained in the program, and $1/\alpha$ represents the size of a fault, that is, the rate at which it causes a failure. (Faults are assumed to be removed on observation of a failure, and each time the rate of occurrence of failures is reduced by $\phi = 1/\alpha$.) The unknown parameters of the model, N and α, must somehow be estimated; this estimation can be done using, for example, Maximum Likelihood Estimation after observing a number of failure times.

In this example, we cannot discuss only the Jelinski–Moranda model, we must specify a prediction system as well. In this case, the prediction system supplements the model with a statistical inference procedure for determining the model parameters, and a prediction procedure for combining the model and the parameter estimates to make statements about future reliability.

3.3.6 Structural and Complexity Metrics

Many high-level quality attributes (i.e., external product attributes) are notoriously difficult to measure. For reasons to be made clear in Chapter 8, we are often forced to consider measures of internal attributes of products as weaker substitutes. The Halstead measures are examples of this situation; defined on the source code, the internal measures suggest what the external measures might be. Numerous other measures have been proposed that are defined on graphical models of either the source code or design. By measuring specific internal attributes like control-flow structure, information flow, and number of paths of various types, one may attempt to quantify those aspects of the code that make the code difficult to understand. Unfortunately, much work in this area has been obfuscated by suggestions that such measures capture directly the external attribute of complexity. In other words, many software engineers claim that a single internal attribute representing complexity can be an accurate predictor of many external quality attributes, as well as of process attributes like cost and effort. We discuss this problem and suggest a rigorous approach to defining structural attributes in Chapter 9.

3.3.7 Management by Metrics

Managers often use metrics to set targets for their development projects. These targets are sometimes derived from best practice, as found in a sample of existing projects.

EXAMPLE 3.14

The US Department of Defense analyzed US government and industry performance, grouping projects into low, median, and best-in-class ratings. From their findings, the analysts recommended targets for Defense Department-contracted software projects, along with indicated levels of performance constituting management malpractice, shown in Table 3.4 (NetFocus 1995). That is, if a project demonstrates that an attribute is below the malpractice level, then something is seriously wrong with the project.

The management metrics are a mixture of measurement types. Defect removal efficiency is a process measures, while defect density, requirements creep (i.e., an increase or change from the original set of requirements) and program documentation are product measures. Cost overrun and staff turnover capture resource attributes.

TABLE 3.4 Quantitative Targets for Managing US Defense Projects

Item	Target	Malpractice Level
Defect removal efficiency	>95%	<70%
Original defect density	<4 per function point	>7 per function point
Slip or cost overrun in excess of risk reserve	0%	≥10%
Total requirements creep (function points or equivalent)	<1% per month average	≥50%
Total program documentation	<3 pages per function point	>6 pages per function point
Staff turnover	1–3% per year	>5% per year

Notice that the measures in Example 3.14 are a mixture of measurement types. Moreover, none of the measures is external. That is, none of the measures reflects the actual product performance as experienced by the user. This situation illustrates the more general need for managers to understand likely product quality as early in the development process as possible. But, as we shall see in the remainder of the book, unless the relationship between internal and external attributes is well understood, such management metrics can be misleading.

3.3.8 Evaluation of Methods and Tools

Often, organizations consider investing in a new method or tool but hesitate because of uncertainty about cost, utility, or effectiveness. Sometimes the proposed tool or method is tried first on a small project, and the results are evaluated to determine whether further investment and broader implementation are in order. For example, we saw in Example 3.8 that AT&T used goals and questions to decide what to measure in evaluating the effectiveness of its inspections.

EXAMPLE 3.15

The results of the AT&T inspection evaluation are summarized in Table 3.5 (Barnard and Price 1994).

For the first sample project, the researchers found that 41% of the inspections were conducted at a rate faster than the recommended rate of 150 lines of code per hour. In the second project, the inspections with rates below 125 found an average of 46% more faults per KLOC than those with faster rates. This finding means either that more faults can be found when inspection rates are slower, or that finding more faults causes the inspection rate to slow. Here, the metrics used to support this analysis are primarily process metrics.

TABLE 3.5   Code Inspection Statistics from AT&T

Metric	First Sample Project	Second Sample Project
Number of inspections in sample	27	55
Total KLOC inspected	9.3	22.5
Average LOC inspected (module size)	343	409
Average preparation rate (LOC/h)	194	121.9
Average inspection rate (LOC/h)	172	154.8
Total faults detected (observable and non-observable) per KLOC	106	89.7
Percentage of re-inspections	11	0.5

These examples show us that it takes all types of attributes and metrics—process, product, and resource—to understand and evaluate software development.

3.4 SOFTWARE MEASUREMENT VALIDATION

The very large number of software measures in the literature aims to capture information about a wide range of attributes. Even when you know which entity and attribute you want to assess, there are many measures from which to choose. Finding the best measure for your purpose can be difficult, as candidates measure or predict the same attribute (such as cost, size or complexity) in very different ways. So it is not surprising when managers are confused by measurement: they see different measures for the same thing, and sometimes the implications of one measure lead to a management decision opposite to the implications of another! One of the roots of this confusion is the lack of software measurement validation. That is, we do not always stop to ensure that the measures we use actually capture the attribute information we seek.

The formal framework presented in Chapter 2 and in this chapter leads to a formal approach for validating software measures. The validation approach depends on distinguishing measurement from prediction, as discussed in Chapter 2. That is, we must separate our concerns about two types of measuring:

1. *Measures or measurement systems* are used to assess an existing entity by numerically characterizing one or more of its attributes.

2. *Prediction systems* are used to predict some attribute of a future entity, involving a mathematical model with associated prediction procedures.

Informally, we say that a *measure* is "valid" if it accurately characterizes the attribute it claims to measure. However, a *prediction system* is "valid" if it makes accurate predictions. So not only are measures different from prediction systems, but the notion of validation is different for each. Thus, to understand why validation is important and how it should be done, we consider measures and prediction systems separately.

3.4.1 Validating Prediction Systems

Validating a prediction system in a given environment is the process of establishing the accuracy of the prediction system by empirical means; that is, by comparing model performance with known data in the given environment.

Thus, validation of prediction systems involves experimentation and hypothesis testing, as we shall see in Chapter 4. Rather than being a mathematical proof, validation involves confirming or refuting a hypothesis.

This type of validation is well accepted by the software engineering community. For example, researchers and practitioners use data sets to validate cost estimation or reliability models. If you want to know whether COCOMO is valid for your type of development project, you can use data that represent that type of project and assess the accuracy of COCOMO in predicting effort and duration.

The degree of accuracy acceptable for validation depends on several things, including the person doing the assessment. We must also consider the difference between *deterministic* prediction systems (we always get the same output for a given input) and *stochastic* prediction systems (the output for a given input will vary probabilistically) with respect to a given model.

Some stochastic prediction systems are more stochastic than others. In other words, the error bounds for some systems are wider than in others. Prediction systems for software cost estimation, effort estimation, schedule estimation and reliability are very stochastic, as their margins of error are large. For example, Boehm has stated that under certain circumstances the COCOMO effort prediction system will be accurate to within 20%; that is, the predicted effort will be within 20% of the actual effort value. An *acceptance range* for a prediction system is a statement of the maximum difference between prediction and actual value. Thus, Boehm specifies 20% as the acceptance range of COCOMO. Some project managers find this range to be too large to be useful for planning, while other project managers find 20% to be acceptable, given the other uncertainties of software development. Where no such range has been specified, you must state in advance what range is acceptable before you use a prediction system.

EXAMPLE 3.16

Sometimes the validity of a complex prediction system may not be much greater than that of a very simple one. For example, Norbert Fuchs points out that if the weather tomorrow in Austria is always predicted to be the same as today's weather, then the predictions are accurate 67% of the time. The use of sophisticated computer models increases this accuracy to just 70%!

In Chapter 10, we present a detailed example of how to validate software reliability prediction systems using empirical data.

3.4.2 Validating Measures

Measures used for assessment are the measures discussed in Chapter 2. We can turn to measurement theory to tell us what validation means in this context:

> *Validating a software measure* is the process of ensuring that the measure is a proper numerical characterization of the claimed attribute by showing that the representation condition is satisfied.

EXAMPLE 3.17

We want to measure the length of a program in a valid way. Here, "program" is the entity and "length" the attribute. The measure we choose must not contradict any intuitive notions about program length. Specifically, we need both a formal model that describes programs (to enable objectivity and repeatability) and a numerical mapping that preserves our intuitive notions of length in relations that describe the programs. For example, if we concatenate two programs P_1 and P_2, we get a program whose length is the combined lengths of P_1 and P_2. Thus, we expect any measure m of length always to satisfy the condition

$$m(P_1, P_2) = m(P_1) + m(P_2)$$

If program P_1 has a greater length than program P_2, then any measure m of length must also satisfy

$$m(P_1) > m(P_2)$$

We can measure program length by counting lines of code (in the carefully defined way we describe in Chapter 8). Since this count preserves these relationships, lines of code is a valid measure of length. We will also describe a more rigorous length measure in Chapter 8, based on a formal model of programs. This form of validation has been applied to measures of *coupling* and *cohesion* of object-oriented software (Briand et al. 1998, 1999), and to the measurement of diagnosability and *vigilance*, which are properties of a software design related to testability (Le Traon et al. 2003; Le Traon et al. 2006).

This type of validation is central to the representational theory of measurement. That is, we want to be sure that the measures we use reflect the behavior of entities in the real world. If we cannot validate the measures, then we cannot be sure that the decisions we make based on those measures will have the effects we expect. In some sense, then, we use validation to make sure that the measures are defined properly and are consistent with the entity's real-world behavior.

3.4.3 A Mathematical Perspective of Metric Validation

In Chapter 2, we discussed the theory of measurement, explaining that we need not use the term "metric" in our exposition. There is another, more formal, reason for using care with the term. In mathematical analysis, a metric has a very specific meaning: it is a rule used to describe how far apart two points are. More formally, a *metric* is a function m defined on pairs of objects x and y such that $m(x,y)$ represents the distance between x and y. Such metrics must satisfy certain properties:

$m(x,x) = 0$ for all x: that is, the distance from a point to itself is 0.

$m(x,y) = m(y,x)$ for all x and y: that is, the distance from x to y is the same as the distance from y to x.

$m(x,z) \leq m(x,y) + m(y,z)$ for all x, y and z: that is, the distance from x to z is no larger than the distance measured by stopping through an intermediate point.

There are numerous examples where we might be interested in "mathematical" metrics in software:

EXAMPLE 3.18

Fault tolerant techniques like N-version programming have been proposed for increasing the reliability of safety-critical systems. The approach involves developing N different versions of the critical software components independently. Theoretically, by having each of the N different teams solving the same problem without knowledge of what the other teams are doing, the probability of all the teams, or even of the majority, making the same error is kept small. When the behavior of the different versions differs, a voting procedure accepts the behavior of the majority of the systems. The assumption, then, is that the correct behavior will always be chosen.

However, there may be problems in assuring genuine design independence, so we may be interested in measuring the level of diversity between two designs, algorithms or programs. We can define a metric m, where $m(P_1, P_2)$ measures the diversity between two programs P_1 and P_2. In this case, the entities being measured are products. Should we use a similar metric to measure the level of diversity between two methods applied during design, we would be measuring attributes of process entities.

EXAMPLE 3.19

We would hope that every program satisfies its specification completely, but this is rarely the case. Thus, we can view program correctness as a measure of the extent to which a program satisfies its specification, and define a metric $m(S,P)$ where the entities S (specification) and P (program) are both products. Then $m(S,P)$ indicates the distance between the specification and a program that implements the specification.

To reconcile these mathematically precise metrics with the framework we have proposed, we can consider pairs of entities as a single entity. For example, having produced two programs satisfying the same specification, we consider the pair of programs to be a single product system, itself having a level of diversity. This approach is consistent with a systems view of N-version programming. Where we have implemented N versions of a program, the diversity of the system may be viewed as an indirect measure of the pairwise program diversity.

3.5 PERFORMING SOFTWARE MEASUREMENT VALIDATION

The software engineering community has always been aware of the need for validation. As new measures are proposed, it is natural to ask whether the measure captures the attribute it claims to describe. But in the past, validation has been a relaxed process, sometimes relying on the credibility of the proposer, rather than on rigorous validation procedures. That is, if someone of stature says that measure X is a good measure of complexity, then practitioners and researchers begin to use X without question. Or software engineers adopt X after reading informal arguments about why X is probably a good measure. This situation can be remedied only by reminding the software community of the need for rigorous validation.

An additional change in attitude is needed, though. Some researchers and practitioners assume that validation of a measure (in the measurement theory sense) is not sufficient assurance that the measure captures the appropriate characteristic of the entity under scrutiny. They expect the validation to demonstrate that the measure is itself part of a valid prediction system.

EXAMPLE 3.20

Many people use lines of code as a measure of software size. The measure has many uses: it is a general size measure in its own right, it can be used to normalize other measures (like number of faults) to enable comparison, and it is input to many derived measures such as productivity (measured as effort divided by lines of code produced). But some software engineers claim that lines of code is not a valid software measure because it is not a good predictor of reliability or maintenance effort.

Thus, a measure must be viewed in the context in which it will be used. Validation must take into account the measurement's purpose; measure X may be valid for some uses but not for others.

3.5.1 A More Stringent Requirement for Validation

As we have seen with lines of code, it is possible for a measure to serve both purposes: to measure an attribute in its own right, and to be a valuable input to a prediction system. But a measure can be one or the other without being both. We should take care not to reject as invalid a reasonable measure just because it is not a predictor. For example, there are many internal product attributes (such as size, structuredness, and modularity) that are useful and practical measures, whether they are also part of a prediction system. If a measure for assessment is valid, then we say that it is *valid in the narrow sense* or is *internally valid*. In designing empirical studies, as shown in Chapter 4, all variables must be internally valid to demonstrate that a study has *construct validity*.

Many process attributes (such as cost), external product attributes (such as reliability), and resource attributes (such as productivity) play a dual role. We say that a measure is *valid in the wide sense* if it is both

1. Internally valid, and

2. A component of a valid prediction system.

Suppose we wish to show that a particular measure is valid in the wide sense. After stating a hypothesis that proposes a specific relationship between our measure and some useful attribute, we must conduct an experiment to test the hypothesis, as described in Chapter 4. Unfortunately, in practice the experimental approach is not often taken. Instead, the measure is claimed to be valid in the wide sense by demonstrating a statistical correlation with another measure. For example, a measure of modularity might be claimed to be valid or invalid on the basis of a comparison with known development costs. This validation is claimed even though there is no demonstrated explicit relationship between modularity and development costs! In other words, our measure cannot be claimed to be a valid measure of modularity simply because of the correlation with development costs. We require a causal model of the relationship between modularity and development costs, showing explicitly all the factors interconnecting each. Only then can we judge whether measuring modularity is the same as measuring development cost.

This type of mistake is common in software engineering, as well as in other disciplines. Engineers forget that statistical correlation does not imply cause and effect. For example, suppose we propose to measure obesity using inches, normally a length measure. We measure the height and weight of each of a large sample of people, and we show that height correlates strongly with weight. Based on that result, we cannot say that height is a valid measure of obesity, since height is only one of several factors that must be taken into account; heredity and body fat percentages play a role, for instance. In other words, just because height correlates with weight does not mean that height is a *valid measure* of weight or obesity. Likewise, there may be a statistical correlation between modularity and development costs, but that does not mean that modularity is the *only* factor determining development cost.

It is sometimes possible to show that a measure is valid in the wide sense without a stated hypothesis and planned experiment. Available data can be examined using techniques such as regression analysis; the measure may be consistently related (as shown by the regression formula) to another variable. However, this approach is subject to the same problems as the correlation approach. It requires a model showing how the various factors interrelate. Given our poor understanding of software, this type of validation appears fraught with difficulty. Nevertheless, many software engineers continue to use this technique as their primary approach to validation.

The many researchers who have taken this approach have not been alone in making mistakes. It is tempting to measure what is available and easy to measure, rather than to build models and capture complex relationships. Indeed, speaking generally about measurement validation, Krantz asserts:

> A recurrent temptation when we need to measure an attribute of interest is to try to avoid the difficult theoretical and empirical issues posed by fundamental measurement by substituting some easily measured physical quantity that is believed to be strongly correlated with the attribute in question: hours of deprivation in lieu of hunger; skin resistance in lieu of anxiety; milliamperes of current in lieu of aversiveness, etc. Doubtless this is a sensible thing to do when no deep analysis is available, and in all likelihood some such indirect measures will one day serve very effectively when the basic attributes are well understood, but to treat them now as objective definitions of unanalyzed concepts is a form of misplaced operationalism.
>
> Little seems possible in the way of careful analysis of an attribute until means are devised to say which of two objects or events exhibits more of the attribute. Once we are able to order the objects in an acceptable way, we need to examine them for additional structure. Then begins the search for qualitative laws satisfied by the ordering and the additional structure.
>
> KRANTZ 1971

Neither we nor Krantz claims that measurement and prediction are completely separate issues. On the contrary, we fully support the observation of Kyburg:

> If you have no viable theory into which X enters, you have very little motivation to generate a measure of X.
>
> KYBURG 1984

However, we are convinced that our initial obligation in proposing measures is to show that they are valid in the narrow sense. Good predictive theories follow only when we have rigorous measures of specific, well-understood attributes.

3.5.2 Validation and Imprecise Definition

In Example 3.20, we noted that LOC is a valid measure of program size. However, LOC has not been shown convincingly to be a valid measure of complexity. Nor has it been shown to be part of an accurate prediction system for complexity. The fault lies not with the LOC measure but with the imprecise definition of complexity. Although complexity is generally described as an attribute that can affect reliability, maintainability, cost, and more, the fuzziness surrounding its definition presents a problem in complexity research. Chapter 9 explores whether complexity can ever be defined and measured precisely.

The problems with complexity do not prevent LOC from being a useful measure for purposes other than size. For example, if there is a stochastic association between a large number of LOC and a large number of unit testing errors, this relationship can be used in choosing a testing strategy and in reducing risks.

Many studies have demonstrated a significant correlation between LOC and the cyclomatic number. The researchers usually suggest that this correlation proves that cyclomatic number increases with size; that is, larger code is more complex code. However, careful interpretation of the measures and their association reveals only that the number of decisions increases with code length, a far less profound conclusion. The cyclomatic number may be just another size measure. Chapter 9 contains more detailed discussion of validation for the McCabe measures.

3.5.3 How Not to Validate

New measures are sometimes validated by showing that they correlate with well-known, existing measures. Li and Cheung present an extensive example of this sort (Li and Cheung 1987). This approach is appealing because:

- It is generally assumed that the well-known existing measure is valid, so that a good correlation means that the new measure must also be valid.

- This type of validation is straightforward, since the well-known measure is often easy to compute; automated tools are often available for computation.

- If the management is familiar with the existing measures, then the proposed measures have more credibility.

Well-known measures used in this way include Halstead's suite of software science measures, McCabe's cyclomatic number, and LOC. However, although these may be valid measures of very specific attributes (such as number of decisions for cyclomatic number, and source code program length for LOC), they have not been shown to be valid measures of attributes such as cognitive complexity, correctness, or maintainability. Thus, we must take great care in validating by comparison with existing measures. We must verify that the qualities associated with the existing measures have been demonstrated in the past.

There is much empirical evidence to suggest that measures such as Halstead's suite, McCabe's cyclomatic number, and LOC are associated with development and maintenance effort and faults. But such correlations do *not* imply that the measures are good predictors of these attributes. The many claims made that Halstead and McCabe measures have been validated are contradicted by studies showing that the correlations with process data are no better than using a simple measure of size, such as LOC (Hamer and Frewin 1982; Fenton and Neil 1999).

There is a more compelling scientific and statistical reason why we must be wary of the correlate-against-existing-measures approach. Unstructured correlation studies run the risk of identifying spurious associations. Using the 0.05 significance level, we can expect a significant but spurious correlation in 1 of 20 times by chance. Thus, if you have 5 independent variables and look at the 10 possible pairwise correlations, there is a 0.5 (1 in 2) chance of getting a spurious correlation. In situations like this, if we have no hypothesis about the reason for a relationship, we can have no real confidence that the relationship is not spurious. Courtney and Gustafson examine this problem in detail (Courtney and Gustafson 1993).

3.5.4 Choosing Appropriate Prediction Systems

To help us formulate hypotheses necessary for validating the predictive capabilities of measures, we divide prediction systems into the following classes:

Class 1. Using internal attribute measures of early life-cycle products to predict measures of internal attributes of later life-cycle products. For example, measures of size, modularity, and reuse of a specification are used to predict size and structuredness of the final code.

Class 2. Using early life-cycle process attribute measures and resource attribute measures to predict measures of attributes of later life-cycle processes and resources. For example, the number of faults found during formal design review is used to predict cost of implementation.

Class 3. Using internal product attribute measures to predict process attributes. For example, measures of structuredness are used to predict time to perform some maintenance task, or number of faults found during unit testing.

Class 4. Using process measures to predict later process measures. For example, measures of failures during one operational period are used to predict likely failure occurrences in a subsequent operational period. In examples like this, where an external product attribute (reliability) is effectively defined in terms of process attributes (operational failures), we may also think of this class of prediction systems as using process measures to predict later external product measures.

Class 5. Using internal *structural* attributes to predict external and process attributes. Examples of these prediction systems include using module coupling or other structural measures to predict the fault-proneness of a component, where fault-proneness is based on failures during testing or operation that are traced to module faults. These prediction systems tend to work only on the specific systems where the prediction system parameters are determined. We doubt that Class 5 prediction systems will work effectively in general. In theory, it is always possible to construct products that appear to exhibit identical external attributes but which in fact vary greatly internally.

However, we usually assume that certain internal attributes that result from modern software engineering techniques (such as modularization, low coupling, control and data structuredness, information hiding and reuse) will generally lead to products exhibiting a high degree of desirable external attributes like reliability and maintainability. Thus, programs and modules that have poor values of desirable internal attributes (such as large, unstructured modules) are likely (but not certain) to have more faults and take longer to produce and maintain. We return to this issue in Chapter 9.

3.6 SUMMARY

This chapter presents a framework to help us to discuss what to measure and how to use the measures appropriately. The key points to remember are the following:

- All entities of interest in software can be classified as either processes, products, or resources. Anything we may wish to measure is an identifiable attribute of these.

- Attributes are either internal or external. Although external attributes (such as reliability of products, stability of processes, or productivity of resources) tend to be the ones we are most interested in measuring, we can rarely do so directly. We are generally forced to use derived measures of internal attributes.

- The Goal-Question-Metric paradigm is a useful approach for deciding what to measure. Since managers and practitioners have little time to measure everything, the GQM approach allows them to choose those measures that relate to the most important goals or the most pressing problems.

- The GQM approach creates a hierarchy of goals to be addressed (perhaps decomposed into subgoals), questions that should be answered in order to know if the goals have been met, and measurements that must be made in order to answer the questions. The technique considers the perspective of the people needing the information and the context in which the measurements will be used.

- Process maturity must also be considered when deciding what to measure. If an entity is not visible in the development process, then it cannot be measured. Five levels of maturity, ranging from *initial* to *optimizing*, can be associated with the types of measurements that can be made.

- GQM and process maturity must work hand-in-hand. By using GQM to decide what to measure and then assessing the visibility of the entity, software engineers can measure an increasingly richer set of attributes.

- Many misunderstandings and misapplications of software measurement would be avoided if people thought carefully about the above framework. Moreover, this framework highlights the relationships among apparently diverse software measurement activities.

- Measurement is concerned not only with assessment (of an attribute of some entity that already exists) but also with prediction (of an attribute of some future entity). We want to be able to predict attributes like the cost and effort of processes, as well as the reliability and maintainability of products. Effective prediction requires a prediction system: a model supplemented with a set of prediction procedures for determining the model parameters and applying the results.

- The validation approach described in this chapter guides you in determining precisely which entities and attributes have to be considered for measurement. In this sense, it supports a goal-oriented approach.

- Software measures and prediction systems will be neither widely used nor respected without a proper demonstration of their validity. However, commonly accepted ideas and approaches to validating software measures bear little relation to the rigorous requirements for measurement validation in other disciplines. In particular, formal validation requirements must first be addressed before we can tackle informal notions such as usefulness and practicality.

- The formal requirement for validating a measure involves demonstrating that it characterizes the stated attribute in the sense of measurement theory.

- To validate a prediction system formally, you must first decide how stochastic it is (i.e., determine an acceptable error range), and then compare performance of the prediction system with known data points. This comparison will involve experiments, such as those described in Chapter 4.

- Software measures that characterize specific attributes do not have to be shown to be part of a valid prediction system in order to be valid measures. A claim that a measure is valid because it is a good predictor of some interesting attribute can be justified only by formulating a hypothesis about the relationship and then testing the hypothesis.

EXERCISES

1. For each entity listed in Table 3.1, find at least one way in which environmental considerations will influence the relevant external attributes.

2. What different product, process, or resource attributes might be the following measure?

   a. The number of faults found in program *P* using a set of test data created specifically for *P*.

   b. The number of faults found in program *P* using a standard in-house set of test data.

   c. The number of faults found in program *P* by programmer *A* during one hour.

   d. The number of faults found in program *P* during a code review by review team *T*.

3. Check the dictionary definitions of the following measures as a basis for validation: (i) candlepower, (ii) decibel, (iii) horsepower, (iv) light-year, (v) span.

4. To which of the above five classes do the following prediction systems belong: (i) the COCOMO model (Example 3.11), (ii) the Jelinski-Moranda reliability model (Example 3.13), (iii) stochastic systems, (iv) coupling as a predictor of program errors.

5. Explain briefly the idea behind the GQM paradigm. Is it always the right approach for suggesting what to measure? Suppose you are managing a software development project for which reliability is a major concern. A continual stream of anomalies is discovered in the software during the testing phase, and you suspect that the software will not be of sufficient quality by the shipping deadline. Construct a GQM tree that helps you to make an informed decision about when to ship the software.

6. Suppose that a software producer considers software quality to consist of a number of attributes, including reliability, maintainability, and usability. Construct a simple GQM tree corresponding to the producer's goal of improving the quality of the software.

7. Your department manager asks you to help improve the maintainability of the software that your department develops. Construct a GQM tree to identify an appropriate set of measures to assist you in this task.

8. Suppose your development team has as its goal: Improve effectiveness of testing. Use the GQM approach to suggest several relevant questions and measures that will enable you to determine if you have met your goal.

9. Your department manager asks you to *shorten the required system testing time* of software products without a loss in quality of the delivered product. Use the GQM approach to derive relevant questions and identify measures to determine whether you have met your goal. Clearly indicate which measures help to answer each question.

10. Our confidence in a prediction system depends in part on how well we feel we understand the attributes involved. After how many successful predictions would you consider the following validated?

  a. A mathematical model to predict the return time of Halley's comet to the nearest hour.

  b. A timetable for a new air service between London and Paris.

  c. A system that predicts the outcome of the spinning of a roulette wheel.

  d. A chart showing how long a bricklayer will take to erect a wall of a given area.

FURTHER READING

The COCOMO model is interesting to study, because it is a well-documented example of measurement that has evolved as goals and technology have changed. The original COCOMO approach is described in

Barry W. Boehm, *Software Engineering Economics*, Prentice-Hall, Englewood Cliffs, New Jersey, 1981.

An overview of the revision to COCOMO is in

Barry W. Boehm and Chris Abts, A. Winsor Brown, Sunita Chulani, Bradford K. Clark, Ellis Horowitz, Ray Madachy, Donald J. Reifer, and Bert Steece, *Software Cost Estimation with COCOMO II*, Prentice-Hall, Upper Saddle River, New Jersey, 2000.

The first paper to mention GQM is

Victor R. Basili and David M. Weiss, "A method for collecting valid software engineering data," *IEEE Transactions on Software Engineering*, 10(6), November 1984, pp. 728–738.

More detail and examples are provided in

Basili V.R. and Rombach H.D., "The TAME project: Towards improvement-oriented software environments," *IEEE Transactions on Software Engineering*, 14(6), pp 758–773, 1988.

van Solingen R. and Berghout E. *The Goal/Question/Metric Method: A Practical Guide for Quality Improvement of Software Development*, McGraw-Hill, London, 1999.

Despite its widespread appeal, GQM is not without its critics. These texts argue strongly against GQM on grounds of practicality.

Bache R. and Neil M., "Introducing metrics into industry: A perspective on GQM," in *Software Quality Assurance and Metrics: A Worldwide Perspective* (Eds: Fenton N.E., Whitty R.W., Iizuka Y.), International Thomson Press, pp. 59–68, 1995.

Hetzel, William C., *Making Software Measurement Work: Building an Effective Software Measurement Program*, QED Publishing Group, Wellesley, Massachusetts, London, 1993.

One extension to GQM is soft systems methodology, which includes stakeholders in the goal setting process and is described in the following paper:

Fredreiksen H.D. and Mathiassen L., "Information-centric assessment of software metrics practices," *IEEE Trans. Engineering Management*, 52(3), August 2005, pp. 350–362.

Much information is available at the CMMI web site: cmmiinstitute.com Researchers at IBM have produced a good analysis of fault categorization. The structure and its application are described in

Chillarege, R., Inderpal S. Bhandari, Jarir K. Chaar, Michael J. Halliday, Diane S. Moebus, Bonnie K. Ray and Man-Yuen Wong, "Orthogonal Defect Classification: A Concept for In-Process Measurements," *IEEE Transactions on Software Engineering*, 18(11), November 1992, pp. 943–956.

Correlation is often used to "validate" software measures. The paper by Courtney and Gustafson shows that use of correlation analysis on many metrics will inevitably throw up spurious "significant" results.

Courtney, R. E. and Gustafson D. A., "Shotgun correlations in software measures," *Software Engineering Journal*, 8(1), pp. 5–13, 1993.

Empirical Investigation*

S OFTWARE ENGINEERS ARE ALWAYS looking for ways to improve the quality of software products and the development process. Thus, they want to use improved tools and techniques in all aspects of development. They want better processes, tools, and languages to gather and analyze requirements, model designs, develop, test, and evolve applications. But how can you determine whether a new tool or technique will really provide the desired benefits? For example, analysts want to use the best techniques for gathering complete requirements and insuring that the requirements are consistent. Software architects want to use the design tools that will help in creating a software design that is relatively easy to implement, adapt, and maintain. Software developers can use languages and tools to build programs that accurately implement a design and are maintainable. Software testers want to use the best tools for generating and running tests that can find program faults.

How can you tell whether you are using appropriate tools and techniques? Often, decisions are made relying on the advice, experience, and opinions of others, and are often not based on objective criteria (Fenton et al. 1994). Many once popular tools and techniques were touted by recognized software "gurus," but have since been abandoned. Generally, you cannot rely on testimonials or evaluations of descriptions or specifications of a tool or technique. Selecting the best or optimal tool or technique requires some empirical support—evidence that the tools and techniques have been effective on a problem and in environments that are similar

---

* Including contributions by Barbara Kitchenham and Shari Lawrence Pfleeger from previous editions.

to the ones where you plan to apply them. The previous chapters demonstrate that measurement plays a crucial role in assessing scientifically the current situation and in determining the magnitude of change when manipulating the environment. In this chapter, we explain the assessment techniques available and provide guidelines for deciding which technique is appropriate in a given situation.

Suppose you are a software practitioner that needs to evaluate a technique, method, or tool. You can either rely on evaluations that have been conducted by others, possibly published in a research journal or conference proceedings, or conduct your own evaluation. In either case, you need to know what kind of study is appropriate, and what are the key elements involved in designing and conducting empirical studies. Otherwise, you might rely on the results of a poorly designed experiment, or conduct an inappropriate study yourself. This chapter begins by introducing key concepts relevant to empirical studies, and the limitations and challenges of empirical studies related to software development tools and techniques. Next, it describes key concepts and issues related to empirical study design. That is, the concepts will help you to determine if a previous study is appropriate for your evaluation, or to design your own study and derive meaningful results from it. Finally, we review the types of studies that are relevant to the problems faced by software developers and provide meaningful results.

4.1 PRINCIPLES OF EMPIRICAL STUDIES

Our basic goal here is to apply the scientific method to software engineering problems. As Marshall Walker explains, "the scientific method is merely a formalization of learning by experience" (Walker 1963, p. 14). Walker further explains that the "purpose of scientific thought is to make correct predictions of events in nature" (Walker 1963, p. 1). Our purpose is to make predictions about events related to software development. For example, would a new testing tool help to reveal program faults sooner and more completely than the old tool? In science, a theory is developed to explain a phenomenon and predict some consequences. Empirical studies are conducted to test the theory. Empirical studies do not prove that a theory is true. Rather, they provide further evidence to support or refute the theory. An empirical study examines some specific sample or observation of all of the possible values of the variables involved in a cause–effect relationship. Figure 4.1 shows the relationship between a theory about the true cause–effect relationships, and what we observe in an empirical study.

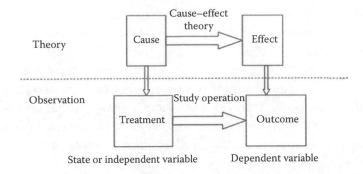

FIGURE 4.1 Empirical theory. (Adapted from Wohlin C. et al. *Experimentation in Software Engineering: An Introduction,* Kluwer Academic Publishers, Norwell, MA, USA, 2000, Figure 7.)

The key principles involved in designing empirical studies include

1. The level of control of study variables that determines the appropriate type of study

2. Study goals and hypotheses

3. Maintaining control of variables

4. Threats to validity

5. The use of human subjects

4.1.1 Control of Variables and Study Type

Ideally, we would like to use controlled experiments to evaluate software engineering tools and techniques. Controlled experiments involve the testing of well-defined hypotheses concerning postulated effects of independent variables on dependent variables in a setting that minimizes other factors that might affect the outcome. However, often, scientists (and software engineers) want to understand entities and processes that involve variables that they cannot control. For example, astrophysicists cannot run experiments that control all variables in order to understand the formation of black holes. Rather, they observe the behavior of existing black holes and try to understand the relationships between variables in order to evaluate hypotheses. Similarly, geologists cannot run controlled experiments to test key ideas about plate tectonics.

Often, ethical and legal issues prevent scientists from conducting controlled experiments that involve humans. For example, a social scientist cannot evaluate the effects of impoverishment on academic achievement by randomly assigning babies to impoverished and affluent families, and then evaluating their academic success. Rather, the social scientist will look for relationships between relative affluence and academic success, without using random assignment to control for the effects of other factors. Empirical studies that involve observations where potential confounding variables cannot be controlled and/or subjects cannot be assigned to treatment or control groups are called *observational studies, natural experiments*, and/or *quasi-experiments*.

Controlled experiments tend to be conducted in academia or research labs. They are rarely conducted inside software development organizations because it is usually not possible to control all potential confounding factors. Application domains, environments, tools, processes, the capabilities of developers, and many other factors can vary widely and are hard to control. Rather, many, but not all, empirical studies in software engineering are case studies that involve the use of a tool or technique on projects without random assignment of subjects to projects and control of all other variables. Thus, these software engineering case studies are really quasi-experiments.

Some questions can be answered by analyzing the data, even though you cannot control key variables. Often, surveys are a useful investigative technique. A *survey* is a retrospective study of a situation to try to document relationships and outcomes. Thus, a survey is done after an event has occurred. You are probably most familiar with social science surveys, where attitudes are polled to determine how a population feels about a particular set of issues, or a demographer surveys a population to determine trends and relationships. Software engineering surveys are similar, in that they poll a set of data from an event that has occurred to determine how the population reacted to a particular method, tool, or technique, or to determine trends or relationships.

EXAMPLE 4.1

Your organization may have used an agile software development process for the first time. After the project is complete, you may perform a survey to capture the size of the code produced, the effort involved, the number of faults and failures, and the project duration. Then, you may compare these figures with those from projects using the prior process to see if the agile process led to improvements over the prior process.

When performing a survey, you have no control over the activity that is under study. That is, because it is a retrospective study, you can record a situation and compare it with similar ones. But you cannot manipulate variables as you do with case studies and controlled experiments.

Controlled experiments must be planned in advance, while case studies may be planned or retrospective. To conduct a controlled experiment, you decide in advance what you want to investigate and then plan how to capture data to support your investigation. A *case study* is a quasi-experiment where you identify key factors that may affect the outcome of an activity and then document the activity inputs, constraints, resources, and outputs. In contrast, a *controlled experiment* is a rigorous, *controlled* investigation of an activity, where the key factors are identified and manipulated to document their effects on the outcome. Figure 4.2 illustrates that controlled experiments, case studies, and surveys are the three types of empirical investigation in software engineering. (They are not the only types of investigation. Other methods, including feature analysis, are not addressed in this book.)

Differences among the research methods are also reflected in their scale. Since experiments require a great deal of control, they tend to involve small numbers of people or events. We can think of experiments as "research in the small." Case studies usually look at a typical project, rather than trying to capture information about all possible cases. Such case studies can be thought of as "research in the typical." And surveys try to poll what is happening broadly over large groups of projects and are thus *research in the large*.

General guidelines can help you decide whether to perform a survey, case study, or a controlled experiment. The first step is deciding whether your investigation is retrospective or not. If the activity you are

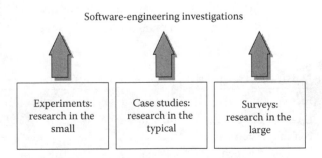

FIGURE 4.2   Three types of investigation.

investigating has already occurred, you must perform a survey or case study. If the activity has yet to occur, you may choose between a case study and a controlled experiment.

The central factor in this choice is the level of control needed for an experiment. If you have a high level of control over the variables that can affect the outcome, then you can consider an experiment. If you do not have that control, an experiment is not possible; a case study is the preferred technique. But the level of control satisfies the technical concerns; you must also address the practical concerns of research. It may be possible but very difficult to control the variables, either because of the high cost of doing so or the degree of risk involved. For example, safety-critical systems may entail a high degree of risk in experimentation, and a case study may be more feasible.

The other key aspect to consider is the degree to which you can replicate the basic situation you are investigating. For instance, suppose you want to investigate the effects of language on the resulting software. Can you develop the same project multiple times using a different language each time? If replication is not possible, then you cannot do a controlled experiment. However, even when replication is possible, the cost of replication may be prohibitive. Table 4.1 summarizes these concerns. The table is read across and up. For example, if the study you want to do has a low replication cost, then an experiment is more appropriate than a case study. Similarly, if you have no control (i.e., the difficulty of control is high), then you should consider a case study.

A controlled experiment is especially useful for investigating alternative methods of performing a particular, self-standing task. For instance, you can perform an experiment to compare the use of Alloy, Object Constraint Language (OCL), and Java Modeling Language (JML) for specifying a set of requirements. Here, the self-standing task can be isolated from the rest of the development process, but the task is still embedded in the usual way when the code is developed. Also, the

TABLE 4.1   Factors Relating to Choice of Research Technique

Factor	Experiments	Case Studies
Level of control	High	Low
Difficulty of control	Low	High
Level of replication	High	Low
Cost of replication	Low	High

self-standing task can be judged immediately, so that the experiment does not delay project completion. On the other hand, a case study may be preferable to a controlled experiment if the process changes caused by the independent variables are wide-ranging, requiring the effects to be measured at a high level and across too many dependent variables to control and measure.

All the three research techniques involve careful measurement. In this chapter, we focus primarily on controlled experiments and case studies. However, we encourage you to become familiar with all the three types of investigation, as each makes a valuable contribution to the body of software engineering knowledge.

4.1.2 Study Goals and Hypotheses

The first step in your investigation is deciding *what* you want to investigate. That goal helps you to decide which type of research technique is most appropriate for your situation. The goal for your research can be expressed as a hypothesis that you want to test. That is, you must specify what it is that you want to know. The *hypothesis* is the tentative idea that you think explains the behavior you want to explore. For example, your hypothesis may be, "Using Scrum produces better quality software than using the Extreme Programming (XP) method." Whether you survey developers to assess what happened when a particular group used each method, evaluate activities of an organization during or after using Scrum (a case study), or conduct a carefully controlled comparison of those using Scrum with those using XP (a controlled experiment), you are testing to see if the data you collect will confirm or refute the hypothesis you have stated.

Wherever possible, you should try to state your hypothesis in quantifiable terms and in terms of independent and dependent variables, so that it is easy to tell whether the hypothesis is confirmed or refuted.

EXAMPLE 4.2   THE STATEMENT

Using the Scrum method produces better quality software than using the XP method

is a hypothesis. However, it is not testable because the notion of quality is not given in a measureable way. You can define "quality" in terms of the defects found and restate the hypothesis as

> If using the Scrum method produces better quality software than using the XP method then code produced using Scrum will have fewer defects per thousand lines of code than code produced using the XP method.
>
> Now the hypothesis is testable. It clearly describes what is meant by better quality, and, if supported empirically, could be used to make predictions.

Quantifying the hypothesis often leads to the use of surrogate measures. That is, in order to identify a quantity with a factor or aspect you want to measure (e.g., quality), you must measure the factor indirectly using something associated with that factor (e.g., defects). Since the surrogate is a derived measure, there is danger that a change in the surrogate is not the same as a change in the original factor. For example, defects (or lack thereof) may not accurately reflect the quality of the software: finding a large number of defects during testing may mean that testing was very thorough and the resulting product is nearly defect-free, or it may mean that development was sloppy and there are likely to be many more defects left in the product (we will return to this issue in Chapter 11). Similarly, the delivered lines of code may not accurately reflect the amount of effort required to complete the product, since not all lines of code are equally difficult to implement. Therefore, along with a quantifiable hypothesis, you should document the relationship between the measures and the factors they intend to reflect. In particular, you should strive for quantitative terms that are as direct and unambiguous at possible. An inappropriate choice of quality measure is an example of a threat to the *construct validity* of a study and is discussed further in Section 4.1.4.

Scientists are skeptics by nature. When deciding whether to accept or reject a hypothesis, they assume that the hypothesis is not true unless the evidence is very strong. The method used is to first evaluate a *null hypothesis*, which states that the proposed relationship does not hold. The null hypothesis relevant to Example 4.2 is the following:

> Hyp_0: There is no difference between the quality of software produced by the Scrum method and the quality of the software produced by using the XP method as indicated by defects per thousand lines of code.

The first objective in analyzing the data collected from an empirical study is to see if you can reject the null hypothesis. Generally, you reject the null hypothesis only with very strong evidence. For example, a common criterion is that you reject the null hypothesis only if there is less than a 5% chance that there is no difference between the two groups. Only after your analysis shows conclusively that you can reject the null hypothesis, you can evaluate *alternative hypotheses*:

Hyp$_{A1}$: The code produced using Scrum will have fewer defects per thousand lines of code than the code produced using the XP method.

Hyp$_{A2}$: The code produced using XP will have fewer defects per thousand lines of code than the code produced using the Scrum method.

Chapter 6 will describe the traditional methods for evaluating hypotheses, and Chapter 7 will present an innovative Bayesian approach to testing hypotheses with incomplete information.

4.1.3 Maintaining Control over Variables

Once you have an explicit hypothesis, you must decide on the variables that can affect its truth. Then, for each variable identified, you must decide how much control you have over it. Knowing what you control is essential, since the key discriminator between experiments and case studies is the degree of control over behavioral events and the variables they represent. A case study is preferable when you are examining events where relevant behaviors cannot be manipulated.

EXAMPLE 4.3

If you are investigating the effect of a design method on the quality of the resulting software, but you have no control over who is using which design method (called subject selection), then you probably want to do a case study to document the results. Experiments are done when you can manipulate behavior directly, precisely, and systematically. Thus, if you can control which subjects use the Scrum method, and which subjects use XP, and when and where they are used, then an experiment is possible. This type of manipulation can be done in a "toy" situation, where events are organized to simulate their appearance in the real world, or in a "field" situation, where events are monitored as they actually happen.

The treatment of state variables differs in case studies and controlled experiments. A *state variable* is a factor that can characterize your project and influence your evaluation results. Sometimes, state variables are called *independent variables*, because they can be manipulated to affect the outcome. The outcome, in turn, is evidenced by the values of the dependent variable; that is, a *dependent variable* is one whose value is affected by changing one or more independent variables.

Examples of state variables include the application area, the system type, or the developers' experience with the language, tool, or method. In a controlled experiment, you can identify the variables and sample *over* them. This phrase means that you select projects exhibiting a variety of characteristics possible for your organization and design your research so that more than one value will be taken for each characteristic. You aim for coverage, so that you have an instance for each possibility or possible combination.

EXAMPLE 4.4

Suppose your hypothesis involves the effect of programming language on the quality of the resulting code. "Language" is a state variable, and an ideal experiment would involve projects where many different languages would be used. You would design your experiment so that the projects studied employed as many languages as possible.

In a case study, you sample *from* the state variable, rather than over it. This phrase means that you select a value of the variable that is typical for your organization and its projects. In Example 4.4, a case study might involve choosing a language that is usually used on most of your projects, rather than trying to choose a set of projects to cover as many languages as possible.

Thus, a state variable is used to distinguish the *control* situation from the *treatment* in a controlled experiment. When you cannot differentiate control from treatment, you must do a case study instead of a controlled experiment. You may consider your current situation to be the control, and your new situation to be the treatment; a state variable tells you how the treatment differs from the control.

EXAMPLE 4.5

Suppose you want to determine whether a change in programming language can affect the productivity of your project. Then language is a state variable. If you currently use C to write your programs and you want to investigate the effects of changing to Java, then you can designate C to be the control language and Java to be the experimental one. The values of all other state variables should stay the same (e.g., application experience, programming environment, and type of problem). Then you can be sure that any difference in productivity is attributable to the change in language. This type of experiment was carried out at NASA's Goddard Space Flight Center to support the decision to move from the FORTRAN language to Ada. The experiments revealed that productivity in Ada did not exceed FORTRAN until the third use of Ada, when programmers began to use Ada appropriately.

4.1.4 Threats to Validity

We have all seen reports about studies with new results that contradict results reported from the past. A new study might show that a medical treatment is not effective, even though prior studies supported the effectiveness of the treatment.

No study is perfect; there are many ways that a study can provide misleading results. Potential problems with empirical studies are classified as categories of *threats to validity*. Wohlin et al. (2000) describe the following four categories of threats to validity:

1. *Conclusion validity.* Conclusion validity refers to the statistical relationship between independent and dependent variables. A study has conclusion validity if the results are statistically significant using appropriate statistical tests. Threats to conclusion validity include using the wrong statistical tests, having too small a sample, and searching for relationships between too many variables, which increases the odds of finding a spurious correlation.

2. *Construct validity.* A study with construct validity uses measures that are relevant to the study and meaningful. Using the measure *faults per KLOC* as a measure of code quality has some threats to validity since its value depends in part on when the measure is taken, for example, during testing (faults found during testing) or after release (faults found by customers). The key is to use meaningful measures that have been validated in the narrow sense, as described in Chapter 2. That is, the measures give values that are consistent with our intuition about the attribute that they purport to quantify.

3. *Internal validity.* Internal validity refers to the cause–effect relationship between independent and dependent variables. A study has internal validity if the treatment actually caused the effect shown in

the dependent variables. Specific threats include the effects of other, possibly unidentified, variables. To have internal validity, there must be a *causal theory*—an *a priori* rationale for why the independent variable would affect the dependent variable. In addition, there should be *temporal precedence* between independent and dependent variables—measurements of independent variables must have been taken before the dependent variables are measured. Otherwise, instead of A causing B, B might cause A.

4. *External validity.* External validity refers to how well you can generalize from the results of one study to the wider world. The ability to generalize depends on how similar the study environment is to the environment used in actual practice. It depends on how similar the study subjects (i.e., undergraduate students, graduate students, novice, and experienced developers) are to software developers in practice. A study results will tend to be more relevant to environments that are most similar to the study environment.

You should design a study to minimize these threats rather than evaluate the threats to a study's validity after the study is completed.

4.1.5 Human Subjects

Empirical studies in software engineering often make use of human subjects. A survey may involve asking developers to respond to questions about their use of tools or other experiences. A case study may involve the study of a programming team and their activities as they develop a product. A controlled experiment may involve assigning students to treatment and control groups to evaluate a new technique.

The US Government uses the following definition of *human subjects*:

> *Human subject* means a living individual about whom an investigator (whether professional or student) conducting research obtains
>
> 1. Data through intervention or interaction with the individual
> 2. Identifiable private information

U.S. DEPARTMENT OF HEALTH AND
HUMAN SERVICES 2009

Generally, government and university regulations require that research involving human subjects meet specified standards. The standards aim to ensure that subjects give informed consent to take part in the study; the study is designed to minimize any damage or injury to subjects; and the study is designed in a way that it is likely to provide useful information. The term *informed consent* refers to a human subject who explicitly agrees to take part in a study after being informed of the nature of the study and the potential risks. In the United States, research organizations use an *institutional review board* (IRB) to evaluate and approve research projects that involve human studies. Other countries have similar procedures.

Software engineering studies rarely involve risks of physical harm to human subjects. The major risks are due to privacy issues. For example, a subject's reputation may be affected by the public release of a subject's participation as a member of the group with a higher defect rate. Other risks include the release of private information that was collected in a study. The public release of a subject's opinion about a project can have negative consequences to the subject. To be approved by an IRB, the investigators must provide specific procedures on how such privacy risks will be protected. The protections might include encrypting all identifying information and destroying identifying information when the study is released. An IRB review will include a review of the study procedures in order to determine that the results are likely to have some impact. This is another reason to carefully plan an empirical study.

4.2 PLANNING EXPERIMENTS

Suppose you have decided that a controlled experiment is the best investigative technique for the questions you want to answer. Controlled experiments, like software development itself, require a great deal of care and planning if they are to provide meaningful and useful results. This section discusses the planning needed to define and run a controlled experiment, including consideration of several key characteristics of the experiment. Many of the concepts in this section apply to case studies. The major difference is that investigators conducting case studies have more limited control over study variables. The discussion should also help you to evaluate the published results from controlled experiments and case studies.

4.2.1 A Process Model for Performing Experiments

Controlled experiments (and case studies) need to be carried out in phases to ensure that all important issues and concerns are addressed and that the results will be useful. We suggest a six-phase process:

1. Conception

2. Design

3. Preparation

4. Execution

5. Analysis

6. Dissemination and decision-making

The process is a sequential. However, it is a good idea to test your experimental design with a small pilot study, involving a small number of subjects before running a full-scale study. You will discover glitches in your design that you can correct before investing the time and effort in a larger study.

4.2.1.1 Conception
The first phase is to define the goals of your experiment. The conception stage includes analysis described earlier to ensure that a controlled experiment, case study, and/or survey are appropriate. Next, you must state clearly and precisely the objective of your study. The objective may include showing the benefits and costs of particular method or tool compared to another method or tool. An alternative goal is to determine, for a particular method or tool, how differences in environmental conditions or quality of resources can affect the use or output of the method or tool. For both experiments and case studies, the objective must be stated so that it can be clearly evaluated at the end of the experiment. That is, it should be stated as a question you want to answer. The next phase is to design an empirical study that will provide the answer.

4.2.1.2 Design
Once your objective is clearly stated, you must translate the objective into a set of hypotheses, as described in the previous section. You should describe two or more hypotheses: one or more null hypotheses and one or more alternative hypotheses for each null hypothesis. A null hypothesis assumes that there is no significant difference between the two treatments (i.e., between two methods, tools, techniques, environments, or other conditions whose effects you are measuring) with respect to the dependent variable you are measuring (such as productivity, quality, or cost). An alternative hypothesis posits that there is a significant difference between the two treatments.

Hypothesis definition is followed by the generation of a formal design to test the hypothesis. The experimental design is a complete plan for applying differing experimental conditions to your experimental subjects so that you can determine how the conditions affect the behavior or result of some activity. In particular, you want to plan how the application of these conditions will help you to test your hypothesis and answer your objective question.

To see why a formal plan or design is needed, consider Example 4.6:

EXAMPLE 4.6

You are managing a software development organization. Your research team has proposed the following objective for an experiment:

> We want to determine the effect of using the Python language on the quality of the resulting code.

The problem as stated is far too general to be useful. You must ask specific questions, such as

1. How quality is to be measured?
2. How is the use of Python to be measured?
3. What are the factors that influence the characteristics to be analyzed? For example, will experience, tools, design techniques, or testing techniques make a difference?
4. Which of these factors will be studied in the investigation?
5. How many times should the experiment be performed, and under what conditions?
6. In what environment will the use of Python be investigated?
7. How should the results be analyzed?
8. How large a difference in quality will be considered important?

These are just a few of the questions that must be answered before the experiment can begin.

There is a formal terminology for describing the components of your experiment or case study. This terminology will help when you consult with a statistician, and encourage you to consider all aspects of the experiment. The new method or tool you wish to evaluate (compared with an existing or different method or tool) is called the *treatment*. You want to determine if the treatment is beneficial in certain circumstances. That is, you want to determine if the treatment produces results that are in some way different. For example, you may want to find out whether a new tool

increases productivity compared with your existing tool and its productivity. Or, you may want to choose between two techniques, depending on
their effect on the quality of the resulting product.

Your experiment will consist of a series of tests of your methods or
tools, and the experimental design describes how these tests will be organized and run. In any individual test run, only one treatment is used. An
individual test of this sort is sometimes called a *trial*, and the *experiment*
is formally defined as the set of trials. Your experiment can involve more
than one treatment, and you will want to compare and contrast the differing results from the different treatments. The *experimental objects* or
experimental units are the objects to which the treatment is being applied.
Thus, a development or maintenance project can be your experimental
object, and aspects of the project's process or organization can be changed
to affect the outcome. Or, the experimental objects can be programs or
modules, and different methods or tools can be used on those objects.

EXAMPLE 4.7

If you are investigating the degree to which a design-related treatment results
in reusable code components, you may consider design components as the
experimental objects.

At the same time, you must identify *who* is applying the treatment;
these people are called the *experimental subjects*. The characteristics of
the experimental subjects must be clearly defined, so that the effects of
differences among subjects can be evaluated in terms of the observed
results.

When you are comparing using the treatment with not using it, you
must establish a *control object*, which is an object not used or being
affected by the treatment. The control provides a baseline of information
that enables you to make comparisons. In a case study, the control is the
environment in which the study is being run or another similar project
that did not use the treatment. In a controlled experiment, the control situation must be defined explicitly and carefully, so that all the differences
between the control object and the experimental object are understood.

The *response variables* (or *dependent variables*) are those factors that
are expected to change or differ as a result of applying the treatment. In
Example 4.6, quality may be considered as a composite of several attributes: the number of defects per thousand lines of code, the number of

failures per thousand hours of execution time, and the number of hours of staff time required to maintain the code after deployment, for example. Each of these is considered a response variable. In contrast, as we noted earlier, state variables (or independent variables) are those factors that may influence the application of a treatment and thus indirectly influence the result of the experiment. We have seen that state variables usually describe the characteristics of the developers, the products, or the processes used to produce or maintain the code. It is important to define and characterize state variables so that their impact on the response variables can be investigated. But state variables can also be useful in defining the scope of the experiment and in choosing the projects that will participate.

EXAMPLE 4.8

Use of a particular design technique may be a state variable. You may decide to limit the scope of the experiment only to those projects that use Java along with a design expressed in the UML, rather than investigating Java on projects using any design technique. Or you may choose to limit the experiment to projects in a particular application domain, rather than considering all possible projects.

Finally, as we saw earlier, state variables (and the control you have over them) help to distinguish case studies from controlled experiments.

The number of and relationships among subjects, objects, and variables must be carefully described in the experimental plan. The more the subjects, objects, and variables, the more complex the experimental design becomes and often the more difficult the analysis. Thus, it is very important to invest a great deal of time and care in designing your experiment, rather than rush to administer trials and collect data. In the remainder of this chapter, we address in more detail the types of issues that must be identified and planned in a formal experimental design. In many cases, the advice in this section should be supplemented with the advice of a statistician, especially when many subjects and objects are involved. Thus, you should seek advice from a statistician during the design phase.

Once the design is complete, you will know what experimental factors (i.e., response and state variables) are involved, how many subjects will be needed, from what population they will be drawn, and to what conditions or treatments each subject will be exposed. In addition, if more than one treatment is involved, the order of presentation or exposure will be laid out. The criteria for measuring and judging effects will be defined, as well as the methods for obtaining the measures.

4.2.1.3 Preparation
Preparation involves readying the subjects for application of the treatment. For example, preparation for your experiment may involve purchasing tools, training staff, or configuring hardware in a certain way. Instructions must be written out or recorded properly. A pilot study—a dry run of the experiment on a small set of people—is very useful. You will surely discover that portions of the plan are incomplete and the instructions need improvement.

4.2.1.4 Execution
During this phase, you conduct the experiment. Following the steps laid out in the plan, and measuring attributes as prescribed by the plan, you apply the treatment to the experimental objects. You must be careful that items are measured and treatments are applied consistently, so that comparison of results is sensible.

4.2.1.5 Analysis
The analysis phase has two parts. First, you must review all the measurements taken to make sure that they are valid and useful. You organize the measurements into sets of data that will be examined as part of the hypothesis-testing process. Second, you analyze the sets of data according to the statistical principles described in Chapter 6. These statistical tests, when properly administered, tell you if the hypotheses are supported or refuted by the results of the experiment. That is, the statistical analysis gives you an answer to the original question addressed by the research.

4.2.1.6 Dissemination and Decision-Making
At the end of the analysis phase, you would have reached a conclusion about how the different characteristics you examined affected the outcome. It is important to document your conclusions in a way that will allow your colleagues to duplicate your experiment and confirm your conclusions in a similar setting. Thus, you must document all of the key aspects of the research: the objectives, the hypotheses, the experimental subjects and objects, the response and state variables, the treatments, and the resulting data. Any other relevant documentation should be included: instructions, tool or method characteristics (e.g., version, platform, and vendor), training manuals, and

more. A report on an experiment should state conclusions clearly, making sure to address any problems experienced during the running of the experiment. For example, you must report any staff changes during project development, and tool upgrades. The report should clearly describe the risks to the validity of the study and how you controlled these risks.

The experimental results may be used in several ways. You may use them to support decisions about how you will develop or maintain software in the future: what tools or methods you will use, and in what situations. Others may use your results to suggest changes to their own development environment. In addition, others may replicate your experiment to confirm the results. Finally, you and others may perform similar experiments with variations in experimental subjects or state variables. These new experiments will help you and others to understand how the results are affected by carefully controlled changes. For example, if your experiment demonstrates a positive change in quality by using Python, others may test to see if the quality can be improved still further by using Python in concert with a particular Python-related tool or in a particular application domain.

4.2.2 Key Experimental Design Concepts

Useful results depend on careful, rigorous, and complete experimental design. In this section, we examine the key concepts that you must consider in designing your experiment. Each key design concept addresses the need for simplicity and for maximizing information. Simple designs help to make the experiment practical, minimizing the use of time, money, personnel, and experimental resources. An added benefit is that simple designs are easier to analyze (and thus are more economical) than complex designs. Maximizing information gives you a complete understanding of your experimental conditions and the results as much as possible, enabling you to generalize your results to the widest possible situations. Following these design concepts will reduce risks to the validity of an experiment.

An experimental design deals with experimental units and experimental error. As noted above, an *experimental unit* is the experimental object to which a single treatment is applied. Usually, you apply the treatment more than once. At the very least, you apply it to the control group as well as at least one other group that differs from the control by a state variable. In many cases, you apply the treatment many times to many groups. In each case, you examine the results to see what the differences are in applying the treatments. However, even when you keep the conditions the same from one trial to another, the results can turn out to be slightly different.

EXAMPLE 4.9

You are investigating the time it takes for a programmer to recognize faults in a program. You have seeded a collection of programs with a set of known faults, and you have asked a programmer to find as many faults as possible. Today, you give the same programmer a different but equivalent program with different but equivalent seeded faults, but the programmer takes more time today to find the set of faults as he or she took yesterday.

To what is this variation attributable? *Experimental error* describes the failure of two identically treated experimental units to yield identical results. The error can reflect a host of problems:

- Errors of experimentation.

- Errors of observation.

- Errors of measurement.

- The variation in experimental resources.

- The combined effects of all extraneous factors that can influence the characteristics under study but which have not been singled out for attention in the investigation.

Thus, in Example 4.9, the differences may be due to things such as

- The programmer's mind wandered during the experiment.

- The timer measured elapsed time inexactly.

- The programmer was distracted by loud noises from another room.

- The programmer found the faults in a different sequence today than yesterday.

- Faults that were thought to be equivalent are not.

- Programs that were thought to be equivalent are not.

The aim of a good experimental design is to control as many variables as possible, both to minimize variability among participants and to minimize the effects of irrelevant variables (such as noise in the next room or the order of presentation of the experiment). Ideally, we would like to eliminate the effects of other variables so that only the effects of the independent variables are reflected in the values of the dependent variable. That is, we would like to eliminate the experimental error. Realistically, complete elimination is rarely possible. Instead, we try to design the experiment so that the effects of irrelevant variables are distributed equally across all the experimental conditions, rather than allowing them to inflate artificially (or bias) the results of a particular condition. In fact, statisticians like, whenever possible, to measure the extent of the variability under "normal circumstances."

The three key experimental design concepts—replication, randomization, and local control—address this problem of variability by giving us guidance on forming experimental units so as to minimize the experimental error.

4.2.2.1 Replication

Replication involves repeating an experiment under identical conditions, rather than repeating measurements on the same experimental unit. This repetition is desirable for several reasons. First, replication (with associated statistical techniques) provides an estimate of experimental error that acts as a basis for assessing the importance of observed differences in an independent variable. That is, replication can help us to know how much confidence we can place in the results of the experiment. Second, replication enables us to estimate the mean effect of any experimental factor.

It is important to ensure that replication does not introduce the confounding of effects. Two or more variables are *confounded* if it is impossible to separate their effects when the subsequent analysis is performed.

EXAMPLE 4.10

Suppose you want to compare the use of a new tool with your existing tool. You set up an experiment where programmer *A* uses the new tool in your development environment, while programmer *B* uses the existing tool. When you compare measures of quality in the resulting code, you cannot say how much of the difference is due to the tools because you have not accounted for the difference in the skills of the programmers. That is, the effects of the tools (one variable) and the programmers' skills (another variable) are confounded. This confounding is introduced with the replication when the repetition of the experiment does not control for other variables (like programmer skills).

EXAMPLE 4.11

Similarly, consider the comparison of two testing techniques. A test team is trained in test technique X and asked to test a set of modules. The number of defects discovered is the chosen measure of the technique's effectiveness. Then, the test team is trained in test technique Y, after which they test the same modules. A comparison of the number of defects found with X and with Y may be confounded with the similarities between techniques or a learning curve in going from X to Y. Here, the sequence of the repetition is the source of the confounding.

Confounding effects are a serious threat to the internal validity of an experiment. For this reason, the experimental design must describe in detail the number and kinds of replications of the experiments. It must identify the conditions under which each experiment is run (including the order of experimentation), and the measures to be made for each replicate.

4.2.2.2 Randomization

Replication makes it possible to test the statistical significance of the results. But it does not ensure the validity of the results. That is, we want to be sure that the experimental results clearly follow from the treatments that were applied, rather than from other variables. Some aspect of the experimental design must organize the experimental trials in a way that distributes the observations independently, so that the results of the experiment are valid. *Randomization* is the random assignment of subjects to groups or of treatments to experimental units, so that we can assume independence (and thus conclusion validity and internal validity) of results. Randomization does not guarantee independence, but it allows us to assume that the correlation on any comparison of treatments is as small as possible. In other words, by randomly assigning treatments to experimental units, you can try to keep some treatment results from being biased by sources of variation over which you have no control.

For example, sometimes the results of an experimental trial can be affected by the time, the place, or unknown characteristics of the participants. These uncontrollable factors can have effects that hide or skew the results of the controllable variables. To spread and diffuse the effects of these uncontrollable or unknown factors, you can assign the order of trials randomly, assign the participants to each trial randomly, or assign the location of each trial randomly, whenever possible.

EXAMPLE 4.12

Consider again the situation in Example 4.11. Suppose each of 30 program-mers is trained to perform both testing techniques. We can randomly assign 15 of the programmers to use technique X on a given program, and the remaining 15 can use technique Y on the same program.

A key aspect of randomization involves the assignment of subjects to groups and treatments. If we use the same subjects in all experimental condi-tions, we say that we have a *related within-subjects* design; this is the situation described in Example 4.11. However, if we use different subjects in different experimental conditions, we have an *unrelated between-subjects* design, as in Example 4.12; different programmers used X than the ones who used Y to test the program. If there is more than one independent variable in the experiment, we can consider the use of same or different subjects separately for each of the variables. (We will describe this issue in more detail later on.)

Earlier, we mentioned the need for protecting subjects from knowing the goals of the experiment; we suggested the use of double-blind experi-ments to keep the subjects' actions from being affected by variable val-ues. This technique is also related to randomization. It is important that researchers do not know who is to produce what kind of result, lest the information bias the experiment.

Thus, an experimental design should include details about the plan to randomize assignment of subjects to groups or of treatments to experi-mental units, and how you plan to keep the design information from bias-ing the analytical results. In cases where complete randomization is not possible, you should document the areas where lack of randomization may affect the validity of the results. In later sections, we shall see the examples of different designs and how they involve randomization.

4.2.2.3 Local Control
As noted earlier, one of the key factors that distinguish a controlled exper-iment from a case study is the degree of control. *Local control* refers to the control that you have over the placement of subjects in experimental units and the organization of those units. Whereas replication and random-ization ensure a valid test of significance, local control makes the design more efficient by reducing the magnitude of the experimental error. Local control is usually discussed in terms of two characteristics of the design: blocking and balancing the units.

Blocking means allocating experimental units to blocks or groups so that the units within a block are relatively homogeneous. The blocks are designed so that the predictable variation among units has been confounded with the effects of the blocks. That is, the experimental design captures the anticipated variation in the blocks by grouping like varieties, so that the variation does not contribute to the experimental error.

EXAMPLE 4.13

You are investigating the comparative effects of three design techniques on the quality of the resulting code. The experiment involves teaching the techniques to 12 developers and measuring the number of defects found per thousand lines of code to assess the code quality. It is possible that the 12 developers graduated from three different universities. It is possible that the universities trained the developers in very different ways, so that the effect of being from a particular university can affect the way in which the design technique is understood or used. To eliminate this possibility, three blocks can be defined so that the first block contains all developers from university *X*, the second block from university *Y*, and the third block from university *Z*. Then, the treatments are assigned at random to the developers from each block. If the first block has six developers, you would expect two to be assigned to design method *A*, two to *B*, and two to *C*, for instance.

Balancing is the blocking and assignment of treatments so that an equal number of subjects is assigned to each treatment, whenever possible. Balancing is desirable because it simplifies the statistical analysis, but it is not necessary. Designs can range from being completely balanced to having little or no balance.

In experiments investigating only one factor, blocking and balancing play important roles. If the design includes no blocks, then it must be completely randomized. That is, subjects must be assigned at random to each treatment. A balanced design, with equal numbers of subjects per treatment, is preferable but not necessary. If one blocking factor is used, subjects are divided into blocks and then randomly assigned to each treatment. In such a design, called a *randomized block design*, balancing is essential for analysis. Thus, this type of design is sometimes called a *complete balanced block design*. Sometimes, units are blocked with respect to two different variables (e.g., staff experience and program type) and then assigned at random to treatments so that each blocking variable combination is assigned to each treatment an equal number of times. In this case, called a *Latin Square*, balancing is mandatory for correct analysis.

Your experimental design should include a description of the blocks defined and the allocation of treatments to each. This part of the design will assist the analysts in understanding what statistical techniques apply to the data that results from the experiments. More details about the analysis can be found in Chapter 6.

4.2.3 Types of Experimental Designs

There are many types of experimental designs. It is useful to know and understand the several types of designs that you are likely to use in software engineering research, since the type of design constrains the type of analysis that can be performed and therefore the types of conclusions that can be drawn. For example, the measurement scale of the variables constrains the analysis. Nominal scales simply divide data into categories and can be analyzed by using statistical tests such as the Sign test (which looks at the direction of a score or measurement); on the other hand, ordinal scales permit rank ordering and can be investigated with more powerful tests such as Wilcoxon (looking at the size of the differences in rank orderings). Parametric tests such as analysis of variance can be used only on data that is at least of an interval scale.

The sampling also enforces the design and constrains the analysis, as we will see in Chapter 6. For example, the amount of random variance should be equally distributed among the different experimental conditions if parametric tests are to be applied to the resulting data. Not only does the degree of randomization make a difference to the analysis, but also the distribution of the resulting data. If the experimental data are normally or near-normally distributed—exhibit a bell-shaped curve, as shown in Figure 4.3 then you can use parametric tests. However, if the data are not normally distributed, or if you do not know what the distribution is, nonparametric methods are preferable; examples of distributions that are not normal are shown in Figure 4.4.

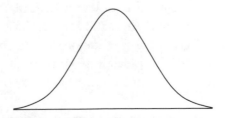

FIGURE 4.3   A normal distribution.

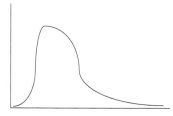 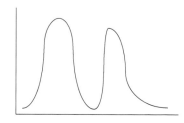

FIGURE 4.4   Example of nonnormal distributions. The distribution on the left is skewed, while that on the right is bimodal.

Many investigations involve more than one independent variable. In addition, the experiment invokes changes in the dependent variable as one or more of the independent variables changes. An independent variable is called a *factor* in the experimental design.

EXAMPLE 4.14

A study to determine the effect of experience and language on the productivity of programmers has two factors: experience and language. The dependent variable is productivity.

Various values or classifications for each factor are called the *levels* of the factor. Levels can be continuous or discrete, quantitative or qualitative. In Example 4.14, we may choose to measure experience in years of experience as a programmer; then each integer number of years can be considered a level. If the most experienced programmer in the study has 8 years of experience, and if there are five languages in the study, then the first factor has eight levels and the second factor five.

There are several types of factors, reflecting things such as treatments, replications, blocking, and grouping. This chapter does not tell you what factors should be included in your design. Neither does it prescribe the number of factors nor the number of levels. Instead, it explains how the factors can be related to each other, and how the levels of one factor are combined with the levels of another factor to form the treatment combinations. The remainder of this section explains how to derive a design from the number of factors and levels you want to consider in your investigation.

Most designs in software engineering research are based on two simple relations between factors: crossing and nesting; each is discussed separately.

		Factor B		
		Level 1	Level 2	Level 3
Factor A	Level 1	a_1b_1	a_1b_2	a_1b_3
	Level 2	a_2b_1	a_2b_2	a_2b_3

FIGURE 4.5   Example of a crossed design.

4.2.3.1 Crossing

The design of an experiment can be expressed in a notation that reflects the number of factors and how they relate to the different treatments. Expressing the design in terms of factors, called the *factorial design*, tells you how many different treatment combinations are required. Two factors, A and B, in a design are *crossed* if each level of each factor appears with each level of the other factor. This relationship is denoted as $A \times B$. The design itself is illustrated with three factors in Figure 4.5, where a_i represents the levels of factor A and b_j the levels of factor B. The figure's first row indicates that you must have a treatment for level 1 of A occurring with level 1 of B, for level 1 of A occurring with level 2 of B, and for level 1 of A with level 3 of B. The first column shows that you must have a treatment for level 1 of B occurring with each of the two levels of A. A crossed design with n levels of the first factor and m levels of the second factor will have $n \times m$ cells, with each cell representing a particular situation. Thus, Figure 4.5 has two levels for A and three for B, yielding six possible treatments in all. In the previous example, the effects of language and experience on productivity can be written as an 8×5 crossed design, requiring 40 different treatment combinations. This design means that your experiment must include treatments for each possible combination of language and experience. For three factors, A, B, and C, the design $A \times B \times C$ means that all combinations of all the levels occur.

4.2.3.2 Nesting

Factor B is *nested* within factor A if each *meaningful* level of B occurs in conjunction with only one level of factor A. The relationship is depicted as $B(A)$, where B is the nested factor and A is the nest factor. A two-factor nested design is depicted in Figure 4.6, where, like Figure 4.5, there are two levels of factor A and three levels of factor B. Now B is dependent on A, and each level of B occurs with only one level of A. In this example, levels 1 and 2 of B occur only with level 1 of A, and level 3 of B only occurs with level 2 of A. Thus, B is nested within A. By nesting we have reduced the number of treatment combinations from 6 to 3.

Factor A		
Level 1		Level 2
Factor B		Factor B
Level 1	Level 2	Level 3
a_1b_1	a_1b_2	a_2b_3

FIGURE 4.6   Example of a nested design.

To understand nesting, and to see how crossing differs from nesting, consider again the effects of language and experience on productivity. In this case, let factor A be the language, and B be the years of experience with a particular language. Suppose A has two levels, PHP and Java, and B separates the programmers into those with less than 2 years of experience with the language and those who have more. For a crossed design, we would have four cells for each language. That is, we would have the following categories:

1. PHP programmers with less than 2 years of experience with PHP (a_1b_1)

2. PHP programmers with less than 2 years of experience with Java (a_1b_3)

3. PHP programmers with at least 2 years of experience with PHP (a_1b_2)

4. PHP programmers with at least 2 years of experience with Java (a_1b_4)

5. Java programmers with less than 2 years of experience with PHP (a_2b_1)

6. Java programmers with less than 2 years of experience with Java (a_2b_3)

7. Java programmers with at least 2 years of experience with PHP (a_2b_2)

8. Java programmers with at least 2 years of experience with Java (a_2b_4)

The above eight categories are generated because we are examining experience with each language type, rather than experience in general. However, with a nested design, we can take advantage of the fact that the two factors are related. As depicted in Figure 4.7, we need to consider only four treatments with a nested design, rather than eight; treatments a_1b_3, a_1b_4, a_2b_1, and a_2b_2 are not used. Thus, nested designs take

Factor A: Language			
Level 1: PHP		Level 2: Java	
<2 years experience with PHP	≥2 years experience with PHP	<2 years experience with Java	≥2 years experience with Java
a_1b_1	a_1b_2	a_2b_3	a_2b_4

FIGURE 4.7   Nested design for language and productivity example.

advantage of factor dependencies, and they reduce the number of cases to be considered.

Nesting can involve more than two factors. For example, three factors can be nested as $C(B(A))$. In addition, more complex designs can be created as nesting and crossing are combined.

There are several advantages to expressing a design in terms of factorials:

1. Factorials ensure that resources are used most efficiently.

2. Information obtained in the experiment is complete and reflects the various possible interactions among variables. Consequently, the experimental results and the conclusions drawn from them are applicable over a wider range of conditions than they might otherwise be.

3. The factorial design involves an implicit replication, yielding the related benefits in terms of reduced experimental error.

On the other hand, the preparation, administration, and analysis of a complete factorial design are more complex and time-consuming than a simple comparison. With a large number of treatment combinations, the selection of homogeneous experimental units is difficult and can be costly. Also, some of the combinations may be impossible or of little interest to you, wasting valuable resources. For these reasons, the remainder of this section explains how to choose an appropriate experimental design for your situation.

4.2.4 Selecting an Experimental Design

As seen above, there are many choices for how to design your experiment. The ultimate choice depends on two things: the goals of your investigation and the availability of resources. The remainder of this section explains how to decide which design is right for your situation.

4.2.4.1 Choosing the Number of Factors

Many experiments involve only one variable or factor. These experiments may be quite complex, in that there may be many levels of the variable that are compared (e.g., the effects of many types of languages, or of several different tools). One-variable experiments are relatively simple to analyze, since the effects of the single factor are isolated from other variables that may affect the outcome. However, it is not always possible to eliminate the effects of other variables. Instead, we strive to minimize the effects, or at least distribute the effects equally across all the possible conditions we are examining. For example, techniques such as randomization aim to prevent variability among people from biasing the results.

But sometimes the absence of a second variable affects the performance of the first variable. That is, people act differently in different circumstances, and you may be interested in the variable interactions as well as in individual variables.

EXAMPLE 4.15

You are considering the effects of a new design tool on productivity. The design tool may be used differently by designers who are well versed in object-oriented design from those who are new to object-oriented design. If you were to design a one-factor experiment by eliminating the effects of experience with object-oriented design, you would get an incomplete (and probably incorrect) view of the effects of the tool. It is better to design a two-factor experiment that incorporates both use of the tool and designer experience. That is, by looking at the effects of several independent variables, you can assess not only the individual effects of each variable (known as the *main effect* of each variable) but also any possible interactions among the variables.

To see what we mean by interaction, consider the reuse of existing code. Suppose your organization has a repository of code modules that is made available to some of the programmers but not all. You design an experiment to measure the time it takes to code a module, distinguishing small modules from large. When the experiment is complete, you plot the results, separating the times for those who reused code from the times of those who did not. This experiment has two factors: module size and reuse. Each factor has two levels; module size is either small or large, and reuse is either present or absent. If the results of your experiment resemble Figure 4.8, then you can claim that there is no interaction between the two factors.

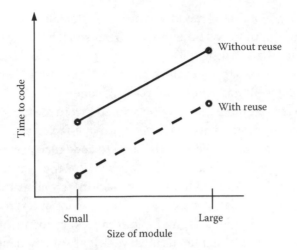

FIGURE 4.8   No interaction between factors.

The lines in Figure 4.8 are parallel, which indicates that the two variables have no effect on each other. Thus, an individual line shows the main effect of the variable it represents. In this case, each line shows that the time to code a module increases with the size of the module. In comparing the two lines, we see that reuse reduces the time to code a module, but the parallel lines indicate that size and degree of reuse do not change the overall trend. However, if the results resemble Figure 4.9, then there is indeed an interaction between the variables, since the lines are not parallel. Such a graph may result if there is considerable time spent searching through

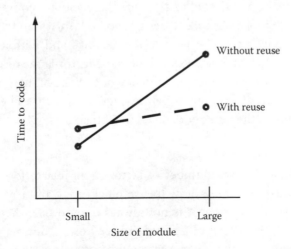

FIGURE 4.9   Interaction between factors.

the repository. For a small module, it may actually take more time to scan the repository than to code the module from scratch. For large modules, reuse is better than writing the entire module, but there is still significant time eaten up in working with the repository. Thus, there is an interaction between the size of the module and reuse of code.

Thus, there is far more information available from the two-factor experiment than there would have been from two one-factor experiments. The latter would have confirmed that the time to code increases with the size of the module, both with and without reuse. But the two-factor experiment shows the relationship between the factors as well as the single-factor results. In particular, it shows that, for small modules, reuse may not be warranted. In other words, multiple-factor experiments offer multiple views of the data and enlighten us in ways that are not evident from a collection of single-factor experiments.

Another way of thinking about whether to use one factor or more is to decide what kind of comparison you want to make. If you are examining a set of competing treatments, you can use a single-factor experiment. For example, you may want to investigate three design methods and their effects on quality. That is, you apply design methods A, B, and C to see which yields the highest quality design or code. Here, you do not have other variables that may interact with the design method.

On the other hand, you may want to investigate treatment combinations, rather than competing treatments. For example, instead of looking just at design methods, you want to analyze the effects of design methods in conjunction with tool assistance. Thus, you are comparing design method A with tool assistance to design method A without tool assistance, as well as design method B with tool assistance and without tool assistance. You have a two-factor experiment: one factor is the design method, and the other is the absence or presence of tool assistance. Your experiment tests $n_1 + n_2$ different treatments, where n_1 is the number of levels in factor 1, and n_2 is the number of levels in factor 2.

4.2.4.2 Factors versus Blocks

Once you decide on the number of factors appropriate for your experiment, you must determine how to use blocking to improve the experiment's precision. However, it is not always easy to tell when something should be a block instead of a factor. To see how to decide, we continue to use the example of staff experience with software design, described in Example 4.15.

In many experiments, we suspect that the experience of the subjects will affect the outcome. One option in the experimental design is to treat experience as a blocking factor, as described earlier. To do this, we can assess the experience of the designers in terms of the number of years each has had experience with design. We can match staff with similar experience backgrounds and then assign staff randomly to the different treatments. Thus, if we are investigating design methods A and B, each block will have at least two subjects of approximately equal experience; within each block, the subjects are assigned randomly to methods A and B.

On the other hand, if we treat experience as a factor, we must define levels of experience and assign subjects in each level randomly to the alternative levels of the other factor. In the design example, we can classify designers as having high and low experience (the two levels of experience); then, within each group, subjects are assigned at random to design method A or B.

To determine which approach (factor or block) is best, consider the basic hypothesis. If we are interested in whether design A is better than design B, then experience should be treated as a blocking variable. However, if we are interested in whether the results of using design methods A and B are influenced by staff experience, then experience should be treated as a factor. Thus, as described before, if we are not interested in interactions, then blocking will suffice; if interactions are important, then multiple factors are needed.

In general, then, we offer the following guidelines about blocking:

- If you are deciding between two methods or tools, then you should identify state variables that are likely to affect the results and sample over those variables using blocks to ensure an unbiased assignment of experimental units to the alternative methods or tools.

- If you are deciding among methods or tools in a variety of circumstances, then you should identify state variables that define the different circumstances and treat each variable as a factor.

In other words, use blocks to eliminate bias; use factors to distinguish cases or circumstances.

4.2.4.3 Choosing between Nested and Crossed Designs

Once you have decided on the appropriate number of factors for your experiment, you must select a structure that supports the investigation and answers the questions you have. As we shall see, this decision is often more complicated in software engineering than in other disciplines, because assigning a group not to use a factor may not be sensible or even possible. That is, there are hidden effects that must be made explicit, and there are built-in biases that must be addressed by the structure of the experiment. In addition, other issues can complicate this choice.

Suppose that a company wants to test the effectiveness of two design methods, A and B, on the quality of the resulting design, with and without tool support. The company identifies 12 projects to participate in the experiment. For this experiment, we have two factors: design method and tool usage. The first factor has two levels, A and B, and the second factor also has two levels, use of the tool and lack of use. A crossed design makes use of every possible treatment combination, and it would appear that a crossed design could be used for this experiment.

As shown in Figure 4.10, the 12 projects are organized so that three projects are assigned at random to each treatment in the design. Consider the implications of the design as shown. Any project has been assigned to any treatment. However, unless the tools used to support the method A are exactly the same as the tools used to support the method B, the factor levels for tool usage are not comparable within the two methods. In other words, with a crossed design such as this, we must be able to make sense of the analysis in terms of interaction effects. We should be able to investigate down columns (in this example, does tool usage make a difference for a given method?) as well as across rows (in this example, does method make a difference with the use of a given tool?). With the design in Figure 4.10, the interaction between method and tool usage (across rows) is not really meaningful. The crossed design yields four different treatments based on method and tool usage that allow us to identify which treatment produces the best result. But the design does not allow us to make statements about the interaction between tool usage and method type.

We can look at the problem in a different way by using a nested design, as shown in Figure 4.11. Although the treatment cells appear to be the same when looking at the figures, the nested design is analyzed differently

Crossed		Design Method	
		Method A	Method B
Tool usage	Not used	Projects 1, 2, and 3	Projects 7, 8, and 9
	Used	Projects 4, 5, and 6	Projects 10, 11, and 12

FIGURE 4.10 Crossed design for design methods and tool usage.

Design Method			
Method A		Method B	
Tool usage		Tool usage	
Not used	Used	Not used	Used
Projects 1, 2, and 3	Projects 4, 5, and 6	Projects 7, 8, and 9	Projects 10, 11, and 12

FIGURE 4.11 Nested design for design methods and tool usage.

from the crossed design (a one-way analysis of variance, as opposed to a two-way analysis of variance), so there is no risk of meaningless interaction effects, as there was with the crossed design. This difference in analysis approach will be discussed in more detail in Chapter 6.

Thus, a nested design is useful for investigating one factor with two or more conditions, while a crossed design is useful for looking at two factors, each with two or more conditions. This rule of thumb can be extended to situations with more than two factors. However, the more the factors, the more complex the resulting analysis. For the remainder of this chapter, we focus on at most two factors, as most situations in software engineering research will involve only one or two factors, with blocking and randomization used to ameliorate the effects of other state variables.

Figure 4.12 summarizes some of the considerations explained so far. Its flowchart helps you to decide on the number of factors, whether to use blocks, and whether to consider a crossed or nested design.

However, there are other, more subtle issues to consider when selecting a design. Let us examine two more examples to see what kinds of problems may be hidden in an experimental design. Consider first the crossed design described by Figure 4.13. The design shows an experiment to investigate two factors: staff experience and design method type. There are two levels of experience, high and low, and two types of design method. The staff can be assigned to a project after the project's status is determined by a randomization procedure. Then, the project can be assigned to a treatment combination. This example illustrates the need to randomize in several ways, as well as the importance of assigning subjects and treatments in an order that makes sense to the design and the goals of the experiment.

Figure 4.14 is similar to Figure 4.13, except that it is examining the method usage, as opposed to method type. In this case, it is important to define exactly what is meant by "not used." Unlike medicine and agriculture, where "not used" means the use of a placebo or the lack of treatment

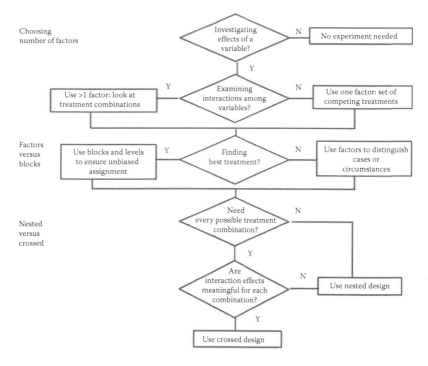

FIGURE 4.12 Flowchart for choosing design.

Crossed		Design method	
		Method A	Method B
Staff experience	Low	Projects 1, 2, and 3	Projects 7, 8, and 9
	High	Projects 4, 5, and 6	Projects 10, 11, and 12

FIGURE 4.13 Crossed design for method types and staff experience.

Crossed		Design method	
		Used	Not used
Staff experience	Low	Projects 1, 2, and 3	Projects 7, 8, and 9
	High	Projects 4, 5, and 6	Projects 10, 11, and 12

FIGURE 4.14 Crossed design for method usage and staff experience.

with a chemical, "not used" in software engineering may be difficult or impossible to control. If we tell designers not to use a particular method, they are likely to use an alternative method, rather than no method at all. The alternative method may be hidden, based on how they were trained or what experience they have, rather than an explicitly defined and well-documented another method. In this case, the design is inappropriate for the goals of the experiment. However, if the goal of the experiment is to assess the benefit of a tool to support the given method, then the design is sufficient.

4.2.4.4 Fixed and Random Effects

Some factors allow us to have complete control over them. For example, we may be able to control what language is used to develop a system, or what processor the system is developed on. But other factors are not easy to control, or are predetermined; staff experience is an example of this type of factor. The degree of control over factor levels is an important consideration in choosing an experimental design. A *fixed-effects model* has factor levels or blocks that are controlled. A *random-effects model* has factor levels or blocks that are random samples from a population of values.

EXAMPLE 4.16

If staff experience is used as a blocking factor to match subjects of similar experience prior to assigning them to a treatment, then the actual blocks are a sample of all possible blocks, and we have a random-effects model. However, if staff experience is defined as two levels, low and high, the model is a fixed-effects model.

The difference between fixed- and random-effects models affects the way the resulting data are analyzed. For completely randomized experiments, there is no difference in analysis. But for more complex designs, the difference affects the statistical methods needed to assess the results. If you are not using a completely randomized experiment, you should consult a statistician to verify that you are planning to use techniques appropriate to the type of effects in your model.

The degree of randomization also affects the type of design that is used in your experiment. You can choose a crossed design when subjects can be assigned to all levels (for each factor) at random. For example, you may be comparing the use of a tool (in two levels: using the tool and not using the tool) with the use of a computing platform (using Linux, Windows, or a Mac OS, for instance). Since you can assign developers to each level at random, your crossed design allows you to look for interaction between the tool and the platform. On the other hand, if you are comparing tool usage and experience (low level of experience versus high level of experience), then you cannot assign people at random to the experience category; a nested design is more appropriate here.

4.2.4.5 Matched- or Same-Subject Designs

Sometimes, economy or reality prevents us from using different subjects for each type of treatment in our experimental design. For instance, we may not find enough programmers to participate in an experiment, or we do not have enough funds to pay for a very large experiment. We can use the same subjects for different treatments, or we can try to match subjects according to their characteristics in order to reduce the scale and cost of the experiments. For example, we can ask the same programmer to use tool *A* in one situation and then tool *B* in another situation. The design of matched- or same-subject experiments allows variation among staff to be assessed and accounts for the effects of staff differences in analysis. This type of design usually increases the precision of an experiment, but it complicates the analysis.

Thus, when designing your experiment, you should decide how many and what type of subjects you want to use. For experiments with one factor, you can consider testing the levels of the factor with the same subjects or with different subjects. For two or more variables, you can consider the question of same-or-different separately for each variable. To see how, suppose you have an experimental design with four different treatments, generated by a crossed design with two factors. If different subjects are used for each treatment (i.e., for each of both variables), then you have a completely unrelated between-subjects design. Alternatively, you could use the same subjects (or subjects matched for similar values of each level) and subject them to all four treatments; this is a completely related within-subjects design. Finally, you can use the same subjects for one factor but different subjects for the other factor to yield a mixed between- and within-subjects design.

4.2.4.6 Repeated Measurements

In many experiments, one measurement is made for each item of interest. However, it can be useful to repeat measurements in certain situations. Repeating a measurement can be helpful in validating it, by assessing the error associated with the measurement process. We explain the added value of repeated measurements by describing an example.

Figure 4.15 depicts the results of an experiment involving one product and three developers. Each developer was asked to calculate the number

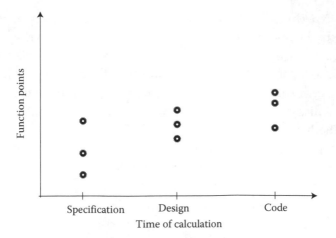

FIGURE 4.15   Repeated measurements on function point calculations.

of function points in the product at each of three different times during development: after the specification was completed, after the design was finished, and after the code was done. Thus, in the figure, there are three points marked at each of the three estimation times. For example, at specification, the three developers produced different function point estimates, so there are three distinct points indicated above "specification" on the *x*-axis. The figure shows that there were two kinds of variation in the data that resulted. The horizontal variation indicates the variation over time, while the vertical differences at each measurement time indicates the variation due to the differences among the developers. Clearly, these repeated measurements add value to the results of the experiment, but at the cost of the more complex analysis required. The horizontal variation helps us to understand the error about the line connecting the means at each measurement time, and the vertical error helps us to understand observational error.

As you can see, there are many issues to consider when choosing a design for your experiment. In the following section, we look at case studies, to see how their design differs from experiments.

4.3 PLANNING CASE STUDIES AS QUASI-EXPERIMENTS

Case studies are often the only practical way to evaluate new methods so that the results will scale up for use in industrial-sized development projects. Many of the issues involved in planning case studies are the same as those relevant to controlled experiments. A case study should be treated as a quasi-experiment and be designed, with limitations, in a similar manner to an experiment. The differences are primarily on the level of control of variables, and the ability to study multiple treatments. In this section, we examine some of the differences and describe the steps to follow.

Every case study requires conception, design, preparation, execution, analysis, dissemination, and decision-making, just as with an experiment. The hypothesis setting is particularly important, as it guides what you measure and how you analyze the results. The projects you select for inclusion in a study must be chosen carefully, to represent what is typical in your organization or company.

A case study usually compares one situation with another: the results of using one method or tool with the results of using another, for example. To avoid bias and make sure that you are testing the relationship you hypothesize, you can organize your study in one of three ways: sister project, baseline, or random selection.

4.3.1 Sister Projects

Suppose your organization is interested in modifying the way it performs code inspections. You decide to perform a case study to assess the effects of using a new inspection technique. To perform such a study, you select two projects, called *sister projects,* each of which is typical of the organization and has similar values for the state variables that you have planned to measure. For instance, the projects may be similar in terms of application domain, implementation language, specification technique, and design method. Then, you perform inspections the current way on the first project, and the new way on the second project. By selecting projects that are as similar as possible, you are controlling as much as you can. This situation allows you to attribute any differences in result to the difference in inspection technique.

4.3.2 Baselines

If you are unable to find two projects similar enough to be sister projects, you can compare your new inspection technique with a general *baseline*. Here, your company or organization gathers data from its various projects, regardless of how different one project is from another. In addition to the variable information mentioned above, the data can include descriptive measures, such as product size, effort expended, number of faults discovered, and so on. Then, you can calculate measures of central tendency and dispersion on the data in the database, so you have some idea of the "average" situation that is typical in your company. Your case study involves completing a project using the new inspection technique, and then comparing the results with the baseline. In some cases, you may be able to select from the organizational database a subset of projects that is similar to the one using the new inspection technique; again, the subset adds a degree of control to your study, giving you more confidence that any differences in result are caused by the difference in inspection technique.

4.3.3 Partitioned Project

Sometimes, it is possible to partition a single project into parts, where one part uses the new technique while the other does not. If possible, you can randomly assign the code components to either the old inspection technique or the new. Then, the case study resembles a controlled experiment, because you are taking advantage of randomization and replication in performing your analysis. It is not a controlled experiment, however, because the project was not selected at random from among the others in the company or organization. As with an experiment, the randomization helps to reduce the experimental error and balance out the confounding factors. It is often not possible to use random selection to partition the project due to pragmatic concerns, especially in a retrospective study.

4.3.4 Retrospective Case Study

You can also conduct a case study on past data from projects. Then you must identify projects using the new method (treatment projects) and similar projects that do not use the new method (control projects). You can also conduct a case study involving portions of a system, by studying version histories and error logs. Such retrospective case studies are limited to available data, and you have limited or no control of variables. However, this allows you to study the evolution of systems over multiple versions and potentially over many years. A number of retrospective studies have used software repositories as sources for case study data.

4.4 RELEVANT AND MEANINGFUL STUDIES

There are many areas of software engineering that can be analyzed using surveys, case studies, and experiments. One key motivator for using a controlled experiment rather than a case study or survey is that the results of an experiment are usually more generalizable. That is, if you use a survey or case study to understand what is happening in a certain organization, the results apply only to that organization (and perhaps to organizations that are very similar). But because an experiment is carefully controlled and involves different values of the controlled variables, its results are generally applicable to a wider community and across many organizations. However, experiments may be limited to studying just a few variables and often do not scale up to an industrial-sized project. In the examples below, we will see how each type of empirical study can help to answer a variety of questions.

4.4.1 Confirming Theories and "Conventional Wisdom"

Many techniques and methods are used in software engineering because "conventional wisdom" suggests that they are the best approaches. Indeed, many corporate, national, and international standards are based on conventional wisdom. For example, many organizations use standard limits on structural measures or rules of thumb about module size to "assure" the quality of their software; they insist that the cyclomatic number be 10 or less, or they restrict module size to 200 lines of code. However, there is very little quantitative evidence to support claims of effectiveness or utility of these and many other standards, methods, or tools. Case studies and surveys can be used to confirm the wisdom of these claims in a single organization, while controlled experiments can investigate the situations in which the claims are generally valid. Thus, empirical studies can be used to provide a context in which certain standards, methods, and tools are recommended for use.

EXAMPLE 4.17

According to the "conventional wisdom" of many software design experts, a good design for an object-oriented system should make use of delegated rather than centralized control. That is, the responsibility for managing actions should be distributed between classes rather than in a few control classes. The experts, as well as most object-oriented design textbooks, claim that a system that uses delegated control will be easier to maintain than one with centralized control (Larman 2004; Fowler 1999).

Arisholm and Sjoberg conducted a controlled experiment to compare the maintainability of delegated control to that of centralized control (Arisholm and Sjoberg 2004). The experimental objects were two implementations of a simple application—a coffee vending machine. One implementation used centralized control, while the other used delegated control. Arisholm and Sjoberg converted the "conventional wisdom" into hypotheses concerning the effect of control style on the time required to perform a set of change tasks and the correctness of the tasks (the dependent variables). The 158 subjects included undergraduate and graduate students, as well as professional Java developers with three levels of experience. The classification of the subjects is the primary independent variable. The experiment assigned subjects using both randomization and blocking to perform the change tasks using either centralized or delegated control.

The results showed that only senior professional developers could complete the maintenance task faster on the implementations that used delegated control. There were more errors on the systems using delegated control for all groups, except the senior professional developers (which had similar error rates for both delegated and centralized designs). Only the senior developers benefited from delegated control.

Often, senior developers determine the overall design and initial implementation of a system, and they are likely to see benefits from using delegated control. Junior developers often perform the maintenance activities. Unfortunately, the experiment suggests that junior developers will need more time and will make more mistakes when modifying software that uses a delegated control.

EXAMPLE 4.18

A number of studies of open-source software projects report that the defect density (defects per thousand lines of code) is higher in small modules than in large modules. This relationship is called the *theory of relative defect proneness*. In a retrospective case study, Koru, Liu, Zang, and El Emam provide evidence that the theory of relative defect proneness also holds for closed-source software projects. The study examined data from ten closed-source systems from NASA, IBM, and other companies (Koru et al. 2010).

4.4.2 Exploring Relationships

Software practitioners are interested in the relationships among various attributes of resources and software products. For example

- How does the project team's experience with the application area affect the quality of the resulting code?

- How does the requirements quality affect the productivity of the designers?

- How does the design structure affect the maintainability of the code?

The relationship can be suggested by a case study or survey, and further explored in follow-up studies. For instance, a survey of completed projects may reveal that software written in Java has fewer faults than projects written in other languages. Clearly, understanding and verifying these relationships is crucial to the success of any future project.

EXAMPLE 4.19

Suppose you want to explore the relationship between programming language and productivity. Your hypothesis may state that certain types of programming languages (e.g., object-oriented languages) make programmers more productive than other types (e.g., procedural languages). A careful experiment would involve measuring not only the type of language and the resulting productivity but also controlling other variables that might affect the outcome. That is, a good experimental design would ensure that factors such as programmer experience, application type, or development environment were controlled so that they would not confuse the results. After analyzing the outcome, you would be able to conclude whether or not programming language affects programmer's productivity.

EXAMPLE 4.20

Identifying the design properties that are common in the most error-prone modules will help us to develop more reliable software. Briand, Wüst, Ikonomovski, and Lounis conducted a case study to explore the relationship between module coupling and cohesion, and module fault proneness in an industrial software system. The study applied 28 coupling measures and 10 cohesion measures. Faults were based on those reported by users over a 1-year period. The study revealed that coupling, particularly import coupling, had the strongest relationship with fault proneness (Briand et al. 1999).

EXAMPLE 4.21

There are many claimed benefits of using an agile software development process. Petersen and Wohlin used both a survey and case study methods to evaluate the perceived effects of migrating from a "plan-driven" software process to one that employed agile methods at Ericsson AB in Sweden (Petersen and Wohlin 2010). The survey involved interviews with 33 developers to collect qualitative information about perceptions of negative process "issues" such as "bottlenecks, unnecessary work, and rework." In addition, Petersen and Wohlin collected quantitative data on "unnecessary work ... and rework." They found more negative issues associated with the plan-driven process, and several benefits from agile methods. They also identified the need for further support for testing and team coordination when applying agile methods.

4.4.3 Evaluating the Accuracy of Prediction Models

Models (as part of prediction systems) are often used to predict the outcome of an activity or to guide the use of a method or tool. For example, cost models predict how much the development or maintenance effort is likely to cost. Capability maturity models guide the use of techniques such as configuration management or the introduction of testing tools and methods. Empirical studies can confirm or refute the accuracy and dependability of these models and their generality by comparing the predictions with the actual values in a carefully controlled environment.

Prediction models present a particularly difficult problem when designing an experiment or case study, because their predictions often affect the outcome. That is, the predictions become goals, and the developers strive to meet the goal, intentionally or not. This effect is common when cost and schedule models are used, and project managers turn the predictions into targets for completion. For this reason, experiments evaluating models can be designed as "double-blind" experiments, where the participants do not know what the prediction is until after the experiment is done. On the other hand, some models, such as reliability models, do not influence the outcome, since reliability measured as mean time to failure cannot be evaluated until the software is ready for use in the field. Thus, the time between consecutive failures cannot be "managed" in the same way that project schedules and budgets are managed.

EXAMPLE 4.22

Kpodjedo, Ricca, Galinier, Gueheneuc, and Antoniol developed a prediction model that uses design evolution measures—changes to class attributes, methods, and relationships between classes (associations and generalizations)—to predict the most fault-prone classes. They developed the prediction model via a case study involving multiple versions of three open-source software systems. The changes were identified from the source code and the faults from both Bugzilla records and Subversion commit messages. The case study suggests that the design evolution measures are better predictors of fault proneness than more traditional static design measures (Kpodjedo et al. 2011).

4.4.4 Validating Measures

Many software measures have been proposed to capture the value of a particular attribute. For example, several measures claim to measure the complexity of code. As we saw in Chapter 3, a measure is said to be *valid* if it reflects the characteristics of an attribute under differing conditions. Thus, suppose code module X has complexity measure C. The code is augmented with new code, and the resulting module X' is now perceived to be much more difficult to understand. If the complexity measure is valid, then the complexity measure C' of X' should be larger than C. In general, a study can be conducted to test whether a given measure appropriately reflects changes in the attribute it is supposed to capture.

As we have seen, validating measures is fraught with problems. Often, validation is performed by correlating one measure with another. But surrogate measures used in this correlation can mislead the evaluator. It is very important to validate using a second measure that is a direct and valid measure of the factor it reflects. Such measures are not always available or easy to measure. Moreover, the measures used must conform to human notions of the factor being measured. For example, if system A is perceived to be more reliable than system B, then the measure of reliability of A should be larger than that for system B; that is, the perception of "more" should be preserved in the mathematics of the measure. This preservation of relationship means that the measure must be objective and subjective at the same time: objective in that it does not vary with the measurer, but subjective in that it reflects the intuition of the measurer (as we saw in Chapter 2).

EXAMPLE 4.23

In order to identify and quantify software security attributes, Manadhata and Wing introduced the notion of a system's *attack surface,* which is "the subset of its resources that an attacker can use to attack the system" (Manadhata and Wing, 2011). They define an empirical relation system to compare the relative attack surface of two systems in terms of the systems' externally accessible channels, methods, and files along with weighting factors to account for the relative damage potential and attack effort for each surface element. They use the empirical relation system and a survey of software developers and system administrators to derive an attack surface metric and validate the metric in the narrow sense. This validation reduces risks to construct validity of studies that use the metric.

Manadhata and Wing validate the attack surface metric in the wide sense through a quasi-experiment (or case study) using Microsoft security bulletins as data. The study confirms the relevance of the dimensions of the metric for predicting attacks and shows that the metric can predict the severity of attacks.

Experiments, quasi-experiments, case studies, and surveys produce data to be analyzed. Chapter 5 provides insights on data collection and Chapter 6 explains how to select appropriate analysis techniques.

4.5 SUMMARY

We have described the key activities necessary for designing or evaluating an empirical study in software engineering. We began by explaining the basic principles of empirical studies. In particular, we explained how the ability to control variables could determine the appropriate study type. We showed the relationship between study goals and hypotheses and provided further insight over maintaining control over study variables.

All empirical studies have limitations. Thus, we described common threats to study validity, so that you can identify these threats when you evaluate published studies and so that you can reduce these threats in your own studies. Human subjects are necessary in many empirical studies. Thus, we describe the ethical standards that you must meet in order to use human subjects in your research.

Next, we explained how to plan a controlled experimentation or case study. We listed six process phases for conducting an experiment or case study: conception, design, preparation, execution, analysis, and dissemination. We discussed experimental design options in some detail. In particular, we pointed out that you must consider the need for replication, randomization, and local control in any experiment that you plan to perform. We showed you how you can think of experimental design in terms of crossing and nesting between factors, as well as other design issues. We looked at some special design issues relevant to case studies: sister projects, organizational baselines, partitioned projects, and retrospective case studies.

Finally, we reviewed examples of relevant and meaningful empirical studies that confirm theories, explore relationships, evaluate prediction models, and validate measures.

EXERCISES

1. Why it is not always possible to apply the principle of random assignment in experiments involving people?

2. How might the principle of blocking be used in an experiment studying the effectiveness of a course of instruction on the productivity of software programmers?

3. Your company is not a software company, but it is beginning to acknowledge the role software plays in the different company products. If you were to create a company baseline for use in case studies, what variable information would you try to capture?

4. Your company is about to start a new project to develop a new software system. This project will use (for the first time in the company) test-driven development (also called test-first development) making use of the JUnit test framework to develop the project. The company management has asked you to assess the use of test-driven development on this project and make a recommendation on whether or not to use it in other projects. Explain how you might assess the costs and benefits of using test-driven development, and what considerations you should make in designing the assessment exercise. Make sure that you describe and justify (a) the type of study that you propose (experiment, case study, survey, or combination of methods), (b) any hypotheses, (c) study variables, and (d) study plan/design.

5. Design a controlled experiment to compare the two programming languages C++ and Java in terms of (1) the productivity of programmers using these languages, (2) reliability, and (3) maintainability of software systems developed using these languages. Use a course project from a senior-level software engineering course as the objects of your study and students as subjects. Make sure that your design includes precise descriptions of the following: (a) all hypotheses, and null hypotheses, the treatment and control object, study variables (state, independent, and dependent), and your plan for carrying out the experiment.

6. You are about to begin a large project that uses new tools, techniques, and languages for building a mission-critical product. Your company president wants to know if the new tools, techniques, and languages should become company standards if the product is a success. Explain how you might assess the product's success, and what considerations you should make in designing the assessment exercise.

7. Assume that an empirical study results allow you to conclusively reject the null hypothesis, and the data exhibit a strong and significant correlation (using appropriate statistical tools) between independent variable x and dependent variable y. What else you need to do to show that the study has internal validity before you accept the alternative hypothesis that x causes y?

FURTHER READING

Curtis was probably the first to address the need for measurement and experimentation in software engineering.

Curtis B. Measurement and experimentation in software engineering, *Proceedings of the IEEE*, 68(9), 1144–1157, September 1980.

Basili and his colleagues produced one of the first papers to point out the need for a rigorous experimentation framework in software engineering.

Basili V.R., Selby R.W., and Hutchens D.H., Experimentation in software engineering, *IEEE Transactions on Software Engineering*, 12(7), 733–743, July 1986.

Wohlin, Runeson, Höst, Ohlsson, Regnell, and Wesslén published one of the most comprehensive books focused on empirical methods in software engineering.

Wohlin C., Runeson P., Höst, M., Ohlsson, M.C., Regnell B., and Wesslén, A., *Experimentation in Software Engineering: An Introduction*, Kluwer Academic Publishers, Norwell, MA, USA, 2000.

One journal exclusively publishes empirical software engineering papers.

Empirical Software Engineering, published by Springer.

A number of journals include empirical studies in all areas of software engineering.
IEEE Transactions on Software Engineering.
ACM Transactions on Software Engineering and Methodology.
Information and Software Technology, published by Elsevier.
Journal of Systems and Software, published by Elsevier.
Software: Practice and Experience, published by Wiley.

Several journals publish empirical studies on focused topics in software engineering.

Software Testing Verification and Reliability, published by Wiley.
Software Quality Journal, published by Springer.
Automated Software Engineering, published by Springer.
Journal of Software Maintenance and Evolution: Research and Practice, published by Wiley.

REFERENCES

Arisholm E. and Sjoberg D.I.K., Evaluating the effect of a delegated versus centralized control style on the maintainability of object-oriented software, *IEEE Transactions on Software Engineering*, 30(8), 521–534, August 2004.

Briand L.C., Wüst, J., Ikonomovski, S.V., and Lounis H., Investigating quality factors in object-oriented designs; An industrial case study. *Proceedings of International Conference on Software Engineering (ICSE)*, 345–354, 1999.

Fowler M., *Refactoring: Improving the Design of Existing Code*. Addison-Wesley, Reading, MA, 1999.

Koru G., Liu H., Zang, D., and Emam E., Testing the theory of relative defect proneness for closed-source software, *Empirical Software Engineering*, 15, 577–598, 2010.

Kpodjedo S., Ricca, F., Galinier, P., Gueheneuc, Y.-G., Antoniol, G., Design evolution metrics for defect prediction in object oriented systems, *Empirical Software Engineering*, 16, 141–175, 2011.

Larman C., *Applying UML and Patterns: an Introduction to Object-Oriented Analysis and Design and Iterative Development*, 3rd Edition, Prentice-Hall PTR, Upper Saddle River, NJ, 2004.

Manadhata P.K. and Wing J.M., An Attack Surface Metric, *IEEE Transactions on Software Engineering*, 37(3), 371–386, May/June 2011.

Petersen K. and Wohlin C. The effect of moving from a plan-driven to an incremental software development approach with agile practices, *Empirical Software Engineering*, 15, 654–693, 2010.

U.S. Department of Health and Human Services, Code of Federal Regulations (CRF), TITLE 45 PUBLIC WELFARE, PART 46 PROTECTION OF HUMAN SUBJECTS, Page 4 (on the pdf version of the regulation), 2009. http://www.hhs.gov/ohrp/humansubjects/guidance/45cfr46.html, August 7, 2014.

There are several good books addressing experimentation in general. Cook, Campbell, and Day discuss many of the issues involved in doing experiments where it is difficult to control all the variables. The other books address statistics and experimental design.

Campbell D.T. and Stanley J., *Experimental and Quasi-Experimental Designs for Research*, Rand McNally, Chicago, 1966.

Cochran W.G., *Sampling Techniques*, 2nd Edition, John Wiley and Sons, New York, 1963.

Cook, T.D., Campbell D.T., and Day A., *Quasi-Experimentation: Design and Analysis Issues for Field Settings*, Houghton-Mifflin, Boston, Massachusetts, 1979.

Green J. and d'Oliveira M., *Units 16 & 21 Methodology Handbook (Part 2)*, Open University, Milton Keynes, England, 1990.

Lee W., *Experimental Design and Analysis*, W.H. Freeman and Company, San Francisco, California, 1975.

Ostle B. and Malone L.C., *Statistics in Research*, 4th Edition, Iowa State University Press, Ames, Iowa, 1988.

Walker M., *The Nature of Scientific Thought*, Prentice-Hall, Inc. Englewood Cliffs, NJ, 1963.

CHAPTER 5

Software Metrics Data Collection

W E SAW IN CHAPTER 3 that measurements should be tied to an organization, project, product and process goals, so that we know what questions we are trying to answer with our measurement program. Chapter 4 showed us how to frame questions as hypotheses, so that we can perform measurement-based investigations to help us understand the answers. But having the right measures is only part of a measurement program. Software measurement is only as good as the data that are collected and analyzed. In other words, we cannot make good decisions with bad data.

> Data should be collected with a clear purpose in mind. Not only a clear purpose but also a clear idea as to the precise way in which they will be analysed so as to yield the desired information. ... It is astonishing that men, who in other respects are clear-sighted, will collect absolute hotchpotches of data in the blithe and uncritical belief that analysis can get something out of it.
>
> MORONEY 1962, P. 120

In this chapter, we consider what constitutes good data, and we present guidelines and examples to show how data collection supports decision making. In particular, we focus on the terminology and organization of data related to software quality, including faults and failures. We also discuss issues related to collecting data on software changes, effort, and productivity.

5.1 DEFINING GOOD DATA

It is very important to assess the quality of data and data collection before data collection begins. Your measurement program must specify not only what metrics to use, but what precision is required, what activities and time periods are to be associated with data collection, and what rules govern the data collection (such as whether a particular tool will be used to capture the data). Of critical importance is the definition of the metric. Terminology must be clear and detailed, so that all involved understand what the metric is and how to collect it.

There are two kinds of data with which we are concerned. As illustrated by Figure 5.1, there is raw data that results from the initial measurement of process, product, or resource. But there is also a refinement process, extracting essential data elements from the raw data so that analysts can derive values about attributes.

To see the difference, consider the measurement of developer effort. The raw effort data may consist of weekly time sheets for each staff member working on a project. To measure the effort expended on the design so far, we must select all relevant time sheets and add up the figures. This refined data are a direct measurement of effort. But we may derive other measures as part of our analysis: for example, average effort per staff member, or effort per design component.

Deciding what to measure is the first step. We must specify which direct measures are needed, and also measures that may be derived from the direct ones. Sometimes, we begin with the derived measures. From a goal, question, metric (GQM) analysis, we understand which derived measures we want to know; from those, we must determine which direct measures are required to calculate them.

Most organizations are different, not only in terms of their business goals but also in terms of their corporate cultures, development preferences, staff skills, and more. So, a GQM analysis of apparently similar projects may result in different metrics in different companies. That is exactly why GQM is preferable to adopting a one-size-fits-all standard measurement set. Nevertheless, most organizations share similar problems. Each

FIGURE 5.1　The role of data collection in software measurement.

is interested in software quality, cost, and schedule. As a result, most developers collect metrics information about quality, cost, and duration. In the next section, we shall learn about the importance of data definition by examining possible definitions for attributes and measures related to the reporting of problems that occur in software development.

5.2 DATA COLLECTION FOR INCIDENT REPORTS

No software developer consistently produces perfect software the first time. Thus, it is important for developers to measure aspects of software quality. Such information can be useful for determining

- How many problems have been found with a product?

- How effective are the prevention, detection, and removal processes?

- Whether the product is ready for release to the next development stage or to the customer?

- How the current version of a product compares in quality with previous or competing versions?

The terminology used to support this investigation and analysis must be precise, allowing us to understand the causes as well as the effects of quality assessment and improvement efforts. However, the use of terms varies widely among software professionals, and terms such as "error," "fault," "failure," and so forth are used inconsistently. For example, there is disagreement with the definition of a software *failure*.

From a formal computer science perspective, a failure is any deviation from *specified behavior*. That is, a failure occurs whenever the output does not match the specified behavior. Using the formal perspective, there is no failure when program behavior matches an incorrect specification even if the behavior does not make sense.

From an engineering perspective, a failure is any deviation from the required or *expected* behavior. Thus, if a program receives an unspecified input it should produce a "sensible" output appropriate for the circumstances.

EXAMPLE 5.1

Consider the following specification for a program:

When a non-negative integer *n* is input into the factorial program, it shall produce *n*! as an integer string for values of *n* up to and including 149.

Table 5.1 gives interpretations of failure events caused by running the program. From a formal perspective, if the program input is an unspecified value (151 or "fred"), it is not a failure for the system to crash or wipe the file system. From an engineering perspective, these behaviors are deemed failures since such output is not sensible and can have catastrophic consequences.

5.2.1 The Problem with Problems

Figure 5.2 depicts some of the components of a problem's cause and symptoms, expressed in terms consistent with Institute of Electrical and Electronic Engineers (IEEE) standard 610.12 (IEEE 610.12-1990) and IEEE standard 1044-2009 (IEEE 1044-2009). A *fault* in a software product occurs due to a human error or mistake. That is, the fault is the encoding of the human error. For example, a developer might misunderstand a user-interface requirement, and therefore create a design that includes the misunderstanding. The design fault can also result in an incorrect code, as well as incorrect instructions in the user manual. Thus, a single error can result in one or more faults, and a fault can reside in any of the products of development.

However, a *failure* is the departure of a system from its required behavior. Failures can be discovered both before and after system

TABLE 5.1　Formal versus Engineering Interpretations of Failure Events for a Factorial Program

Input	Output	Formal Failure?	Engineering Failure?
150	"Too large a number"	No	No
151	System crash	No	Yes
"fred"	Delete all files	No	Yes

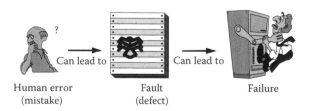

Human error　　　　　Fault　　　　　Failure
(mistake)　　　　　(defect)

FIGURE 5.2　Software quality terminology. Terminology from the IEEE standard 1044-2009 is in parentheses when it differs from ours. (IEEE Standard 1044-2009: *Standard Classification for Software Anomalies*, IEEE Computer Society Press, 2009.)

delivery, as they can occur in testing as well as in operation. Since we take an engineering perspective to identify failures, we compare actual system behavior, rather than with specified behavior. Thus, faults in the requirements documents can result in failures, too.

In some sense, you can think of faults and failures as inside and outside views of the system. Faults represent problems that the developer sees, while failures are problems that the user sees. Not every fault corresponds to a failure, since the conditions under which a fault results in system failure may never be met.

EXAMPLE 5.2

The software in an actual nuclear reactor failed due to a fault that was caused by the process described below:

1. *Human error:* Failure to distinguish signed and absolute value numbers in an algorithm resulted in the following fault.
2. *Fault:* '*X*: = *Y*' is coded instead of '*X*: = *ABS(Y)*' which in turn led to the following failure.
3. *Failure:* Nuclear reactor shut down because it was wrongly determined that a meltdown was likely.

A failure may or may not occur depending on the sequence of inputs processed by the system—a fault may reside in a program segment that is never executed, thus preventing the processing error from occurring.

We need to introduce another stage to this failure process that of the software error. Ammann and Offutt define the term *software error* as "an incorrect internal state that is the manifestation of some fault" (Ammann and Offutt, 2008). This notion of error is commonly used in the software testing community, which is concerned with identifying software errors. In general, the triggering of a fault may not lead to instantaneous failure; rather, an erroneous state arises within the system which at some later time leads to failure. We can think of software errors as intermediate error states which when propagated through the system ultimately lead to failure. One example is an operating system fault that is triggered by a particular input that results in a state error that causes another state error in the word processor which then crashes. Thus, the fault starts a chain of software errors that ultimately result in a failure.

This notion of software error is also highly relevant for software *fault tolerance*, which is concerned with how to prevent failures in the presence of software errors. If an error state is detected and countered before it propagates to the output, the failure can be prevented. Preventing error states from propagating is the goal of software fault tolerance. Researchers continue to search for methods that enable software to operate dependably even when it contains faults that have been triggered. Software errors do not necessarily cause failures in fault-tolerant software.

Unfortunately, the terminology used to describe software problems is not uniform. If an organization measures software quality in terms of faults per thousand lines of code, it may be impossible to compare the result with the competition if the meaning of "fault" is not the same. The software engineering literature is rife with differing meanings for the same terms. Below are just a few examples of how researchers and practitioners differ in their usage of terminology.

- In many organizations, *errors* can mean faults. This meaning contrasts with the notion of *software error* as used in the software testing community—an invalid system state that results when a fault is triggered and may or may not lead to a failure.

- *Anomalies* usually mean a class of faults that are unlikely to cause failures in themselves but may nevertheless eventually lead to failures indirectly. In this sense, an anomaly is a deviation from the usual, but it is not necessarily wrong. For example, deviations from accepted standards of good programming practice (such as use of nonmeaningful names) are often regarded as anomalies. Note that the *IEEE Standard Glossary of Software Engineering Terminology* defines an anomaly as "anything observed in the documentation or operation of software that deviates from expectations" (IEEE 610.12-1990). Also the IEEE Standard Classification for Software Anomalies addresses both faults (defects) and failures (IEEE 1044-2009).

- *Defects* often refer collectively to faults and failures. However, the IEEE Standard 1044-2009 uses the term "defect" to refer only to faults (IEEE 1044-2009).

- *Bugs* refer to faults occurring in the code, but in some cases are used to describe failures.

- *Crashes* are a special type of failure, where the system ceases to function.

Since the terminology used varies widely, it is important for you to define your terms clearly, so that they are understood by all who must supply, collect, analyze, and use the data. Often, differences in meaning are acceptable, as long as the data can be translated from one framework to another.

We also need a good, clear way of describing what we do in reaction to problems. For example, if an investigation of a failure results in the detection of a fault, then we make a *change* to the product to remove it. A change can also be made if a fault is detected during a review or inspection process. In fact, one fault can result in multiple changes to one product (such as changing several sections of a piece of code) or multiple changes to multiple products (such as a change to requirements, design, code, and test plans).

In this book, we describe the observations of development, testing, system operation, and maintenance problems in terms of failures, faults, and changes. Whenever a problem is observed, we want to record its key elements, so that we can investigate the causes and cures. In particular, we want to know the following:

1. *Location*: Where did the problem occur?

2. *Timing*: When did it occur?

3. *Symptom*: What was observed?

4. *End result*: Which consequences resulted?

5. *Mechanism*: How did it occur?

6. *Cause*: Why did it occur?

7. *Severity:* How much was the user affected?

8. *Cost:* How much did it cost?

These eight attributes are similar to the failure attributes and defect (fault) attributes in the IEEE Standard Classification for Software Anomalies (IEEE 1044-2009). However, the IEEE standard for failure attributes does not separate *symptom* from *end result*; both are included in the general attribute called "description." The IEEE standard for defect (fault) attributes includes *symptom* under the "description" attribute, and *end result* under the "effect" attribute.

We use the above eight problem attributes because they are (as far as possible) mutually independent, so that proposed measurement of one does not affect measurement of another*; this characteristic of the attributes is called *orthogonality*. Orthogonality can also refer to a classification scheme within a particular category. For example, cost can be recorded as one of the several predefined categories, such as *low* (under $100,000), *medium* (between $100,000 and $500,000), and *high* (more than $500,000). However, in practice, attempts to over-simplify the set of attributes sometimes result in nonorthogonal classifications. When this happens, the integrity of the data collection and metrics program can be undermined, because the observer does not know in which category to record a given piece of information.

EXAMPLE 5.3

Some organizations try to provide a single classification for software faults, rather than using the eight components suggested above. Here, one attribute is sometimes regarded as most important than the others, and the classification is a proposed ordinal scale measure for this single attribute. Consider the following classification for faults, based on severity:

- Major
- Minor
- Negligible
- Documentation
- Unknown

This classification is not orthogonal; for example, a documentation fault could also be a major problem, so there is more than one category in which the fault can be placed. The nonorthogonality results from confusing severity with location, since the described fault is located in the documentation but has a severe effect.

When the scheme is not orthogonal, the developer must choose which category is most appropriate. It is easy to see how valuable information is lost or misrepresented in the recording process.

---

* This mutual independence applies only to the initial measurement. Data analysis may reveal, for example, that severity of effects correlates with the location of faults within the product. However, this correlation is discovered after classification, and cannot be assumed when faults are classified.

EXAMPLE 5.4

Riley described the data collection used in the analysis of the control system software for the Eurostar train (the high-speed train used to travel from Britain to France and Belgium via the Channel tunnel) (Riley 1995). In the Eurostar software problem-reporting scheme, faults are classified according to only two attributes, cause and category, as shown in Table 5.2. Note that "cause" includes notions of timing and location. For example, an error in software implementation could also be a deviation from functional specification, while an error in test procedure could also be a clerical error. Hence, Eurostar's scheme is not orthogonal and can lead to data loss or corruption.

On the surface, our eight-category report template should suffice for all types of problems. However, as we shall see, the questions are answered very differently, depending on whether you are interested in faults, failures, or changes.

5.2.2 Failures

A failure report focuses on the external problems of the system: the installation, the chain of events leading up to the failure, the effect on the user or other systems, and the cost to the user as well as the developer. Thus, a typical failure report addresses each of the eight attributes in the following way:

TABLE 5.2   Fault Classifications Used in Eurostar Control System

Cause	Category
Error in software design	Category not applicable
Error in software implementation	Initialization
Error in test procedure	Logic/control structure
Deviation from functional specification	Interface (external)
Hardware not configured as specified	Interface (internal)
Change or correction-induced error	Data definition
Clerical error	Data handling
Other (specify)	Computation
	Timing
	Other (specify)

Source: IEEE Standard 1044-2009: *Standard Classification for Software Anomolies*, IEEE Computer Society Press, 2009.

Failure Report

Location: Such as installation where failure was observed

Timing: CPU time, clock time, or some temporal measure

Symptom: Type of error message or indication of failure

End result: Description of failure, such as "operating system crash," "services degraded," "loss of data," "wrong output," and "no output"

Mechanism: Chain of events, including keyboard commands and state data, leading to failure

Cause: Reference to possible fault(s) leading to failure

Severity: Reference to a well-defined scale, such as "critical," "major," and "minor"

Cost: Cost to fix plus cost of lost potential business

Let us examine each of these categories more closely.

Location is usually a code (e.g., hardware model and serial number, or site and hardware platform) that uniquely identifies the installation and platform on which the failure was observed. The installation description must be interpreted according to the type of system involved. For example, if software is embedded in an automatic braking system, the "installation" may move about. Similarly, if the system is distributed, then the terminal at which a failure is observed must be identified, as well as the server to which it was online.

Timing has two, equally important aspects: real time of occurrence (measured on an interval scale), and execution time, up to the occurrence of failure (measured on a ratio scale).

The *symptom* category explains what was observed as distinct from the *end result*, which is a measure of the consequences. For example, the symptom of a failure may record that, the screen displayed a number that was one greater than the number entered by the operator; if the larger number resulted in an item's being listed as "unavailable" in the inventory (even though one was still left), that symptom belongs to the "end result" category.

End result refers to the consequence of the failure. Generally, the "end result" requires a (nominal scale) classification that depends on the type of system and application. For instance, the end result of a failure may be any of the following:

- Operating system crash

- Application program aborted

- Service degraded

- Loss of data

- Wrong output

- No output

Mechanism describes how the failure came about. This application-dependent classification details the causal sequence leading from the activation of the source to the symptoms eventually observed. This category may also characterize what function the system was performing or how heavy the workload was when the failure occurred. Unraveling the chain of events is a part of diagnosis, so often this category is not completed at the time the failure is observed.

Cause is also part of the diagnosis (and as such is more important for the fault form associated with the failure). Cause involves two aspects: the type of trigger and the type of source (i.e., the fault that caused the problem). The trigger can be one of several things, such as

- Physical hardware failure

- Operating conditions

- Malicious action

- User error

- Erroneous report

While the actual source can be faults such as these:

- Physical hardware fault

- Unintentional design fault

- Intentional design fault

- Usability problem

For instance, the cause can involve a *switch* or *case* statement with no code to handle the drop-through condition (the source), plus a situation (the trigger) where the listed cases were not satisfied.

The cause category is often cross-referenced to fault and change reports, so that the collection of reports paints a complete picture of what happened (the failure report), what caused it (the fault reports), and what was done to correct it (the change reports). Then, post-mortem analysis can identify the *root cause* of a fault. The analysis will focus on how to prevent such problems in the future, or at least catch them earlier in the development process.

Severity describes how serious the failure's end result was for the service required from the system. For example, failures in safety-critical systems are often classified for severity as follows (Riley 1995):

- *Catastrophic* failures involve the loss of one or more lives, or injuries causing serious and permanent incapacity.

- *Critical* failures cause serious and permanent injury to a single person but would not normally result in loss of life to a person of good health. This category also includes failures causing environmental damage.

- *Significant* failures cause light injuries with no permanent or long-term effects.

- *Minor* failures result neither in personal injury nor a reduction in the level of safety provided by the system.

Severity may also be measured in terms of cost to the user.

Cost to the system provider is recorded in terms of how much effort and other resources were needed to diagnose and respond to the failure. This information may be a part of diagnosis and therefore supplied after the failure occurs.

Sometimes, a failure occurs several times before it is recognized and recorded. In this case, an optional ninth category, *count*, can be useful. *Count* captures the number of failures that occurred within a stated time interval.

EXAMPLE 5.5

Early on December 31, 2008, Microsoft's first-generation Zune portable media players hung. This event was widely reported in the media since there were many thousands of affected users (Robertson 2009). The failure was traced to a fault in the program code related to date calculations—2008 is a leap year with 366 days, and the code used 365 days as the length of all years. The workaround was to wait until January 1, let the battery drain completely, and then restart the device (this workaround left the fault in place). Using our failure attributes, a possible failure report might look like the following:

- *Location*: Many first-generation Zune 30 media players in use around the world.
- *Timing*: December 31, 2008, starting early in the morning.
- *Symptom*: The device froze.
- *End result*: The device became unusable, even after a restart.
- *Mechanism*: Upon startup, the loading bar indicates "full," and then the device hangs.
- *Cause* (1): (Trigger) Starting the device on December 31, 2011.
- *Cause* (2): (Source type) Coding fault related to date calculation.
- *Severity*: Serious, as it made the device unusable until the inconvenient workaround was communicated to the large and diverse user community.
- *Cost*: Effort to diagnose the problem, develop and publicize a workaround, and repair the fault. Perhaps, the greatest cost was damage to the company reputation.

EXAMPLE 5.6

In the 1980s, problems with a radiation therapy machine were discovered, and the description of the problems and their software causes are detailed in an article by Leveson and Turner (1993). Mellor used the above framework to analyze the first of these failures, in which patients died as a result of a critical design failure of the Therac 25 radiation therapy machine (Mellor 1992). The Therac administered two types of radiation therapy: x-ray and electron. In x-ray mode, a high-intensity beam struck a tungsten target, which absorbed much of the beam's energy and produced x-rays. In electron mode, the Therac's computer retracted the metal target from the beam path while reducing the intensity of the radiation by a factor of 100. The sequence of a treatment session was programmed by an operator using a screen and keyboard; the Therac software then performed the treatment automatically. In each of the accidents that occurred, the electron beam was supposed to be at a reduced level but instead was applied full strength, without the tungsten target in place. At the same time, the message "Malfunction 54" appeared on the monitor screen. Diagnosis revealed that the use of the up-arrow key to correct a typing mistake by the operator activated a fault in the software, leading to an error in the system which scrambled the two modes of operation.

Applying our failure framework to the 1986 accident which caused the death, six months later, of Mr V. Cox, the failure report might look like this:

- *Location*: East Texas Cancer Center in Tyler, Texas, USA.
- *Timing* (1): March 21, 1986, at whatever the precise time that "Malfunction 54" appeared on the screen.
- *Timing* (2): Total number of treatment hours on all Therac 25 machines up to that particular time.
- *Symptom* (1) "Malfunction 54" appeared on the screen.
- *Symptom* (2) Classification of the particular program of treatment being administered, type of tumor, etc.
- *End result*: Strength of the beam too great by a factor of 100.
- *Mechanism*: Use of the up-arrow key while setting up the machine led to the corruption of a particular internal variable in the software.
- *Cause* (1) (Trigger) Unintentional operator action.
- *Cause* (2) (Source type) Unintentional design fault.
- *Severity*: Critical, as injury to Mr Cox was fatal.*
- *Cost*: Effort or actual expenditure by accident investigators.

In Leveson and Turner's post-mortem analysis, they identified several root causes of the failures including the following:

- "Overconfidence in the software" especially "among nonsoftware professionals."
- "Confusing reliability with safety." The software rarely failed. However, the consequences of failure were tragic.
- "Lack of defensive design."
- "Failure to eliminate root causes" of prior failures.
- "Complacency."
- "Unrealistic risk assessments." Software risk was not treated as seriously as hardware risks.
- "Inadequate investigation or follow-up on accident reports."
- "Inadequate software engineering practices."

Remember that only some of the eight attributes can usually be recorded at the time the failure occurs. These are:

- Location

- Timing

- Symptom

- End result

- Severity

---

* The failure is classed as "critical," rather than "catastrophic," in accordance with several safety guidelines that (unfortunately) require the loss of *several* lives for catastrophic failure.

The others can be completed only after diagnosis, including root-cause analysis. Thus, a data collection form for failures should include at least these five categories.

When a failure is closed, the precipitating fault in the product has usually been identified and recorded. However, sometimes there is no associated fault. Here, great care should be exercised when closing the failure report, so that readers of the report will understand the resolution of the problem. For example, a failure caused by user error might actually be due to a usability problem, requiring no immediate software fix (but perhaps changes to the user manual, or recommendations for enhancement or upgrade). Similarly, a hardware-related failure might reveal that the system is not resilient to hardware failure, but no specific software repair is needed.

Sometimes, a problem is known but not yet fixed when another, similar failure occurs. It is tempting to include a failure category called "known software fault," but such classification is not recommended because it affects the orthogonality of the classification. In particular, it is difficult to establish the correct timing of a failure if one report reflects multiple, independent events; moreover, it is difficult to trace the sequence of events causing the failures. However, it is perfectly acceptable to cross-reference the failures, so the relationships among them are clear.

The need for cross-references highlights the need for forms to be stored in a way that allows pointers from one form to another. The storage system must support changes to stored data. For example, a failure may initially be thought to have one fault as its cause, but subsequent analysis reveals otherwise. In this case, the failure's "type" may require change, as well as the cross-reference to other failures.

The form storage scheme must also permit searching and organizing. For example, we may need to determine the first failure due to each fault for several different samples of trial installations. Because a failure may be a first manifestation in one sample, but a repeat manifestation in another, the storage scheme must be flexible enough to handle this.

5.2.3 Faults

A failure reflects the user's view of the system, but a fault is seen only by the developer. Thus, a fault report is organized much like a failure report but has very different answers to the same questions. It focuses on the internals of the system, looking at the particular module where the fault occurred and the cost to locate and fix it. A typical fault report interprets the eight attributes in the following way:

Fault Report

Location: Within system identifier, such as module or document name

Timing: Phases of development during which fault was created, detected and corrected

Symptom: Type of error message reported, or activity which revealed fault (such as review)

End result: Failure caused by the fault

Mechanism: How source was created, detected, and corrected

Cause: Type of human error that led to fault

Severity: Refer to severity of resulting or potential failure

Cost: Time or effort to locate and correct; can include analysis of cost had fault been identified during an earlier activity

Again, we investigate each of these categories in more detail.

In a fault report, *Location* tells us which product (including both identifier and version) or part of the product (subsystem, module, interface, document) contains the fault. The IEEE Standard Classification for Software Anomalies (IEEE 1044-2009) provides location attributes in four levels:

1. *Asset*, which identifies the application with the fault.

2. *Artifact*, for example, a specific code file, requirements document, design specification, test plan, test case.

3. *Version detected*, which identifies the version where the fault was found.

4. Version corrected.

Timing relates to the three events that define the life of a fault:

1. When the fault is created

2. When the fault is detected

3. When the fault is corrected

Clearly, this part of a fault report will need revision as a causal analysis is performed. It is also useful to record the time taken to detect and correct the fault, so that product maintainability can be assessed.

The *Symptom* classifies what is observed during diagnosis or inspection. The 1993 draft version of the IEEE standard on software anomalies (IEEE 1044-2009) provides a useful and extensive classification that we can use for reporting the symptom. We list the categories in the following table:

Classification of Fault Types

Logic problem
    Forgotten cases or steps
    Duplicate logic
    Extreme conditions neglected
    Unnecessary function
    Misinterpretation
    Missing condition test
    Checking wrong variable
    Iterating loop incorrectly
Computational problem
    Equation insufficient or incorrect
        Missing computation
        Operand in equation incorrect
        Operator in equation incorrect
        Parentheses used incorrectly
    Precision loss
        Rounding or truncation fault
        Mixed modes
    Sign convention fault
Interface/timing problem
    Interrupts handled incorrectly
    I/O timing incorrect
        Timing fault causes data loss
    Subroutine/module mismatch
        Wrong subroutine called
        Incorrectly located subroutine call
        Nonexistent subroutine called
        Inconsistent subroutine arguments
Data-handling problem
    Initialized data incorrectly
    Accessed or stored data incorrectly
        Flag or index set incorrectly
        Packed/unpacked data incorrectly

continued

(continued) Classification of Fault Types

Referenced wrong data variable
Data referenced out of bounds
Scaling or units of data incorrect
Dimensioned data incorrectly
Variable-type incorrect
Subscripted variable incorrectly
Scope of data incorrect
Data problem
Sensor data incorrect or missing
Operator data incorrect or missing
Embedded data in tables incorrect or missing
External data incorrect or missing
Output data incorrect or missing
Input data incorrect or missing
Documentation problem
Ambiguous statement
Incomplete item
Incorrect item
Missing item
Conflicting items
Redundant items
Confusing item
Illogical item
Nonverifiable item
Unachievable item
Document quality problem
Applicable standards not met
Not traceable
Not current
Inconsistencies
Incomplete
No identification
Enhancement
Change in program requirements
Add new capability
Remove unnecessary capability
Update current capability
Improve comments
Improve code efficiency
Implement editorial changes
Improve usability
Software fix of a hardware problem
Other enhancement

(continued) Classification of Fault Types
Failure caused by a previous fix Other problems

Source: IEEE Standard 1044-2009, *Standard Classification for Software Anomolies,* IEEE Computer Society Press, 2009.

The *End result* is the actual failure caused by the fault. If separate failure or incident reports are maintained, then this entry should contain a cross-reference to the appropriate failure or incident reports.

Mechanism describes how the fault was created, detected, and corrected. Creation explains the type of activity that was being carried out when the fault was created (e.g., specification, coding, design, maintenance). Detection classifies the means by which the fault was found (e.g., inspection, unit testing, system testing, integration testing), and correction refers to the steps taken to remove the fault or prevent the fault from causing failures.

Cause explains the human error (mistake) that led to the fault. Although difficult to determine in practice, the cause is classified in terms of the suspected causes such as lost information, requirements misunderstanding, management not taking engineering concerns seriously, and so forth.

Severity assesses the impact of the fault on the user. That is, severity examines whether the fault can actually be evidenced as a failure, and the degree to which that failure would affect the user.

The *Cost* explains the total cost of the fault to the system provider. Much of the time, this entry can be computed only by considering other information about the system and its impact.

The optional *Count* field can include several counts, depending on the purpose of the field. For example, count can report the number of faults found in a given product or subsystem (to gauge inspection efficiency), or the number of faults found during a given period of operation (to assist in reliability modeling).

EXAMPLE 5.7

We reexamine the problem with the Zune media player described in Example 5.5. The fault causing the failure was quickly identified as a coding error (Zuneboards, 2008), and was used as an example by Weimer, Forrest, Le Goues, and Nguyen in a study of automated program repair (Weimer et al. 2010).

- *Location*: Module rtc.c, Convert Days function, lines 249–275.
- *Timing*: Created during coding, detected during operational use.
- *Symptom*: Missing condition test causing a loop to iterate incorrectly (nontermination).
- *End result*: The device froze.
- *Mechanism*: Creation: during code development; Detection: diagnosis of operational failure; Correction: workaround provided, code correction probably done.
- *Cause*: Human mistake in dealing with a special case—leap years.
- *Severity*: Serious, as all of the first-generation Zune devices froze.
- *Cost*: Minimal cost to diagnose, prepare workaround, and repair; however, there was significant cost with respect to the reputation of the company.

EXAMPLE 5.8

We return to the Therac problem described in Example 5.6 (Leveson and Turner 1993). The fault causing the failure that we discussed may be reported in the following way:

- *Location*: Product is the Therac 25; the subsystem is the control software. The version number is an essential part of the location. The particular module within the software is also essential.
- *Timing*: The fault was created at some time during the control software's development cycle.
- *Symptom*: The category and type of software fault that were the root cause of the failure.
- *End result*: "Malfunction 54," together with a radiation beam that was very strong.
- *Mechanism*: Creation: during code development; Detection: diagnosis of operational failure; Correction: the immediate response was to remove the up-arrow key from all machines and tape over the hole!
- *Cause*: This is discussed at length by Leveson and Turner (1993).
- *Severity*: Critical,* in spite of the fact that, not all of the "Malfunction 54" failures led to injury.
- *Cost*: The cost to the Therac manufacturers of all investigations of all the failures, plus corrections.

---

* Again, the failure is not catastrophic, because only one life was lost.

EXAMPLE 5.9

The popular open source problem tracking tool Bugzilla, which describes both failures and faults as "bugs," supports recording the eight problem attributes either directly or indirectly as follows:

- *Location*: Location attributes include product, component, version, and platform.
- *Timing*: Timing attributes include dates for when a bug is reported and when an artifact is modified.
- *Symptom, end result, and mechanism*: These attributes are not separated. They can be recorded under the category "steps to reproduce."
- *Cause*: Cause can be recorded as "additional information."
- *Severity*: Severity is directly supported.
- *Cost*: Cost of fault diagnosis and repair is recorded under "time tracking."

5.2.4 Changes

Once a failure is experienced and its cause determined, the problem is fixed through one or more changes. These changes may include modifications to any or all of the development products, including the specification, design, code, test plans, test data, and documentation. Change reports are used to record the changes and track the products most affected by them. For this reason, change reports are very useful for evaluating the most fault-prone modules, as well as other development products with unusual numbers of defects. A typical change report may look like this:

Change Report

Location: Identifier of document or module changed

Timing: When the change was made

Symptom: Type of change

End result: Success of change, as evidenced by regression or other testing

Mechanism: How and by whom change was performed

Cause: Corrective, adaptive, preventive, or perfective

Severity: Impact on the rest of the system, sometimes as indicated by an ordinal scale

Cost: Time and effort for change implementation and test

The *Location* identifies the product, subsystem, component, module, or subroutine affected by a given change. *Timing* captures when the change was made, while *End result* describes whether the change was successful or not. (Sometimes changes have unexpected effects and have to be redone; these problems are discovered during regression or specialized testing.)

The *IEEE Standard Glossary of Software Engineering Terminology* defines four types of maintenance activities, which are classified by the reasons for the changes (IEEE 1990). A change may be *corrective maintenance*, in that the change corrects a fault that was discovered in one of the software products. It may be *adaptive maintenance*: the underlying system changes in some way (the computing platform, network configuration, or some part of the software is upgraded), and a given product must be adapted to preserve functionality and performance. Developers sometimes make *perfective* changes, refactoring code to make it more adaptable by removing "bad smells" (Fowler 1999), rewriting documentation or comments, and/or renaming a variable or routine to clarify the system structure so that new faults are not likely to be introduced as part of other maintenance activities. Finally, *preventive maintenance* involves removing anomalies that represent potential faults. The *Cause* entry in the change report is used to capture one of these reasons for change: corrective, adaptive, perfective, or preventive or perfective.

The *Cost* entry explains the cost to the system developer of implementing a change. The expense includes not only the time for the developer to find and fix the system but also the cost of doing regression tests, update documentation, and return the system to its normal working state.

A *Count* field may be used to capture the number of changes made in a given time interval, or to a given system component.

5.3 HOW TO COLLECT DATA

Since the production of software is an intellectual activity, the collection of data requires human observation and reporting. Managers, systems analysts, programmers, testers, and users must record raw data on forms. This manual recording is subject to bias (deliberate or unconscious), error, omission, and delay. Automatic data capture is therefore desirable, and sometimes essential, such as in recording the execution time of real-time software.

Unfortunately, in many instances, there is no alternative to manual data collection. To ensure that the data are accurate and complete, we must plan our collection effort before we begin to measure and capture data. Ideally, we should do the following:

- Keep the procedures simple.

- Avoid unnecessary recording.

- Train staff in the need to record data and in the procedures to be used.

- Provide the results of data capture and analysis promptly to the original providers and in a useful form that will assist them in their work.

- Validate all data collected at a central collection point.

The quality of collected fault and failure data can vary. Often key data are missing.

EXAMPLE 5.10

Open source projects are potentially a rich source of data for empirical studies. They often include source code, test cases, and change log files. The GNU project coding standards say that you should "keep a change log to describe all the changes made to program source code" (GNU 2013). Yet, a study of three open-source projects—GNUJSP, GCC-g++, and Jikes—by Chen, Schach, Yu, Offutt, and Heller found that overall 22% of the changes in project files were not recorded in the change logs. In one of the versions of Jikes, 62% of the changes were not recorded (Chen et al. 2004).

In a similar study, Bachmann and Bernstein examined data from five open-source and one closed-source project (Bachmann and Bernstein 2009). They found that most of the fault reports are not linked to an entry in the change logs.

Clearly, we must check the integrity of the data collected when analyzing project data whether it is from an open-source or proprietary projects. This requires careful planning.

Planning for data collection involves several steps. First, you must decide which products to measure, based on your GQM analysis. You may need to measure several products that are used together, or you may measure one part or subsystem of a larger system.

EXAMPLE 5.11

Ultimately, failures of all types have to be traced to some system component, such as a program, function, unit, module, subsystem, or the system itself. If the measurement program is to enable management to take action to prevent problems, rather than waiting for problems to happen, it is vital that these components be identified at the right level of granularity. This ability to focus on the locus of a problem is a critical factor for your measurement program's success. For example, software development work at a large British computer manufacturer was hampered by not having data amalgamated at high levels. The software system had a large number of very small modules, each of which had no faults or very few faults. At this level of granularity, it was impossible to identify trends in the fault locations. However, by combining modules according to some rule of commonality (e.g., similar function, same programmer, or linkage by calling routines), it may have been possible to see patterns not evident at lower levels. In other words, it was of little use to know that each program contained either 0 or 1 as known fault, but it was of great interest to see, for example, a set of 25 programs implementing a single function that had 15 faults, whereas a similar set of programs implementing another function had no faults. Such information suggested to management that the first, more fault-prone function, be subject to greater scrutiny.

In Example 5.11, the level of the granularity of collection was too fine. Of course, lowering the granularity can be useful, too. Suppose, your objective is to monitor the fault density of individual modules. That is, you want to examine the number of faults per thousand lines of code for each of a given set of modules. You will be unable to do this if your data collection forms associate faults only with subsystems, not with individual modules. In this case, the level of granularity in your data collection is too *coarse* for your measurement objectives. Thus, determining the level of granularity is essential to planning your data collection activities.

The next step in planning data collection is making sure that the product is under configuration control. We must know which version(s) of each product we are measuring.

EXAMPLE 5.12

In measuring reliability growth, we must decide what constitutes a "baseline" version of the system. This baseline will be the system to which all others will be compared. To control the measurement and evaluate reliability over time, the changed versions must have a multi-level version numbering scheme, including a "mark" number that changes only when there is a major functional enhancement. Minor version numbers track lesser changes, such as the correction of individual faults.

The GQM analysis suggests which attributes and measures you would like to evaluate. Once you are committed to a measurement program, you must decide exactly which attributes to measure and how derived measures will be derived. This will determine what raw data will be collected, and when.

EXAMPLE 5.13

We saw in Chapter 3 that Barnard and Price used GQM to determine what measures they wanted to investigate in evaluating inspection effectiveness (Barnard and Price 1994). They may have had limited resources, so they may have captured only a subset of the metrics initially, focusing first on the ones supporting the highest-priority goals. Once their metrics were chosen, they defined carefully exactly which direct and derived measures were needed. For example, they decided that defect-removal efficiency is the percentage of coding faults found by code inspections. To calculate this derived metric, they needed to capture two direct metrics: total faults detected at each inspection, and the total coding faults detected overall. Then, they defined an equation to relate the direct measures to the derived one:

$$Defect\_removal\_efficiency = 100 \times \frac{\sum_{i=1}^{N} total\_faults\_detected_i}{total\_coding\_faults\_detected}$$

The direct and derived measures may be related by a measurement model, defining how the metrics relate to one another, equations such as the one in Example 5.13 are useful models. Sometimes, graphs or relationship diagrams are used to depict the ways in which metrics are calculated or related.

Once the set of metrics is clear, and the set of components to be measured has been identified, you must devise a scheme for identifying each entity involved in the measurement process. That is, you must make clear how you will denote products, versions, installations, failures, faults, and more on your data collection forms. This step enables you to proceed to form design, including only the necessary and relevant information on each form. We shall look at forms more closely in the next section.

Finally, you must establish procedures for handling the forms, analyzing the data, and reporting the results. Define who fills in what, when, and where, and describe clearly how the completed forms are to be processed. In particular, set up a central collection point for data forms, and determine who is responsible for the data, each step of the way. If no one person or group has responsibility for a given step, the data collection and analysis process will stop, as each developer assumes that another is handling the data. Analysis and feedback will ensure that the data are used, and useful results will motivate staff to record information.

5.3.1 Data Collection Forms

A data collection form encourages collecting good, useful data. The form should be self-explanatory, and include the data required for analysis and feedback. Regardless of whether the form is to be supplied on paper or computer, the form design should allow the developer to record both fixed-format data and free-format comments and descriptions. Boxes and separators should be used to enforce formats of dates, identifiers and other standard values. By pre-printing data, that is the same on all forms (e.g., a project identifier) will save effort and avoid mistakes. Figure 5.3 shows an actual form used by a British company in reporting problems for air traffic control support system. (Notice that it is clearly designed to capture failure information, but it is labeled as a fault report!)

As we have seen, many measurement programs have objectives that require information to be collected on failures, faults, and changes. The remainder of this section describes the forms used in case study of a project whose specific objective was to monitor software reliability. The project developed several data collection forms, including separate ones for each failure, fault, and change. We leave it as an exercise for you to determine whether the data collection forms have all the attributes we have suggested in this chapter.

Table 5.3 depicts an index for a suggested comprehensive set of forms for collecting data to measure reliability (and other external product attributes). Each form has a three-character mnemonic identifier. For example, "FLT" refers to the fault record, while "CHR" is a change record. These forms are derived from actual usage by a large software development organization. As we describe each set of forms, you can decide if the forms are sufficient for capturing data on a large project.

Some fields are present on all or most of the forms. A coded Project Identifier is used so that all forms relevant to one project can be collected and filed together. The name and organization of the person who completes each form must be identified so that queries can be referred back to the author in case of question or comment. The date of form completion must be recorded and distinguished from the date of any failures or other significant events. The identifier and product version must always be recorded,

CDIS FAULT REPORT		S.P0204.6.10.3016

ORIGINATOR:	Joe Bloggs
BRIEF TITLE:	Exception 1 in dps_c.c line 620 raised by NAS

FULL DESCRIPTION Started NAS endurance and allowed it to run for a few minutes. Disabled the active NAS link (emulator switched to standby link), then re-enabled the disabled link and CDIS exceptioned as above. (I think the re-enabling is a red herring.) (during database load)

ASSIGNED FOR EVALUATION TO: DATE:

CATEGORISATION: 0 ①2 3 Design Spec Docn
SEND COPIES FOR INFORMATION TO:
EVALUATOR: ⟨signature⟩   DATE: 8/7/92

CONFIGURATION ID	ASSIGNED TO	PART
dpo_s.c		

COMMENTS: dpo_s.c appears to try to use an invalid CID, instead of rejecting the message. AWJ

ITEMS CHANGED

CONFIGURATION ID	IMPLEMENTOR/DATE	REVIEWER/DATE	BUILD/ISSUE NUM	INTEGRATOR/DATE
dpo_s.c v.10	AWJ 8/7/92	MAR 8/7/92	6.120	RA 8-7-92

COMMENTS:

CLOSED

FAULT CONTROLLER: ⟨signature⟩   DATE: 9/7/92

FIGURE 5.3   Problem report form used for air traffic control support system.

TABLE 5.3   Data Collection Forms for Software Reliability Evaluation

Identifier	Title
PVD	Product version
MOD	Module version
IND	Installation description
IRP	Incident report
FLT	Fault record
SSD	Subsystem version
DOD	Document issue
LGU	Log of product use
IRS	Incident response
CHR	Change record

and a single project may monitor several products over several successive versions. For example, the Installation Description form may record the delivery and withdrawal of several successive versions of a product at a given installation, while Log of Product Use may be used to record the use of several different products at the same terminal or workstation.

The Product Version, Subsystem Version, Module Version, and Document Issue forms identify a product version to be measured: all its component subsystems, modules, and documents, together with their version or issue numbers. They are used to record when the product or component enters various test phases, trial, and service. Previous versions, if any, may be cross-referenced. Some direct product measures (such as size) are recorded; sometimes, the scales used for some associated process measures are also captured. Each Subsystem Version and Module Version Description form should cross-reference the product version or higher-level subsystem version, to define a hierarchical product structure. Similarly, the Document Issue Description should identify the product version of which the particular issue of the document is part. Together, this set of forms provides a basis for configuration control.

The Installation Document identifies a particular installation on which the product is being used and measured. It records the hardware type and configuration, and the delivery date for each product version.

The Log of Product Use records the amount the product is used on a given installation. Separate records must be kept for each product version, and each record must refer to a particular period of calendar time, identified by a date; for example, the period can be described by the date of the end of the week. Total product use in successive periods on all installations or terminals can then be extracted as discussed above. This form may be used to record product use on a single central installation, or adapted for use at an individual terminal on a distributed system.

The Incident Report form has fields corresponding to the attributes listed in the sample failure report at the beginning of this chapter. Each incident must be uniquely identified. The person who completes the form will usually identify it uniquely among those from a particular installation. When the report is passed to the Central Collection Point, a second identifier may be assigned, uniquely identifying it within the whole project.

The Incident Response report is returned to the installation from the central collection point following investigation of the incident's cause. If the wrong diagnosis is made, there may be a request for further diagnostic information, and several responses may be combined into one report. This report includes the date of response, plus other administrative information. After a response, the incident may be still open (under investigation) or closed (when the investigation has reached a conclusion). A response that closes an incident records the conclusion and refers to the appropriate fault record.

The Fault Record records a fault found when inspecting a product or while investigating an incident report. Each fault has a unique identifier. Note that a given fault may be present in one or more versions of the product, and may cause several incidents on one or more installations. The fields in this report correspond to those in the fault report recommended in Section 5.2.3. Finally, the Change Record captures all the fields described in the recommended change report in Section 5.2.4.

Thus, the collection of 10 forms includes all aspects of product fault, failure, and change information. Most organizations do not implement such an elaborate scheme for recording quality information. Scrutinize these descriptions to determine if any data elements can be removed. At the same time, determine if there are data elements missing. We will return to the need for these measures in Chapter 11, where we shall discuss software reliability in depth.

5.3.2 Data Collection Tools

There are many software tools available that support the recording and tracking of software faults and their attributes. These tools provide data collection forms or frameworks for designing your own forms. We found 98 different commercial and freeware fault-tracking tools listed online.* These tools can make it much easier for developers to monitor faults from their discovery to their resolution. Figure 5.4 gives a screenshot of a tailored form using the Bugzilla open-source freeware tool.

In addition to tools to support tracking of faults, tools are also available to support the collection and analysis of source code, analysis of the social networks involved in developing software, extraction, and analysis of a variety of information from CVS and subversion repository logs including the analysis of the evolution of related components and the overall architecture of a system.†

Tools to support data collection are constantly changing. One source for up-to-date information on data collection tools is the International Software Benchmarking Standards Group (ISBSG). The ISBSG web site

---

* Conduct a web search on terms such as "free software testing tools" or "software test automation."
† Sites with information about software project data collection and analysis tools include http://www.swag.uwaterloo.ca/tools.html and http://tools.libresoft.es/.

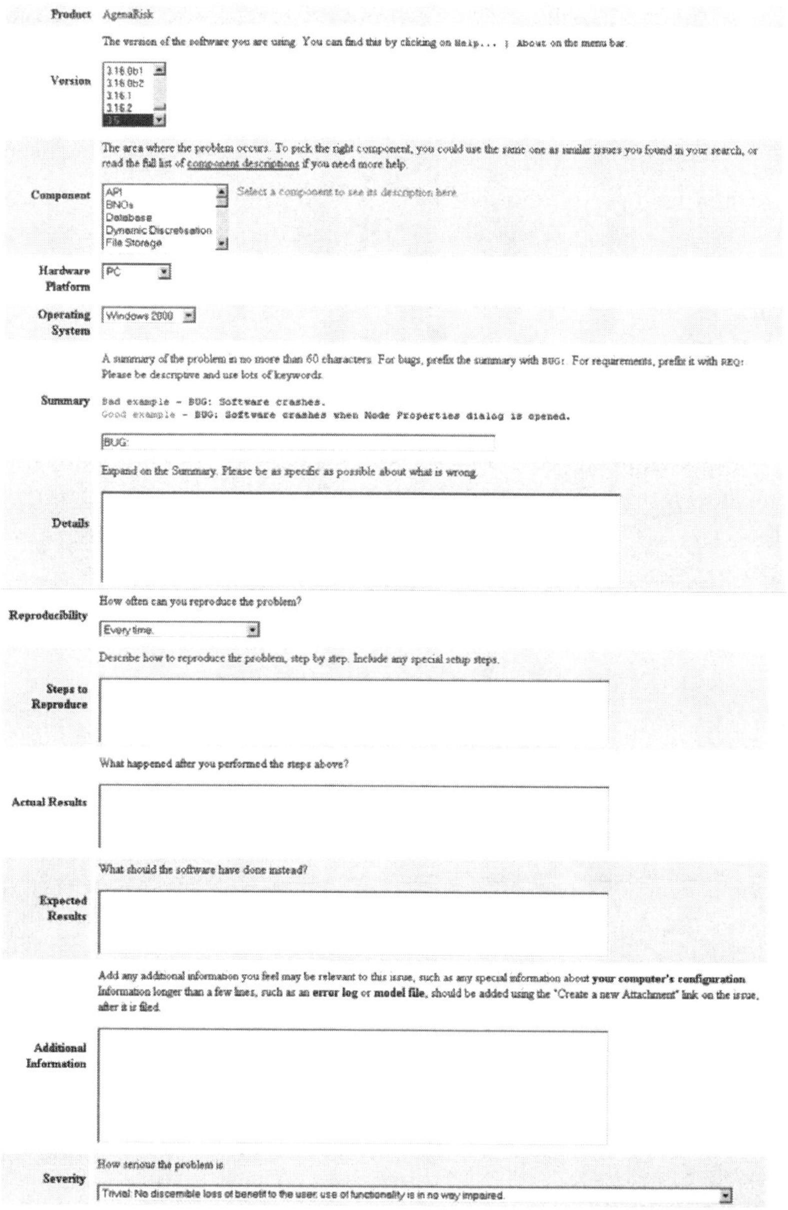

Product AgenaRisk

The version of the software you are using. You can find this by clicking on Help... | About on the menu bar.

Version
```
3.16.0b1
3.16.0b2
3.16.1
3.16.2
```

The area where the problem occurs. To pick the right component, you could use the same one as similar issues you found in your search, or read the full list of component descriptions if you need more help.

Component
```
API
BNOs
Database
Dynamic Discretisation
File Storage
```
Select a component to see its description here

Hardware Platform PC

Operating System Windows 2000

A summary of the problem in no more than 60 characters. For bugs, prefix the summary with BUG: For requirements, prefix it with REQ: Please be descriptive and use lots of keywords.

Summary Bad example - BUG: Software crashes.
Good example - BUG: Software crashes when Node Properties dialog is opened.

```
BUG:
```

Expand on the Summary. Please be as specific as possible about what is wrong.

Details

Reproducibility How often can you reproduce the problem?

Every time.

Steps to Reproduce Describe how to reproduce the problem, step by step. Include any special setup steps.

Actual Results What happened after you performed the steps above?

Expected Results What should the software have done instead?

Additional Information Add any additional information you feel may be relevant to this issue, such as any special information about **your computer's configuration** Information longer than a few lines, such as an **error log** or **model file**, should be added using the "Create a new Attachment" link on the issue, after it is filed.

Severity How serious the problem is.

Trivial: No discernible loss of benefit to the user; use of functionality is in no way impaired.

FIGURE 5.4   Bugzilla screenshot.

(http://www.isbsg.org) includes repositories of software development data from thousands of software projects. Thus it is a great source for industry software project data, and has links to several data collection and analysis tools. These tools focus on project size and cost estimation. Of particular interest are the ISBSG guidelines on what data the group finds to be most effective for improving software processes. ISBSG data are available for both academic and commercial use.

5.4 RELIABILITY OF DATA COLLECTION PROCEDURES

For data collection to be reliable and predictable, it needs to be automated. Data collection technology must also be adaptable so that you can continue to collect data while development environments (both development languages and tools) and the measurement tools evolve (Sillitti et al. 2004). Silliti et al. found that it takes a significant and flexible infrastructure to collect data from projects involving many developers using multiple clients employing different languages and tools. Application domains tend to be unique and require varied data collection support environments and tools. Reports generated from the data need to be customized for an application domain and the context of use.

Data that is not reliable—data that is not appropriate for the application domain or the context of use—represents a threat to the construct validity of results gleaned from the data (as described in Chapter 4).

EXAMPLE 5.14

Lethbridge, Sim, and Singer identify many factors that can affect the reliability of collected data (Lethbridge et al. 2005). One factor to consider is the closeness of the connection between the evaluators or researchers and the software developers that are part of a project being studied. The relative closeness of the connection is classified as *first degree*, *second degree*, or *third degree*:

- *First degree:* Evaluators or researchers directly interact with developers. They may interview developers, observe them working, or otherwise interact directly with developers in their daily activities.
- *Second degree:* Evaluators or researchers indirectly interact with developers. They may instrument software development tools to collect information or may record meetings or other development activities.
- *Third degree:* Evaluators have no interactions with developers. Rather, they study artifacts such as revision control system records, fault reports and responses, testing records, etc. Third-degree studies can be performed retrospectively.

First-degree studies offer researchers the greatest level of control over the quality of the data, since they directly observe and control the data collection. However, first-degree studies generally require the greatest level of resources, as researchers must be present to collect data. Also, there is a potential for the presence of researchers to affect the behavior of developers. Second-degree studies support the direct collection of data, without the presence of researchers. However, second-degree studies require the availability (or development) of appropriate data collection tools. Also, with second-degree studies, it is more difficult to collect data concerning the rationale that developers use to make particular decisions. Third-degree studies do not support the direct collection of data from developers. However, they do allow researchers to mine available software repositories for their studies.

EXAMPLE 5.15

Lincke, Lundberg, and Löwe find different data collection tools can produce measurement values that vary widely (Lincke et al. 2008). They examined the values of nine common measures generated by 10 different measurement tools, and found discrepancies. In particular, they found large discrepancies in the measured values of coupling between objects (CBO), and lack of cohesion between object classes (LCOM). The discrepancies were large enough to affect the ranking of classes in terms of these measures.

The usability of collected data depends on how the data are stored. To be useful over an extended period, data must be stored in a manner that supports future needs. Unfortunately, we may not know now how we will need to use the data in the future. Harrison proposed a flexible design for a metrics repository based on a transformational view of software development (Harrison 2004). The design uses meta-data and their relations and defers the choice of specific metrics to quantify attributes. Such flexible designs promise to support repositories that can remain useful over the long term.

5.5 SUMMARY

We have seen in this chapter that the success or failure of any metrics program depends on its underlying data collection scheme. Chapter 3 showed us how to use measurement goals to precisely specify the data to be collected; there is no single type of data that can be useful for all studies. Thus, knowing if your goals are met requires careful collection of valid, complete, and appropriate data.

Data collection should be simple and nonobtrusive, so that developers and maintainers can concentrate on their primary tasks, with data collection playing a supporting role. Because quality is a universal concern, almost all measurement programs require data collection about software problems and their resolution. We distinguish among, and record information about, several types of problems: *faults*, *failures*, and *changes*. For each class of entity, we consider measuring the following attributes:

Location: Where is the entity?

Timing: When did it occur?

Symptom: What was observed?

End result: Which consequences resulted?

Mechanism: How did it occur?

Cause: Why did it occur?

Cost: How much was incurred by the developer?

Count: How many entities were observed?

Data collection requires a classification scheme, so that every problem is placed in a class, and no two classes overlap. Such a scheme, called an orthogonal classification, allows us to analyze the types and sources of problems, so that we can take action to find and fix problems earlier in the life cycle.

It is important that system components be identified at appropriate levels of granularity. For some projects, this means that we collect subsystem data, while for others it means that we must look at each module carefully. Often, the granularity required means that we are dealing with large amounts of data. For this reason, and for consistency of counting, we should automate our data collection, or at least generate uniform data collection forms. The collected data should be stored in an automated database, and the results should be reported back to the developers as soon as possible, so that they can use the findings to improve product and process.

EXERCISES

1. Define the notions of error, fault, and failure in software.

2. Below is the severity classification for problems discovered in a large air traffic control system. Discuss the quality of the classification; in particular, are there problems with orthogonality?

Category 1: Operational system critical

- Corruption or loss of data in any stage of processing or presentation, including database.

- Inability of any processor, peripheral device, network, or software to meet response times or capacity constraints.

- Unintentional failure, halt, or interruption in the operational service of the system or the network for whatever reason.

- Failure of any processor or hardware item or any common failure mode point within the quoted mean time between failures that is not returned to service within its mean time to repair.

Category 2: System inadequate

- Noncompliance with or omission from the air traffic control operational functions as defined in the functional or system specifications.

- Omission in the hardware, software, or documentation (including testing records and configuration management records) as detected in a physical configuration audit.

- Any Category 1 item that occurs during acceptance testing of the development and training system configurations.

Category 3: System unsatisfactory

- Noncompliance with or omission from the non-air traffic control support and maintenance functions as defined by the functional or system specifications applicable to the operational, development, and training systems.

- Noncompliance with standards for deliverable items including documentation.

- Layout or format errors in data presentation that do not affect the operational integrity of the system.

- Inconsistency or omission in documentation.

3. Your company is developing the software for a telephone switching system for a single client. This very large system is delivered to the customer in phased releases. The customer occasionally observes system failures, such as loss of availability, loss of specific services, or erroneous services. There are two testing phases at which it is possible to gather additional failure data internally: integration testing and system testing. An attempt must be made to fix all failures, whether observed by the user or in test. Devise the necessary forms for a data-collection scheme that has to take account of the following:

- There are rigid reliability requirements for the system.

- There are wide variations in the ability to fix failures occurring in certain parts of the system (one of the project manager's major concerns).

- Your change control procedure has to be auditable.

4. A software development company has a metrics program in which the following specific measures are collected for each separate development project:

- Total effort (in person years)

- Total number of lines of code

- Total number of faults recorded during testing

a. Describe three objectives that could sensibly be addressed by this set of measures. In each case, describe the goal that is to be addressed and how the metrics may enable you to understand and meet your goal. State clearly any limitations or reservations you might have about using this particular set of measures for the stated purposes.

b. Each development project is divided up into modules which, depending on their functionality and criticality, may be specified, designed, and tested using different techniques. For example, some modules are subject to no specific quality assurance techniques, while others may be formally specified, formally reviewed, and subject to several different testing strategies. What modest additions or changes would you make to the metrics program in order to be able to assess the effectiveness of the different software quality assurance methods?

5. Your company has been asked to build a Patient Observation and Control System (POCS), based on the following specification:

Patient Observation and Control System

POCS is intended to improve the efficiency of intensive care, providing a better level of customer service while reducing the cost of trained medical staff required to operate it. Sensors attached to the patient will monitor all vital signs: heartbeat, respiration, blood pressure, temperature, and brain activity. The set of parameters to be recorded will vary from patient to patient. For example, a diabetic patient will require having their blood sugar level measured in addition to the usual signs, and brain activity should not be recorded for an Arsenal football supporter, to avoid false indications of brain death.

Drugs will be administered intravenously by pumps that are controlled by the system. The operation of these pumps and the dosage administered are also recorded. The signals from the sensors and pumps are concentrated at the parameter analysis node (PAN) beside the bed. Complete data is held for the past hour on a local file, and essential information only (deviation of vital signs from normal values) is transmitted via a communications link to the nurse's station, where the breakdown of essential data (BED) software displays the status of each patient on a monitor screen, and sounds an alarm to call the nurse to the bedside if necessary. If an emergency arises, a signal will be sent to the junior doctor's paging device. During quiet periods, a signal (to which the doctor must respond by pressing a button on the device) will be sent to the paging device at random intervals to prevent the doctor's falling asleep. If a cardiac arrest is detected, the cardiac team will automatically be alerted. In the case where the customer is a private patient, the senior consultant will also be paged.

a. What are the three most important dependability attributes of POCS, and why?

b. What data would you record about any failure that occurred during the operation of POCS to enable you to measure the three dependability attributes selected in (a)?

   c. In addition to data about failures, what other types of data would you need in order to measure the attributes chosen in (a), and how would you record them?

   d. Describe five modes of failure of POCS that could arise from the activation of latent software faults. In each case, state to which dependability attribute that mode of failure is relevant, and how its end result and severity are classified.

6. You are a consultant who builds software for the transportation industry. One night in the Edward the Confessor pub (known locally as the "Ted the Grass") in Stevenage, you happen to get into conversation with Big Mick, the owner of a local taxi firm. Over a pint or three, he outlines to you his requirement for a computer system to assist him in his business. The Stevenage Compute-a-cab Automated Management System (SCAMS) will control the running of Mick's taxis. In particular, it will automate the selection of routes and communication with the control room, to minimize voice radio communication. (At the moment, Mick's taxis are controlled by radio, and many of the drivers are inexperienced and do not have "the knowledge" of the layout of the town, particularly in outlying areas. The controllers therefore spend a lot of time guiding drivers to their destinations by referring to the map on the wall.)

   Mick needs a "Control and Management" system (CAM) in the control room, to communicate via a digital radio link with a "Computer on Taxi" system (COT) on-board each taxi. CAM will maintain a database of the location and status of every taxi. (E.g., a taxi may be classified as waiting on rank, on way to pick up, waiting at pick-up location, on way to destination, arrived at destination, returning to base, or taking a tea break.) On receipt of a phone call from a prospective client, a controller will enter the pick-up location in the system, and COM will respond with the identification number of the taxi that can get there most quickly, plus an estimated time of arrival (ETA). If the ETA is acceptable to the client, CAM will signal the appropriate COT. If the driver responds with an "accept" message, the booking is confirmed, and CAM will direct the driver to the pick-up location. After the driver signals "customer picked up," CAM will transmit directions to the destination and, after the "at destination" signal, calculate the fare, inform the driver of the amount received, and record it on the accounts database.

COT will communicate with CAM, and display its directions to the driver. It will transmit back to CAM the status indicated by the driver. If a robbery is attempted, the driver can press a silent alarm, causing the COT to transmit an emergency signal both to CAM and to the police station.

Ideally, the location of each taxi will be determined by a global positioning system (GPS) receiver, but it is also possible for the driver to report location directly. The calculation of journey times by CAM requires a geographical database of the town, available by lease from Stevenage Borough Council, which already uses one for managing road repairs.

The next morning, you vaguely recall that you promised Mick a very dependable software system for SCAMS for a highly competitive price, but unfortunately you cannot remember exactly how dependable you promised the software would be; you also cannot remember what price you quoted. Before you meet Mick again and commit yourself to a written contract, you must think seriously about the dependability requirements for the proposed system.

a. Describe five ways in which the proposed SCAMS system could fail due to a software fault. In each case, imagine that you had just received a report of a failure due to that fault during the operation of SCAMS. State what data you would measure and record for that failure.

b. Accounting for various modes of failure, state which dependability attributes are important for SCAMS, and suggest a reasonable quantitative target. Give reasons for your answer.

c. What data you would collect during a trial of SCAMS in order to be able to measure the dependability attributes you have identified above, and convince Mick that the system was adequately dependable?

7. The IEEE Standard Classification for Software Anomalies (IEEE 1044-2009) lists the following fault attributes (called "defect" in the IEEE standard). Comment on the completeness and utility of the list for use in a data recording form.

- *Defect ID:* Unique identifier for the defect.

- *Description:* What is missing, wrong, or unnecessary?

- *Status:* Current state within defect report life cycle.

- *Asset:* The software asset (product, component, module, etc.) containing the defect.

- *Artifact:* The specific work product (e.g., source file or requirements document) containing the defect.

- *Version detected:* Version ID where the defect was detected.

- *Version corrected:* Version ID of the version where the defect was corrected.

- *Priority:* Ranking for processing.

- *Severity:* Highest failure impact that the defect could (or did) cause.

- *Probability:* Probability of recurring failure caused by this defect.

- *Effect:* The class of requirement (e.g., functionality or usability) that is impacted by a failure caused by the defect.

- *Type:* A categorization based on the class of code or work product where the defect is found (e.g., interface, data, or method body).

- *Mode:* A categorization based on whether the defect is due to incorrect implementation or representation, the addition of something unneeded, or an omission.

- *Insertion activity:* The activity during which the defect was injected/inserted.

- *Detection activity:* The activity during which the defect was detected (i.e., inspection or testing).

- *Failure reference(s):* Identifier of failure(s) caused by the defect.

- *Change reference(s):* Identifier of the change request initiated to correct the defect.

- *Disposition:* Final disposition of the defect report upon closure.

8. Examine the "fault" report form of Figure 5.3. Using the guidelines for data collection discussed in this chapter, suggest changes to the form that would improve data quality and help you to make better decisions about the data. For each suggested change, explain why the change will help.

REFERENCES

Lincke, R., Lundberg, J., and Löwe, W., Comparing software metrics tools, *Proceedings of the International Symposium on Software Testing and Analysis (ISSTA)*, ACM, Seattle, Washington, pp. 131–142, 2008.

Sillitti A., Russo B., Zuliani P., and Succi G., Deploying, updating, and managing tools for collecting software metrics, *Proceedings of the 2004 Workshop on Quantitative Techniques for Software Agile Process*, ACM, pp. 1–4, 2004.

FURTHER READING

The definitions and attributes of failures, faults, and changes in this chapter are consistent with the IEEE Standard Glossary, listed below. The framework is essentially the same as that in the British draft standard, BS 5760, which provides comprehensive coverage concentrating on reliability assessment. The text by Ammann and Offutt gives precise definitions of *failures, faults,* and *software errors* (state errors) and the relations between them.

Ammann, P. and Jeff Offutt, J., *Introduction to Software Testing*, Cambridge University Press, New York, 2008.
IEEE Standard 610.12-1990: *Glossary of Software Engineering Terminology*, IEEE Computer Society Press, New York, 1990.

The 2009 IEEE Standard Classification for Software Anomalies provides a standard set of attributes for both faults (defects) and failures, as well as examples of attribute values. The 1993 draft version of this standard provides a comprehensive set of classifications for software faults, and also contains some useful templates for various data-collection forms.

IEEE Draft Standard 1044-1993, *Draft Standard Classification for Software Anomalies*, IEEE Computer Society Press, New York, 1993.
IEEE Standard 1044-2009, *Standard Classification for Software Anomolies*, IEEE Computer Society Press, New York, 2009.

For a comprehensive analysis of the Therac-25 accidents discussed in Example 5.4, Leveson and Turner's article is compelling reading. Jordan Roberson provided one of the first accounts of the Zune failure, and Weimer et al. use the Zune fault in their research on program repair. See the *Risks Digest* for comprehensive descriptions of failures that involve computer software that have occurred since 1985.

Leveson N.G. and Turner C.S., "An investigation of the Therac-25 accidents," *IEEE Computer*, 18–41, July 1993.

Neumann P.G. (moderator), *The Risks Digest: Forum on Risks to the Public in Computers and Related Systems*, ACM Committee on Computers and Public Policy. http://www.risks.org.

Robertson, J., "Microsoft Zune's New Year Crash," *The Street*. January 1 2009. http://www.thestreet.com/story/10455712/microsoft-zunes-new-year-crash.html.

Weimer, W., Forrest, S., Le Goues, C., and Nguyen, T., "Automatic program repair with evolutionary computation," *Communications of the ACM*, 53(5), 109–116, May 2010.

The following empirical research studies identify inaccuracies in change logs. In particular, they find that many changes are never logged.

Bachmann, A. and Bernstein, A. "Software process data quality and characteristics—A historical view on open and closed source projects," *Proc. Joint ERCIM Workshop on Software Evolution and Int. Workshop on Principles of Software Evolution (IWPSE-Evol'09)*, Amsterdam, Netherlands, August 24–25, 2009, pp. 119–128, 2009.

Chen, K., Schach, S., Yu, L., Offutt, J., and Heller, G. "Open-source change logs," *Empirical Software Engineering*, 9, 197–210, 2004.

Many tools support reporting and tracking faults and failures. The "Testing FAQS" site lists numerous commercial and freeware tools. One of the most popular fault tracking tools is Bugzilla. The Working Conference on Mining Software Repositories is an established conference series that is devoted to studying, collecting, and analyzing data from both proprietary and open-source repositories. It is a rich source for both research results and analysis tools. Studies involve the analysis of artifacts in repositories as well as the social networks involved in software development. There are also sites with information about various software repository analysis tools:

Bugzilla, www.bugzilla.org.
http://www.faqs.org/faqs/software-eng/testing-faq/

Sites with information about software project data collection and analysis tools related to the MSR community include http://www.swag.uwaterloo.ca/tools.html and http://projects.libresoft.es/projects/dataanalysis.

CHAPTER 6

Analyzing Software Measurement Data*

In Chapter 4, we introduced general techniques for designing empirical studies. Then, in Chapter 5, we explained the importance of good data-gathering to support rigorous investigation. In this chapter, we turn to the analysis of the data.

Data analysis involves several activities and assumptions:

- We have a number of measurements of one or more attributes from a number of software entities (products, processes, or resources). We call the set of measurements a *dataset* or a *batch*.

- We expect the software items to be *comparable* in some way. For example, we may compare modules from the same software product by examining the differences or similarities in the data. Or, we may look at several tasks from the same project to determine what the data tell us about differences in resource requirements. Similarly, we can compare several projects undertaken by the same company to see if there are some general lessons we can learn about quality or productivity.

- We wish to determine the *characteristics* of the attribute values for items of the same type (usually we look at measures of central tendency and measures of dispersion), or the *relationships* between attribute values for software items of the same or different types.

---

* Including contributions by Barbara Kitchenham and Shari Lawrence Pfleeger from previous editions.

To perform the analysis, we use statistical techniques to describe the distribution of attribute values, as well as the relationship between or among attributes. That is, we want to know what a picture of the data looks like: how many very high values, how many very low values, and how the data progress from low to high. And we want to know if we can express relationships among some attribute values in a mathematical way. Thus, the purpose of the data analysis is to make any patterns or relationships more visible, so that we can use the patterns and relationships to make judgments about the attributes we are measuring.

We begin our discussion of analysis by examining the role of statistical distributions and hypothesis testing in data analysis. Then we describe several classical statistical techniques. We explain why some statistical methods must be treated with caution when we apply them to software measurement data. Then, we describe some simple methods for exploring software measurements. In the next section, we turn to the relationships among attributes and how we can use analysis techniques to describe them. Several advanced techniques can be helpful in understanding software, and we discuss three of them: multiattribute utility theory (including the analytical hierarchy process), outranking methods, and the Bayesian approach to evaluating multiple hypotheses. We end this chapter with a reminder of how statistical tests relate to the number of groups you are analyzing.

6.1 STATISTICAL DISTRIBUTIONS AND HYPOTHESIS TESTING

After we have collected data from an empirical study, we will use the distribution of data to evaluate the hypotheses and to make decisions and predictions. We do this in order to better understand the principles of software engineering or to make decisions relevant to a practical software engineering problem. As before, you should use your goal to guide the data analysis process. The collected data is generally a sample, and represents the best available information about the status of the variables involved in the empirical study. There is always uncertainty about the relationship between the collected data and the actual relationships; thus, we use probabilistic reasoning. We examine probability distributions and their role in evaluating hypotheses.

6.1.1 Probability Distributions

Consider the experiment of selecting a contractor to develop a software system. We are interested in the quality of the contractor and therefore consider the set of possible outcomes to be

(a)

Very poor	0.4
Poor	0.3
Average	0.15
Good	0.1
Very good	0.05

(b)

Contractor quality
Very poor ▬ 40%
Poor ▬ 30%
Average ▬ 15%
Good ▬ 10%
Very good ▬ 5%

FIGURE 6.1 Probability distribution. (a) Probability table; (b) Distribution shown as a graph.

{very poor, poor, average, good, very good}

On the basis of our previous experience with contractors, or purely based on subjective judgment, we might assign the probabilities to these outcomes as shown in the table of Figure 6.1a.

Since the numbers are all between 0 and 1 and since they sum to 1, this assignment is a valid probability measure and represents the likelihood of randomly selecting a contractor with each outcome level.

A table like the one in Figure 6.1a, or an alternative graphical representation of it like the one in Figure 6.1b, is called a *probability distribution*. For experiments with a discrete set of outcomes, it is defined as follows:

A probability distribution for an experiment is the assignment of probability values to each of the possible outcomes.

Once we have the probability distribution defined, we can easily calculate the probability of any event. The *probability of an event E* is simply the sum of the probabilities of the individual outcomes that make up the event *E*.

EXAMPLE 6.1

In the above experiment of selecting a contractor, suppose *E* is the event "quality is at least average," then *E* consists of the following set of outcomes:

{average, good, very good}

And so, from the table in Figure 6.1a

$$P(A) = P(\text{average}) + P(\text{good}) + P(\text{very good}) = 0.15 + 0.1 + 0.05 = 0.3$$

The expression $P(A)$ can refer to both the probability of an event A and the probability distribution for an experiment A, depending on what A is. So far we have only used $P(A)$ to denote the probability of an event A. However, if A is itself the experiment, then we will also use $P(A)$ as shorthand for the full probability distribution for the experiment A. In the above experiment A of selecting a contractor, we use $P(A)$ as shorthand for the following probability distribution:

Very poor	0.4
Poor	0.3
Average	0.15
Good	0.1
Very good	0.05

The notation is a legacy from when discussions revolved primarily around experiments that simply had two outcomes yes and no (or equivalently true and false). In such cases, the experiment is simply named after the "yes" outcome, so the following are examples of experiment names:

- "Guilty": This is the experiment whose outcome is "yes" when a person tried is guilty and "no" when not guilty.

- "Disease": This is the experiment whose outcome is "yes" when a person has the disease and "no" when the person does not have the disease.

- "Test positive": This is the experiment whose outcome is "yes" when a person tests positive for the disease and "no" when the person tests negative.

So if we say $P(\text{Guilty}) = 0.9$, then this genuinely has two meanings. On the one hand, this means the probability of the event "guilty = yes" is 0.9. On the other hand, since it tells us that the probability of the event "guilty = yes" is 0.9, it follows that we also know the whole probability distribution, since $P(\text{yes}) = 0.9$ and $P(\text{no}) = 0.1$.

In many situations, assumptions about the underlying experiment enable us to automatically define the whole probability distribution rather than having to assign individual probabilities to each outcome. For example, if we are rolling a die, we can assume that each of the six outcomes is equally likely. In general, if there are n equally likely outcomes, then the probability of each outcome is simply 1/*n*. Such a probability distribution is called the *uniform* distribution.

For experiments with a continuous set of outcomes, such as measuring the height of a person in centimeters, it is meaningless to assign probability values to each possible outcome. Instead, we do one of the following:

1. Divide the continuous range of outcomes into a set of discrete intervals (this process is called *discretization*) and assign probability values to each interval. For example, if the continuous range is 0 to infinity, we might define the discrete set as

   [0, 100), [100, 110), [110, 120), [120, 130), [130, 140), [140, 150), [150, infinity)

2. Use a continuous function whose area under the curve for the range is 1. One common example of this is the *normal distribution*. For example, we could use a normal distribution such as shown in Figure 6.2a for our height experiment (in a tool like AgenaRisk* you simply enter the function as shown in Figure 6.2b; note that the *variance* is simply the square of the standard deviation—it is more usual to specify the variance rather than the standard deviation). Although the graph plotted here is shown only in the range 0–300, the normal distribution extends from minus infinity to plus infinity. Thus, formally, a normal distribution implies that there is a "probability" of negative values, which is really zero for height measurements.

(a)

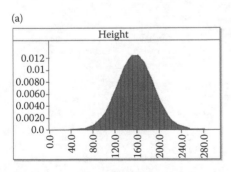

(b)

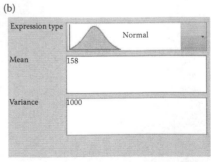

FIGURE 6.2 A continuous probability distribution. (a) Normal distribution (with mean 158 and variance 1000) displayed as a graph; (b) Defining the distribution in AgenaRisk.

---

* http://www.agenarisk.com/.

This is one reason why the normal distribution can only be a crude approximation to the "true" distribution.

If you are using a continuous function as a probability distribution, you need to understand that, unlike discrete probability distributions, it is not meaningful to talk about the probability of a (point) value. So, we cannot talk about the probability that a person's height is 158 cm. Instead, we always need to talk in terms of a range, such as the probability that a person's height is between 157.9 and 158.1 cm. The probability is then the proportion of the curve lying between those two values. This is shown in Example 6.2, which also describes the important example of the continuous uniform distribution.

EXAMPLE 6.2

Suppose that we have modeled the outcomes of measuring people's height using the normal distribution shown in Figure 6.3.

The model is an idealized distribution for the underlying population. We cannot use the model to determine the probability that a person has some exact (point value), but we can use it to calculate the probability that a person's height is within any nonzero length range. We do this by computing the area under the curve between the endpoints of the range. So, in Figure 6.3, the shaded area is the probability that a person is between 178 and 190 cm tall. In general, the area under a curve is calculated by integrating the mathematical function that describes the distribution. Even for experienced mathematicians, this can be very difficult, so they use statistical tables or tools like Excel or AgenaRisk to do it for them.

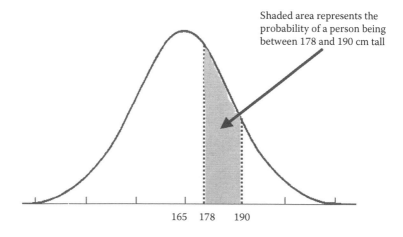

FIGURE 6.3   Probabilities as area under the curve.

6.1.2 Hypothesis Testing Approaches

Empirical studies are generally conducted to evaluate hypotheses. The goal is to determine which one out of a set of competing hypotheses is most plausible. By *plausible*, we mean which hypothesis has the greatest probability of being true given the data that we have gathered. As described in Chapter 4, the traditional approach is to develop a null hypothesis, along with an alternative hypothesis, or multiple alternative hypotheses. A key criterion for testing a hypothesis is a "test of significance," which evaluates the probability that a relationship was due to chance (Fisher 1925).

The classical approaches examine whether or not the null hypothesis can be refuted with some predetermined *confidence level*, often .05. Using the .05 confidence level, we can refute the null hypothesis only if our evidence is so strong that there is only a probability of .05 (5%) that, in spite of an apparent relationship, the null hypothesis is really true.

EXAMPLE 6.3

Consider the following hypotheses:

- H_0: The use of a test-first methodology during unit testing does not affect the fault densities found during system testing.
- H_1: The use of a test-first methodology during unit testing reduces the fault densities found during system testing.

Assume that a study was conducted comparing the use of a test-first methodology to the prior method used for unit testing. The collected data shows that the mean fault densities obtained during system testing are lower for the code developed by teams employing a test-first methodology. Using the .05 confidence level, we reject H_0 only if statistical tests show that there is less than a 5% chance that H_0 is really true, due to the randomness in selecting program units or teams or some other factor relevant to the study. We do not even consider H_1 if the confidence level is greater than .05.

Using a classical approach, we do not evaluate hypotheses in terms of the magnitude of the outcome differences between the treatment and control group. Rather, we accept or reject hypotheses only on the basis of a predetermined confidence level. Suppose the results described in Example 6.3 found that the fault densities of the treatment group (modules developed using a test-first methodology) averaged less than 50% of the fault densities of the control group (modules not developed using test-first methodology), but the statistical confidence level was .07. The results would not be significant and we would not even consider H_1, even though there was a large magnitude difference between the treatment and control groups.

The key issue is that the classical approach focuses exclusively on the precision of the results rather than the magnitude of the differences. According to Ziliak and McClosey, "statistical significance should be a tiny part of an inquiry concerned with the size and importance of a relationship" (Ziliak and McCloskey 2008). This emphasis on precision over magnitude has limited the benefits of empirical research. As a result, critical information that can aid decision-makers is often discarded as insignificant. We will show in Chapter 7 that an alternative approach to data analysis is often more appropriate than the classical approach. This alternate approach employs causal models and Bayesian statistics.

The classical approach remains the predominant way that researchers analyze data. Thus, we present classical techniques in the following sections. Chapter 7 will present data analysis using causal models and Bayesian statistics.

6.2 CLASSICAL DATA ANALYSIS TECHNIQUES

After you have collected the relevant data (based on the framework we presented in Chapters 3 and 5), you must analyze it appropriately. This section describes what you must consider in choosing a classical data analysis technique. We discuss typical situations in which you may be performing a study, and what technique is most appropriate for each situation. Specific statistical techniques are described and used in the discussion. We assume that you understand basic statistics, including the following notions:

- Measures of central tendency

- Measures of dispersion

- Distribution of data

- Student's t-test

- F-statistic

- Kruskal–Wallis test

- Level of significance

- Confidence limits

The other statistical tests are described here in overview, but the details of each statistical approach (including formulae and references to other statistical textbooks) can be found in standard statistical textbooks (Caulcutt 1991; Chatfield 1998; Dobson 2008; Draper 1998; Ott and Longnecker 2010). However, you need not be a statistical expert to read this chapter; you can read about techniques to learn of the issues involved and types of problems addressed. Moreover, many of the commonly used spreadsheet and statistical packages analyze and graph the data automatically; your job is to choose the appropriate test or technique.

There are three major items to consider when choosing analysis techniques: the nature of the data you collected, why you performed the study, and the study design. We consider each of these in turn.

6.2.1 Nature of the Data

In the previous chapters, we have considered data in terms of its measurement scale and its position in our entity-and-attribute framework. To analyze the data, we must also look at the larger population represented by the data, as well as the distribution of that data.

6.2.1.1 Sampling, Population, and Data Distribution

The nature of your data will help you to decide what analysis techniques are available to you. Thus, it is very important for you to understand the data as a sample from a larger population of all the data you could have gathered, given infinite resources. Because you do not have infinite resources, you are using your relatively small sample to generalize to that larger population, so the characteristics of the population are important.

From the sample data, you must decide whether measured differences reflect the effects of the independent variables, or whether you could have obtained the result solely through chance. To make this decision, you use a variety of statistical techniques, based in large degree on the sample size and the data distribution. That is, you must consider the large population from which you could have selected experimental subjects and examine how your smaller sample relates to it.

In an experiment where we measure each subject once, the sample size is simply the number of subjects. In other types of experiments where we measure repeatedly (i.e., the repeated-measures designs discussed in Chapter 4), the sample size is the number of times a measure is applied to a single subject. The larger the sample size, the more confident we can be that observed differences are due to more than just chance variation. As we have seen in Chapter 4, an experimental error can affect our results, so we try to use large samples to minimize that error and increase the likelihood that we are concluding correctly from what we observe. In other words, if the sample size is large enough, we have confidence that the sample of measurements adequately represents the population from which it is drawn.

Thus, we must take care in differentiating what we see in the sample from what we believe about the general population. *Sample statistics* describe and summarize measures obtained from a finite group of subjects, while *population parameters* represent the values that would be obtained if all possible subjects were measured. For example, we can measure the productivity of a group of programmers before and after they have been trained to use a new programming language; an increase in productivity is a sample statistic. But we must examine our sample size, population distribution, experimental design, and other issues before we can conclude that productivity will increase for any programmers using the new language.

As we have seen in Chapter 2, we can describe the population or the sample by measures of central tendency (mean, median, and mode) and measures of dispersion (variance and standard deviation). These characteristics tell us something about how the data are distributed across the population or sample. Many sets of data are distributed normally, or according to a Gaussian distribution, and have a bell-shaped curve similar to the graph shown in Figures 6.2a and 6.3. By definition, the mean, median, and mode of such a distribution are all equal, and 96% of the data occur within three standard deviations of the mean.

For example, the data represented by the histogram of Figure 6.4 is sometimes called "normal" because it resembles the bell-shaped curve. As the sample gets bigger, we would expect its graph to look more and more bell-shaped.

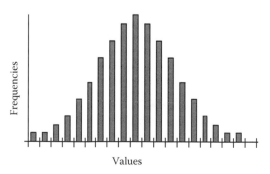

FIGURE 6.4   Data resembling a normal distribution.

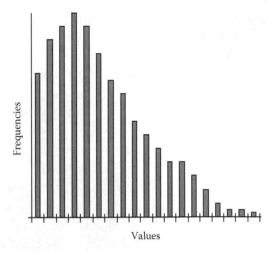

FIGURE 6.5   Distribution where data are skewed to the left.

You can see in Figure 6.4 that the data are evenly distributed about the mean, which is a significant characteristic of the normal distribution. But there are other distributions where the data are skewed, so that there are more data points on one side of the mean than another. For example, as you can see in Figure 6.5, most of the data are on the left-hand side, and we say that the distribution is skewed to the left.

There are also distributions that vary radically from the bell-shaped curve, such as the one shown in Figure 6.6. As we shall see, the type of distribution determines the type of analysis we can perform.

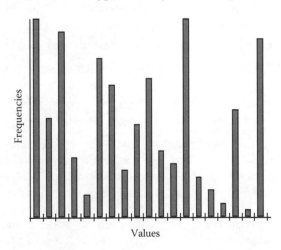

FIGURE 6.6   Nonnormal distribution.

6.2.1.2 Distribution of Software Measurements

Let us see how common software engineering data "measures up" to being normally distributed. In earlier chapters, we mentioned that many common statistical operations and tests are not meaningful for measures that are not on an interval or ratio scale. Unfortunately, many software measures are only ordinal. This scale results from our wanting to use measures that categorize and rank. For example, we create categories of failure severity, so that we can tell most of our system failures are trivial, rather than life threatening. Similarly, we ask our users to rate their satisfaction with the system, or ask designers to assign a quality measure to each requirement before the design begins. Such ordinal measures do not allow us to calculate means and standard deviations, as those are suitable only for interval, ratio, and absolute data. Thus, we must take great care in choosing analysis techniques that are appropriate to the data we are collecting. (And as we saw in Chapter 5, we may collect data based on the type of analysis that we want to do.)

In addition to measurement scale considerations, we must also consider how the data are gathered. As we have seen in Chapter 4, many statistical tests and analysis techniques are based on the assumption that datasets are made up of measurements drawn at random. Indeed, we design our data collection techniques to encourage this randomization, although it is not always possible. Moreover, even when software measurements are on a ratio scale and are selected randomly, the underlying distribution of the general population of data is not always normal. Frequently, the datasets are discrete and nonnegative, skewed (usually toward the left), and usually include a number of very large values.

Let us examine these problems more carefully by looking at examples of two real datasets.

EXAMPLE 6.4

Consider the datasets in Tables 6.1a and 6.1b. Both contain measures of project and product information from commercial software systems. Dataset 1 includes all of the data that was available from a particular environment during a particular time interval; it is typical of the project-level data used for project cost control. Dataset 2 reflects attributes of all the procedures in a particular product subsystem and is typical of component-level data. This second dataset includes various internal product measures that will be described in detail in Chapter 9.

In a normal distribution, the mean, median, and mode of the data are the same. We can use this information to tell us whether the example datasets are normally distributed or not. If, for each attribute of Table 6.1, we compare the mean with the median, we see in Table 6.2 that the median is usually considerably smaller than the mean. Thus, the data are not normally distributed.

TABLE 6.1A   Dataset 1

Project Effort (Months)	Project Duration (Months)	Product Size (Lines of Code)
16.7	23.0	6050
22.6	15.5	8363
32.2	14.0	13,334
3.9	9.2	5942
17.3	13.5	3315
67.7	24.5	38,988
10.1	15.2	38,614
19.3	14.7	12,762
10.6	7.7	13,510
59.5	15.0	26,500

TABLE 6.1B   Dataset 2

Module Size	Module Fan-Out	Module Fan-In	Module Control Flow Paths	Module Faults
29	4	1	4	0
29	4	1	4	2
32	2	2	2	1
33	3	27	4	1
37	7	18	16	1
41	7	1	14	4
55	1	1	12	2
64	6	1	14	0
69	3	1	8	1
101	4	4	12	5
120	3	10	22	6
164	14	10	221	11
205	5	1	59	11
232	4	17	46	11
236	9	1	38	12
270	9	1	80	17
549	11	2	124	16

TABLE 6.2A   Summary Statistics for Dataset 1

Statistic	Effort	Duration	Size
Mean	26.0	15.2	16,742
Median	18.3	14.8	13,048
Standard deviation	21.3	5.1	13,281

TABLE 6.2B   Summary Statistics for Dataset 2

Statistic	Size	Fan-Out	Fan-In	Paths	Faults
Mean	133.3	5.6	5.8	40	5.9
Median	69	4	1	14	4.0
Standard deviation	135.6	3.5	7.9	57.0	5.8

Many statistical references describe other techniques for assessing whether a distribution is normal. Until we know something about our data, we must be very cautious about the use of techniques that assume an underlying normal distribution. When we do not know anything about the distribution, there are a number of approaches dealing with our lack of knowledge:

- We can use robust statistics and nonparametric methods. *Robust statistical methods* are descriptive statistics that are resilient to non-normality. That is, regardless of whether the data are normally distributed or not, robust methods yield meaningful results. On the other hand, *nonparametric statistical techniques* allow us to test various hypotheses about the dataset without relying on the properties of a normal distribution. In particular, nonparametric techniques often use properties of the *ranking* of the data.

- We can attempt to transform our basic measurements into a scale in which the measurements conform more closely to the normal distribution. For example, when investigating relationships between project effort and product size, it is quite common to transform to the logarithmic scale. Whereas the original data are not normally distributed, the logarithms of the data are.

- We can attempt to determine the true underlying distribution of the measurements and use statistical techniques appropriate to that distribution.

EXAMPLE 6.5

Mayer and Sykes looked at the relationship between lines of code (LOC) per module and the number of decisions contained in each module. They found a very good fit for two different datasets using the negative binomial distribution (Mayer and Sykes 1989).

6.2.1.3 Statistical Inference and Classical Hypothesis Testing

Classical hypothesis testing, introduced in Section 6.1.2, makes use of statistical inference. The distribution type plays a big part in how we make inferences from our data. Statistical inference is the process of drawing conclusions about the population from observations about a sample. The process and its techniques depend on the distribution of the data. As we have noted above, parametric statistical techniques apply only when the sample has been selected from a normally distributed population; otherwise, we must use nonparametric techniques. In both cases, the techniques are used to determine if the sample is a good representation of the larger population. For example, if we perform an experiment to determine the increase in productivity of programmers, as described in Example 4.15, we can calculate a mean productivity and standard deviation. Statistical inference tells us whether the average productivity for any programmer using the new language is likely to be the same as the average productivity of our sample.

The logic of statistical inference is based on the two possible outcomes that can result from any statistical comparison:

1. The measured differences observed in the course of the experiment reflect simple chance variation in measurement procedures alone (i.e., there is no real difference between treated subjects and untreated ones).

2. The measured differences indicate the real treatment effects of the independent variable(s).

The first case corresponds to our statement of the null hypothesis; there is no change. Statisticians often denote the null hypothesis as H_0. The second case is the alternative hypothesis, often written as H_1. The purpose of statistical analysis is to see whether the data justify rejecting H_0. The rejection of H_0 does not always mean the acceptance of H_1; there may be several alternative hypotheses, and rejection of H_0 simply means that more experimentation is needed to determine which alternative hypothesis is the best explanation of the observed behavior.

We emphasize that statistical analysis is directed only at whether we can reject the null hypothesis. In this sense, our data can refute the alternative hypothesis in light of empirical evidence (i.e., the data support the null hypothesis because there is no compelling evidence to reject it), but we can never prove it. In many sciences, a large body of empirical data is amassed, wherein each case rejects the same null hypothesis; then, we say loosely that this evidence "confirms," "suggests," or "supports" the alternative hypothesis, but we have not proven the hypothesis to be fact.

The statistical technique applied to the data yields the probability that the sample represents the general population; it provides the confidence we can have in this result and is called the *statistical significance* of the test. Usually referred to as the alpha (α) level, acceptable significance is agreed upon in advance of the test and often α is set at 0.05 or 0.01. That is, the experiment's results are not considered to be significant unless we are sure that there is at least a 0.95 or 0.99 probability that our conclusions are correct.

Of course, there is no guarantee of certainty. Accepting the null hypothesis when it is actually false is called a *Type II error.* Conversely, rejecting the null hypothesis when it is true is a *Type I error.* Viewed in this way, the level is the probability of committing a Type I error, as shown in Table 6.3.

We have seen how a normal distribution is continuous and symmetric about its mean, yet software data are often discrete and not symmetric. If you are not sure whether your data are normal or not, you must assume that they are not, and use techniques for evaluating nonnormal data.

In the examples that follow, we consider both normal and nonnormal cases, depending on the type of data. The examples mention specific statistical tests, descriptions of which are at the end of this chapter, and additional information can be found in standard statistical textbooks. Many of these statistical tests can be computed automatically by spreadsheets,

TABLE 6.3   Results of Hypothesis Testing

State of the World	Decision: Accept H_0	Decision: Reject H_0
H_0 is true	Correct decision	Type I error
	Probability = $1 - \alpha$	Probability = α
H_0 is false	Type II error	Correct decision
	Probability = α	Probability = $1 - \alpha$

so you need not master the underlying theory to use them; it is important only to know when the tests are appropriate for your data.

6.2.2 Purpose of the Experiment

In Chapter 4, we noted two major reasons to conduct a formal investigation, whether it is an experiment, case study, or survey:

- To confirm a theory

- To explore a relationship

Each of these requires analysis carefully designed to meet the stated objective. In particular, the objective is expressed formally in terms of the hypothesis, and the analysis must address the hypothesis directly. We consider each one in turn, mentioning appropriate analysis techniques for each objective; all of the techniques mentioned in this section will be explained in more detail later in this chapter.

6.2.2.1 Confirming a Theory

Your investigation may be designed to explore the truth of a theory. The theory usually states that the use of a certain method, tool, or technique (the treatment) has a particular effect on the subjects. The theoretical effect is to improve the process or product in some way compared to another treatment (usually the existing method, tool, or technique). For example, you may want to investigate the effect of a test-first methodology by comparing it with your existing testing methods. The usual analysis approach for this situation is *analysis of variance* (ANOVA). That is, you consider two populations, the one that uses the old technique and the one that uses the new, and you do a statistical test to see if the difference in treatment results is statistically significant. You analyze the variance between the two sets of data to see if they come from one population (and therefore represent the same phenomenon) or two (and therefore may represent different phenomena). The first case corresponds to accepting the null hypothesis, while the second corresponds to rejecting the null hypothesis. Thus, the theory is not proven; instead, the second case provides empirical evidence that suggests some reason for the difference in behavior.

There are two cases to consider: normal data and nonnormal data. If the data come from a normal distribution and you are comparing two groups, you can use tests such as the *Student's t-test* to analyze the effects of the two treatments. If you have more than two groups to compare, a more general ANOVA test, using the *F statistic*, is appropriate. Both of these are described in most statistics books.

EXAMPLE 6.6

You are investigating the effect of the use of a new tool on productivity. You have two groups that are otherwise equal except for use of the tool: group *A* is using the existing method (without the tool), while group *B* is using the tool to perform the designated task. You are measuring productivity in terms of thousands of delivered source code instructions per month, and the productivity data come from a normal distribution. You can use a Student's *t*-test to compare group *A*'s productivity data with group *B*'s to see if the use of the tool has made a significant change in productivity.

EXAMPLE 6.7

On the other hand, suppose you want to investigate whether the test-first technique yields higher-quality code than your current testing technique. Your null hypothesis is stated as:

> Code developed and tested using the test-first technique has the same number of defects per hundred LOC as code developed using current testing techniques.

You collect data on number of defects per line of code (a measures of defect density) for each of two groups, and you seek an analysis technique that will tell you whether or not the data support the hypothesis. Here, the data on defects per line of code are not normally distributed. You can analyze the defect data by ranking it (e.g., by ranking modules according to their defect density) and using the *Kruskal–Wallis test* to tell you if the mean rank of the test-first modules is lower than that of the nontest-first data.

6.2.2.2 Exploring a Relationship

Often, an investigation is designed to determine the relationship among data points describing one variable or across multiple variables. For example, you may be interested in knowing the normal ranges of productivity or quality on your projects, so that you have a baseline to compare for the future. A case study may be more appropriate for this objective, but you may want to answer this question as part of a larger experiment. Several techniques can help to answer questions about a relationship: box plots, bar charts, control charts, scatter plots (or scatter diagrams), and correlation analysis.

A *box plot* can depict a summary of the range and distribution of a set of data for one variable. It shows where most of the data are clustered and the location of outlier data. A *bar* chart provides an alternative way to display a single variable. Bar charts are especially useful when you are comparing the data from a small number of identified entities. A *control chart* shows the trends of a variable over time, and can help you to spot occurrences of abnormal data values. While box plots, bar charts, and control charts show information about one variable, a *scatter plot* depicts the relationship between two variables. By viewing the relative positions of pairs of data points, you can visually determine the likelihood of an underlying relationship between the variables. You can also identify data points that are atypical, because they are not organized or clustered in the same way as the other data points.

Correlation analysis goes a step further than a scatter diagram by using statistical methods to confirm whether there is a relationship between two attributes. Correlation analysis can be done in two ways: by generating measures of association that indicate the closeness of the behavior of the two variables, or by generating an equation that describes that behavior.

6.2.3 Decision Tree

To help you understand when to use one analysis technique or another, we have constructed a decision tree, shown in Figure 6.7, to take into account the major considerations discussed in this section. The decision tree is to be read from left to right. Beginning with the objective of your investigation, move along the branch that fits your situation until you reach a leaf node with the appropriate analysis technique(s). The next section provides further details concerning these analysis techniques.

6.3 EXAMPLES OF SIMPLE ANALYSIS TECHNIQUES

There are many robust techniques that are useful with software measurement data, regardless of the distribution. You need not be a statistician to understand and use them, and you can implement them using simple spreadsheets or statistical packages. The previous section outlined several approaches, based on the goals of your investigation. In this section, we look at some of the techniques in more detail.

6.3.1 Box Plots

Software measurement datasets are often not normally distributed, and the measurements may not be on a ratio scale. Thus, you should use the

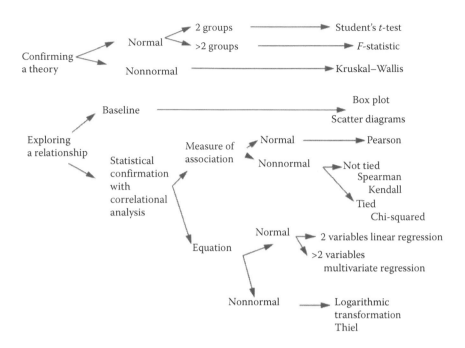

FIGURE 6.7   Decision tree for analysis techniques.

median and quartiles to define the central location and spread of the component values, rather than the more usual mean and variance. These robust statistics can be presented in a visual form called a *box plot*, as shown in Figure 6.8. Box plots are constructed from three summary statistics: the median, the upper quartile, and the lower quartile.

The *median* is the middle-ranked item in the dataset. That is, the median is the value *m* for which half the values in the dataset are larger than *m* and half are smaller than *m*. The *upper quartile u* is the median of the values that are more than *m*, and the *lower quartile l* is the median of the values that are less than *m*. Thus, *l*, *m*, and *u* split the dataset into

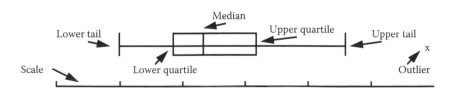

FIGURE 6.8   Drawing a box plot.

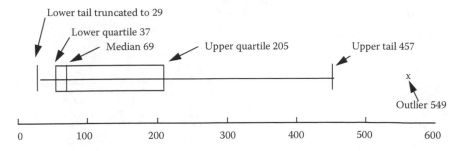

FIGURE 6.9   Box plot of lines of code (17 procedures) for dataset 2 of Table 6.1b.

four parts. We define the *box length*, *d*, to be the distance from the upper quartile to the lower; thus, $d = u - l$. Next, we define the *tails* of the distribution. These points represent the theoretical bounds between which we are likely to find all the data points if the distribution is normal. If the data is on an interval, ratio, or absolute scale, the theoretical upper tail value is the point $u + 1.5d$, and the lower tail value is $l - 1.5d$. These theoretical values must then be truncated to the nearest actual data point to avoid meaningless concepts (such as negative LOC) and to demonstrate the essential asymmetry of skewed datasets. Values outside the upper and lower tails are called *outliers*; they are shown explicitly on the box plot, and they represent data points that are unusual in some way.

The relative positions of the median, the quartiles, and the tails in a box plot can indicate if the dataset is skewed. If the dataset is symmetric about the median, the median will be positioned in the center of the box, and the tail lengths (i.e., the length from the quartile to the tail point) will be equal. However, if the dataset is skewed, the median will be offset from the center of the box and the tail lengths will be unequal.

To see how box plots are used, we apply them to dataset 2 of Table 6.1. Figure 6.9 shows the result.

EXAMPLE 6.8

The LOC values in Table 6.1b are arranged in ascending order. The median is the ninth value: 69. The lower quartile is the fifth value: 37. The upper quartile is the thirteenth value: 205. Thus, the box length is 168. Constructing a box plot is more difficult when there is an even number of observations; in that case, you must take the average of two values when there is no "middle" value. Hoaglin describes this procedure in more detail (Hoaglin et al. 2000). It is clear that the box plot in Figure 6.9 is strongly skewed to the left.

Box plots are simple to compute and draw, and they provide a useful picture of how the data are distributed. The outliers are especially important when you are looking for abnormal behaviors.

EXAMPLE 6.9

Figure 6.10 shows the box plots for measures taken on 17 software systems. Each system provided three measures: thousands of lines of code (KLOC), average module size in LOC (MOD), and the number of faults found per KLOC (fault density, or FD). The top box plot illustrates KLOC, the middle MOD, and the bottom FD. Each identifies outliers with respect to the measure it is depicting.

Notice that MOD and FD have exactly the same outliers! Systems D, L, and A have abnormally high FDs, and they also have unusual module sizes: D and A have abnormally low MOD, while L has abnormally high MOD. Further investigation must identify more clearly the relationship between MOD and FD, but the initial data seem to confirm the widely held belief that a system should be composed of modules that are neither too small nor too large.

The outliers in the KLOC box plot seem to be unrelated to those in fault density.

In general, since the outliers represent unusual behavior, quality assurance staff can use box plots to identify the modules that should be tested or inspected first. In this example, we might want to examine all systems whose MOD values are outliers in the box plot, since these modules are the ones most likely to be fault-prone.

System	KLOC	MOD	FD
A	10	15	36
B	23	43	22
C	26	61	15
D	31	10	33
E	31	43	15
F	40	57	13
G	47	58	22
H	52	65	16
I	54	50	15
J	67	60	18
K	70	50	10
L	75	96	34
M	83	51	16
N	83	61	18
P	100	32	12
Q	110	78	20
R	200	48	21

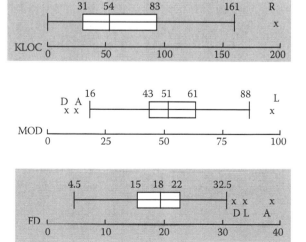

FIGURE 6.10   Box plots for different attributes.

Thus, box plots point us to abnormal or unusual behavior. On the other hand, they can also help us to see what is usual or normal.

EXAMPLE 6.10

We often analyze system test fault density (defined as the number of faults discovered during system test divided by some measure of product size) on past projects to identify the fault density norm for an organization or product line. We can calculate the likely number of system test faults expected for a new product entering system test by multiplying the median value of the fault density by the actual product size measure. Then, we can define an acceptable range to be between the upper and lower quartiles, and use it to monitor the results of system test.

Many developers seek to implement statistical quality control on software development. For this kind of control, it is too stringent to restrict the identification of potential problem components to those that are outliers. A compromise solution is to review quickly those components with values greater than the upper quartile; then, give greater scrutiny to the components with values greater than the upper quartile plus the box length. The components identified by these criteria are considered to be *anomalies* rather than outliers.

These examples show the utility of box plots in providing norms for planning purposes, summarizing actual measurements for the purposes of project monitoring, and identifying software items with unusual values for quality control and process evaluation.

6.3.2 Bar Charts

We saw the utility of box plots in visualizing what is happening in a dataset. The box plot hides most of the expected behavior and shows us instead what is unusual. Another useful depiction of data is a *bar chart*. Here, we simply display all the data ordered to see what the patterns or trends may be. For example, the graph in Figure 6.11 contains a bar for each of the ten projects in dataset 1 of Table 6.1. The x-axis is labeled from 1 to 10 for the project number, the bar height shows the effort, and there is one bar for each measurement in the dataset.

We can see from the bar chart that most of the projects require less than 40 person-months of effort, but two require a great deal more. Such information raises many questions, and often we want to know the relationship between one attribute (such as effort in this example) and others (such as

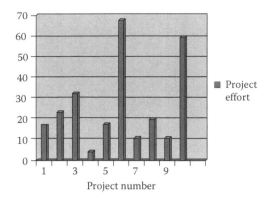

FIGURE 6.11   Bar chart of effort from dataset 1.

size). Unlike box plots, bar charts allow us to readily identify the entity associated with each measured value.

6.3.3 Control Charts

Another helpful technique is a *control chart*, which helps you to see when your data are within acceptable bounds. By watching the data trends over time, you can decide whether to take action to prevent problems before they occur. To see how control charts work, consider first some nonsoftware examples. Many processes have a normal variation in a given attribute. For instance, steel manufacturers rarely make a 1-inch nail that is exactly 1-inch long; instead, they set tolerances, and a 1-inch nail can be a very small fraction above or below an inch in length and still be acceptable. We would expect the actual length values to be randomly distributed about the mean of 1 inch, and 95% of the nails would have lengths falling within two standard deviations of the 1-inch mean. Similarly, the voltage in a 220-V electrical outlet is rarely exactly 220 V; instead, it ranges in a band around 220 V, and appliances are designed to work properly within that small band (but may fail to work if the voltage drops too low or jumps too high).

In the same way, there are parts of the software process that can be expected to behave in a random way, and we can use the control limits to warn us when a value is unusual. We use the values of two standard deviations above and below the mean as guidelines. We want to determine the reasons why any value falls outside these boundaries.

Consider the data in Table 6.4 (Schulmeyer and McManus 1987). Here, we have information about the ratio between preparation hours and inspection hours for a series of design inspections. We calculate the mean and standard

TABLE 6.4   Selected Design Inspection Data

Component Number	Preparation Hours/ Inspection Hours
1	1.5
2	2.4
3	2.2
4	1.0
5	1.5
6	1.3
7	1.0
Mean	1.6
Standard deviation	0.5
Upper control limit (UCL)	2.6
Lower control limit (LCL)	0.4

deviation of the data, and then two control limits. The *upper control limit* is equal to two standard deviations above the mean, while the *lower control limit* is two standard deviations below the mean (or zero, if a negative control limit is meaningless). The control limits act as guidelines for understanding when the data are within random statistical variation and when they represent unusual behavior. In this sense, control charts are similar to box plots.

To visualize the behavior of the data, we construct a graph called a *control chart*. The graph shows the upper control limit, the mean, and the lower control limit. As you can see in Figure 6.12, the data are plotted so that we can see when they are getting close to or exceeding the control limits. In Figure 6.12, the data stay within the control limits. Thus, the preparation hours per hour of inspection are within what we would expect of random

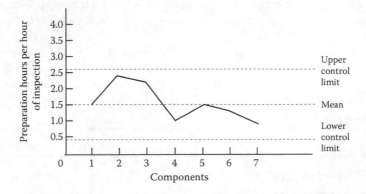

FIGURE 6.12   Inspection control chart showing hours of preparation per hour of inspection.

variation. However, if the data were to exceed the control limits, we would take action to bring them back under control—that is, back within the band between the upper and lower control limits. In this example, exceeding the upper control limit means that too much preparation is being done, or not enough time is being spent during inspection. If the value is below the lower control limit, then not enough preparation is being done, or too much time is being spent during inspection. Even when data do not exceed the control limits, action can be taken when the trend veers toward the edge of the acceptable band, so that behavior is drawn back toward the middle.

6.3.4 Scatter Plots

In Chapter 4, we saw many reasons for wanting to investigate the relationships among attribute values. For instance, understanding relationships is necessary when, for planning purposes, we wish to predict the future value of an attribute from some known value(s) of the same or different attributes. Likewise, quality control requires us to identify components having an unusual combination of attribute values.

When we are interested in the relationship between two attributes, a scatter plot offers a visual assessment of the relationship by representing each *pair* of attribute values as a point in a Cartesian plane.

EXAMPLE 6.11

Figure 6.13 is a scatter plot of effort (measured in person-months) against size (measured in thousands of LOC) for dataset 1 in Table 6.1. Each point represents one project. The x-coordinate indicates project effort and the y-coordinate is the project size. This plot shows us that there appears to be a general relationship between size and effort, namely, that effort increases with the size of the project.

However, there are a few points that do not support this general rule. For instance, two projects required almost 40 person-months to complete, but one was 10 KLOC, while the other was 68. The scatter plot cannot explain this situation, but it can suggest hypotheses that we can then test with further investigation.

EXAMPLE 6.12

Similarly, Figure 6.14 is a scatter plot of the module data shown in dataset 2 of Table 6.1. Here, module size (measured in LOC) is graphed against the number of faults discovered in the module. That is, there is a point in the

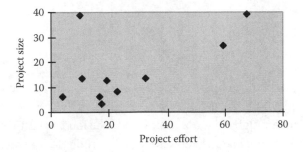

FIGURE 6.13   Scatter plot of project effort against project size for dataset 1.

scatter plot for each module in the system. Again, the plot shows a trend, in that the number of faults usually increases with the module size. However, again we find one point that does not follow the rule: the largest module does not appear to have as many faults as might have been expected.

Thus, scatter plots show us both general trends and atypical situations. If a general relationship seems to be true most of the time, we can investigate the situations that are different and determine what makes the projects or products anomalous. The general relationship can be useful in making predictions about future behavior, and we may want to generate an estimating equation to project likely events. Similarly, we may project likely anomalies to control problems by taking action to avert their behavior or by minimizing their effects.

For obvious reasons, scatter plots are not helpful for more than three variables, unless you evaluate relationships for all possible pairs and

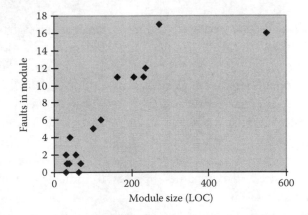

FIGURE 6.14   Scatter plot of module faults against module size for dataset 2.

triples. Such partitioning of the problem does not give you a good over-view of the behavior that interests you.

6.3.5 Measures of Association

Scatter plots depict the behavior of two attributes, and sometimes we can determine that the two attributes are related. But the appearance of a rela-tionship is not enough evidence to draw conclusions. Statistical techniques that can help us evaluate the likelihood that the relationship seen in the past will be seen again in the future. We call these techniques *measures of association*, and the measures are supported by statistical tests that check whether the association is significant.

For normally distributed attribute values, the *Pearson correlation coef-ficient* is a valuable measure of association. Suppose we want to examine the association between two attributes, say x and y. For instance, we saw in Example 6.12 that x could be the size of a module, while y is the number of faults found in the module. If the datasets of x and y values are normally distributed (or nearly), then we can form pairs (x_i, y_i), where there are i software items and we want to measure the association between x and y. The total number of pairs is n, and for each attribute, we calculate the mean and variance. We represent the mean of the x values by m_x, and the mean of the y values by m_y. Likewise, var(x) is the variance of the set of x values, and var(y) is the variance of the y values. Finally, we calculate

$$r = \sum_{i=1}^{n} \frac{(x_i - m_x)(y_i - m_y)}{\sqrt{n \, \text{var}(x) \, \text{var}(y)}}$$

The value of r, called the correlation coefficient, varies from −1 to 1. When r is 1, then x and y have a perfect positive linear relationship; that is, when x increases, then so does y in equal linear steps. Similarly, −1 indicates a perfect negative linear relationship (i.e., when x increases, y decreases linearly), and 0 indicates no relationship between x and y (i.e., when x increases, y is just as likely to increase as to decrease). Statistical tests can be used to check whether a calculated value of r is significantly different from zero at a specified level of significance; in other words, com-putation of r must be accompanied by a test to indicate how much confi-dence we should have in the association.

However, as we have noted before, most software measurements are not normally distributed and usually contain atypical values. In fact, the dataset in Example 6.12 is not from a normal distribution, so the Pearson correlation coefficient is not recommended for it. In these cases, it is preferable to use robust measures of association and nonparametric tests of significance. One approach to the problem is the use of a robust correlation coefficient; other approaches, including contingency tables and the chi-squared test, are discussed in standard statistical textbooks.

6.3.6 Robust Correlation

The most commonly used robust correlation coefficient is *Spearman's rank correlation coefficient*. It is calculated in the same way as the Pearson correlation coefficient, but the x and y values are based on ranks of the attributes, rather than raw values. That is, we place the attribute values in ascending order and assign 1 to the smallest value, 2 to the next smallest, etc. If two or more raw values are equal, each is given the average of the related rank values.

EXAMPLE 6.13

The two smallest modules in dataset 2 have 29 LOC. In a ranking, each module is assigned the rank of 1.5, calculated as the average of ranks 1 and 2.

Kendall's robust correlation coefficient t varies from -1 to 1, as Spearman's, but the underlying theory is different. The Kendall coefficient assumes that, for each two pairs of attribute values, (x_i, y_i) and (x_j, y_j), if there is a positive relationship between the attributes, then when x_i is greater than x_j, then it is likely that y_i is greater than y_j. Similarly, if there is a negative relationship, then when x_i is greater than x_j, it is likely that y_i is less than y_j. Kendall's t is based on assessing all of pairs of vectors (in a dataset of points) and comparing positive with negative indications.

Kendall's correlation coefficient is important because it can be generalized to provide partial correlation coefficients. These partial values are useful when the relationship between two attributes may in fact be due to the relationship between both attributes and a third attribute. (See Seigel and Castellan's classic statistics book for an explanation of partial coefficients and their use (Siegel and Castellan 1988).)

The rank correlation coefficients are intended to be resilient both to atypical values and to nonlinearity of the underlying relationship, as well as not being susceptible to the influence of very large values.

EXAMPLE 6.14

Tables 6.5 and 6.6 contain the values of the Pearson, Spearman, and Kendall correlation coefficients for various attribute pairs from datasets 1 and 2 in Table 6.1. Notice that, in both tables, the robust correlations (i.e., the Spearman and Kendall values) are usually less than the Pearson correlations. This difference in correlation value is caused by the very large values in both datasets; these large values tend to inflate the Pearson correlation coefficient but not the rank correlations.

Notice, too, that we have calculated some of the correlations both with and without some of the atypical values, as identified by scatter plots earlier in this chapter. It appears that the rank correlation coefficients are reasonably resilient to atypical values. Moreover, the relationship between control flow paths and faults in a module is nonlinear, but the rank correlations are not affected by it.

Table 6.5, reflecting dataset 1, shows that, in general, all the correlation coefficients were susceptible to the gross outliers in the data. So, although the rank correlation coefficients are more reliable for software measurement data than the Pearson correlation, they can be misleading without visual inspection of any supposed relationship.

Example 6.14 includes a practice common to data analysis: eliminating atypical data points. Sometimes, a scatter plot or correlational analysis reveals that some data illustrate behavior different from the rest. It is important to try to understand what makes these data different. For example, each point may represent a project, and the unusual behavior may appear only in those projects done for a particular customer, X. With those data points removed, the behavior of the rest can be analyzed and used to predict likely behavior on future projects done for customers other than X. However, it is not acceptable to remove data points only to make the associations look stronger; there must be valid reasons for separating data into categories and analyzing each category separately.

TABLE 6.5 Correlation Coefficients for Dataset 1

Type	Effort vs. Size	Effort vs. Size (Atypical Items Removed)	Effort vs. Duration	Effort vs. Duration (Atypical Items Removed)
Pearson	0.57	0.91	0.57	0.59
Spearman	0.46	0.56	0.48	0.50
Kendall	0.33	0.67	0.38	0.65

TABLE 6.6   Correlation Coefficients for Dataset 2

Type	LOC vs. Faults	LOC vs. Faults (Atypical Removed)	Paths vs. Faults	Paths vs. Faults (Atypical Removed)	FO vs. Faults	FI vs. Faults
Pearson	0.88	0.96	0.68	0.90	0.65	−0.11
Spearman	0.86	0.83	0.80	0.79	0.54	0.02
Kendall	0.71	0.63	0.60	0.62	0.58	0.02

6.3.7 Linear Regression

Having identified a relationship using box plots, scatter plots, or other techniques, and having evaluated the strength of the association between two variables, our next step is expressing the nature of the association in some way. Linear regression is a popular and useful method for expressing an association as a linear formula. The linear regression technique is based on a scatter plot. Each pair of attributes is expressed as a data point, (x_i, y_i), and then the technique calculates the line of best fit among the points. Thus, our goal is to express attribute y, the dependent variable, in terms of attribute x, the independent variable, in an equation of the form

$$y = a + bx$$

To see how this technique works, consider again the scatter plot of Figure 6.10. We can draw a line roughly to estimate the trend we see, showing that effort increases as size increases; this line is superimposed on the scatter plot in Figure 6.15.

The theory behind linear regression is to draw a line from each data point vertically up or down to the trend line, representing the vertical distance from the data point to the trend. In some sense, the length of these

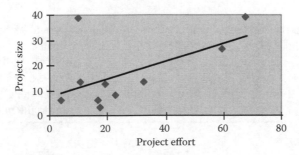

FIGURE 6.15   Plot of effort against size for dataset 1, including line to show trend.

lines represents the discrepancy between the data and the line, and we want to keep this discrepancy as small as possible. Thus, the line of "best fit" is the line that minimizes these distances.

The mathematics required to calculate the slope, b, and intercept, a, of this "best fit" line are straightforward. The discrepancy for each point is called the *residual*, and the formula for generating the linear regression line minimizes the sum of the squares of the residuals. We can express the residual for a given data point as

$$r_i = y_i - a - bx_i$$

Minimizing the sum of squares of the residuals leads to the following equations for a and b:

$$b = \frac{\sum (x_i - m_x)(y_i - m_y)}{\sum (x_i - m_x)^2}$$

$$a = m_y - bm_x$$

The least squares technique makes no assumptions about the normality of the distribution of either attribute. However, to perform any statistical tests relating to the regression parameters (e.g., we may want to determine if the value of b is significantly different from 0), it is necessary to assume that the residuals are distributed normally. In addition, the least squares approach must be used with care when there are many large or atypical values; these data points can distort the estimates of a and b.

EXAMPLE 6.15

Figure 6.16 shows the scatter plot of module size against faults from dataset 2 in Table 6.1. Imposed on the scatter plot is the line generated by linear regression. Notice that there are several values in the dataset that are far from the line, including the point representing the largest module. If we remove the data point for the largest module, the regression line changes dramatically, as shown in Figure 6.17. The remaining data points are much closer to the new line.

After fitting a regression line, it is important to check the residuals to see if any are unusually large. We can plot the residual values against the

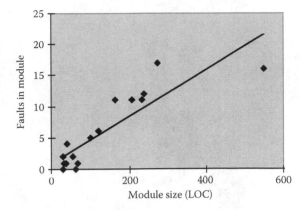

FIGURE 6.16   Module size against faults, including linear regression line.

corresponding dependent variable in a scatter plot. The resulting graph should resemble an ellipsoidal cloud of points, and any outlying point (representing an unusual residual) should be clearly visible, as shown in Figure 6.18.

6.3.8 Robust Regression

There are a number of robust linear regression approaches, and several statistical textbooks (see (Sprent 2007), for example) describe them. For instance, Theil proposed estimating the slope of the regression line as the

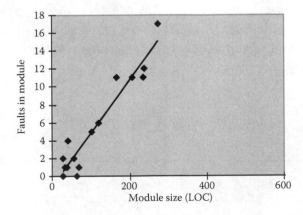

FIGURE 6.17   Module size against faults, using dataset 2 without the largest module.

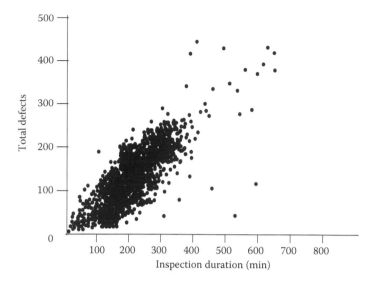

FIGURE 6.18 A scatter plot of data from a major telecommunications company, with residuals noted.

median of the slopes of all lines joining pairs of points with different values. For each pair (x_i, y_i) and (x_j, y_j), the slope is

$$b_{ij} = \frac{y_j - y_i}{x_j - x_i}$$

If there are n data points and all the x_i are different, then there are $n(n - 1)/2$ different values for b_{ij}, and b is estimated by the median value. It is also possible to determine the confidence limits for this estimate.

Theil suggested estimating the intercept, a, as the median of all the values

$$a_i = y_i - bx_i$$

EXAMPLE 6.16

We can apply Theil's method to the relationship between size and effort represented by dataset 1, and we generate the equation

Effort (months) = 8.869 + 1.413 size (KLOC)

This equation is quite similar to the regression line obtained by least squares when the atypical value is removed. However, the confidence limits on the estimate of the slope range from −0.22 to 2.13 (at the 0.05 level of significance), which includes 0. Thus, we must conclude that the robust technique implies that the relationship between effort and size is not significant.

6.3.9 Multivariate Regression

The regression methods we have considered so far focus on determining a linear relationship between two attributes. Each involves the use of the least squares technique. This technique can be extended to investigate a linear relationship between one dependent variable and two or more independent variables; we call this *multivariate linear regression*. However, we must be cautious when considering a large number of attributes, because

- It is not always possible to assess the relationship visually, so we cannot easily detect atypical values.

- Relationships among the dependent variables can result in unstable equations. For example, most size and structure attributes are correlated, so it is dangerous to include several size and structure measures in the same multivariate regression analysis.

- Robust multivariate regression methods can be quite complicated.

For multivariate linear regression, we recommend using least squares analysis but avoiding the use of many correlated-dependent variables. Be sure to analyze the residuals to check for atypical data points (i.e., data points having very large residuals).

6.4 MORE ADVANCED METHODS

There are many other statistical methods for analyzing data. In this section, we consider several advanced methods that can be useful in investigating relationships among attributes: classification tree analysis, transformations, and multivariate data analysis.

6.4.1 Classification Tree Analysis

Many statistical techniques deal with pairs of measures. But often we want to know which measures provide the best information about a particular goal or behavior. That is, we collect data for a large number of measures, and we want to know which ones are the best predictors of the behavior in a given attribute. A statistical technique called *classification tree analysis* can be used to address this problem. This method, applied successfully on data from multiple releases of a large telecommunications system (Khoshgoftaar et al. 2000), allows large sets of metrics data to be analyzed with respect to a particular goal.

EXAMPLE 6.17

Suppose you have collected data on a large number of code modules, and you want to determine which of the metrics are the best predictors of poor quality. You define poor quality in terms of one of your measures; for instance, you may say that a module is of poor quality if it has more than three faults. Then, a classification tree analysis generates a decision tree to show you which of the other measures are related to poor quality. The tree may look like the one in Figure 6.19, which suggests that if a module is between 100 and 300 LOC and has a cyclomatic number of at least 15, then it may have a large number of faults. Likewise, if the module is over 300 LOC, has had no design review, and has been changed at least five times, then it too may have a large number of faults. This technique is useful for shrinking a large number of collected measures to a smaller one, and for suggesting design or coding guidelines. In this example, the tree suggests that, when resources are limited, design reviews may be restricted to large modules, and code reviews may be advisable for large modules that have been changed a great deal.

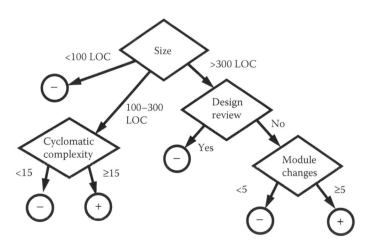

FIGURE 6.19   Classification tree.

6.4.2 Transformations

Sometimes, it is difficult to understand the behavior of data in its original form, but much easier if the data are transformed in some way to make the behavior more visible. We noted earlier that it is sometimes possible to transform nonnormal data into normal data by applying a logarithmic function. In general, a *transformation* is a mathematical function applied to a measurement. The function *transforms* the original dataset into a new dataset. In particular, if a relationship between two variables appears to be nonlinear, it is often convenient to transform one of the attributes in order to make the relationship more linear.

EXAMPLE 6.18

Figure 6.20 is a scatter plot of some of the data from dataset 2 in Table 6.1. Here, we have graphed the structure of each module, as measured by control flow paths, against the number of faults found in the module. The relationship between faults and number of paths appears to be nonlinear and therefore not suitable for analysis using linear regression.

There are many choices for transformation. Tukey suggested a "ladder" from which to decide on an appropriate transformation function (Mosteller and Tukey 1977). The ladder consists of a sequence of transformation functions, where the "top" of the ladder is on the left as shown in Figure 6.21.

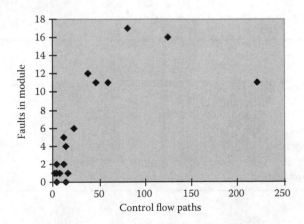

FIGURE 6.20   Structure graphed against faults, from dataset 2.

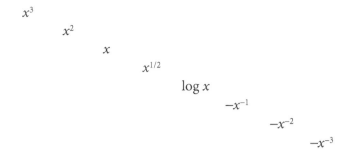

$$x^3$$
$$x^2$$
$$x$$
$$x^{1/2}$$
$$\log x$$
$$-x^{-1}$$
$$-x^{-2}$$
$$-x^{-3}$$

FIGURE 6.21 Tukey's ladder.

We begin by positioning ourselves at the x value of the ladder (i.e., midway on the ladder). If we seem to have a curved rather than linear relationship between two attributes, we use the ladder as follows:

- If the relationship looks positive and convex (as in Figure 6.20), we may transform either the independent variable by going down the ladder (i.e., using a square root transformation), or the dependent variable by going up the ladder (i.e., using a square transformation).

- If the relationship looks positive and concave, we may transform either the independent variable by going up the ladder, or the dependent variable by going down.

If it is important to obtain an equation for predicting the dependent variable, we recommend transforming the independent variable, so that the dependent variable is in the correct scale.

EXAMPLE 6.19

Figure 6.22 is a graph of the data depicted in Figure 6.20, except that the square root of the control flow paths is plotted, rather than the raw data of dataset 2. The relationship of the transformed data to the module faults is clearly more linear for the transformed data than for the raw data. In fact, the relationship obtained using the raw data accounts for only 46% of the variation of the dependent variable, whereas the relationship obtained using the transformed data accounts for 68% of the variation.

There are many other reasons for transforming data. For example, data are often transformed to cope with relationships of the form

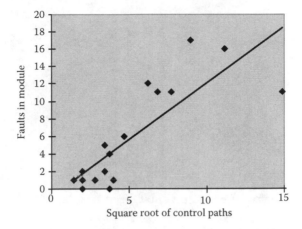

FIGURE 6.22   Module structure (transformed) plotted against faults.

$$y = ax^b$$

where x and y are two attributes, a is a coefficient, and b is an exponent. We can use logarithms to generate a linear relationship of the form

$$\log (y) = \log (a) + b \log (x)$$

Then, $\log(a)$ and b can be estimated by applying the least squares technique to the transformed attribute values. The logarithmic transformation is used frequently to investigate the relationship between project effort and product size, or between project effort and project duration, as we saw in the case of the COCOMO model in Chapter 3 (Example 3.11). However, we must take care in the judgments that we make from such transformed data. Plotting relationships on a log–log scale can give a misleading impression of the variability in the raw data. To predict effort (as opposed to log(effort)), you should plot an appropriate line on the *untransformed* scatter plot.

EXAMPLE 6.20

The relationship between effort and duration for dataset 1 is depicted in Figure 6.23. A linear relationship is not significant for the raw data. However, when we transform the data using the logarithm of each attribute, the result is the scatter plot of Figure 6.24. This figure shows the linear regression line, which is significant after using the log–log transformation.

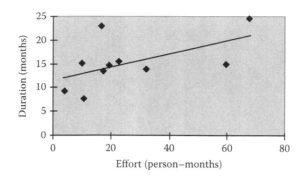

FIGURE 6.23   Effort plotted against duration, from dataset 1.

6.4.3 Multivariate Data Analysis

There are several different techniques that can be applied to data involving many variables. In this section, we look at three related techniques. The *principal component analysis* can simplify a set of data by removing some of the dependencies among variables. The *cluster analysis* uses the principal components that result, allowing us to group modules or projects according to some criterion. Then, the *discriminant analysis* derives an assessment criterion function that distinguishes one set of data from another; for instance, the function can separate data that is "error-prone" from data that is not.

6.4.3.1 Principal Component Analysis

Often, we measure a large set of attributes where some subset of the attributes consists of data that are related to one another. For example, the size of a project affects the amount of effort required to complete it, so size and effort are related. Suppose we want to predict project duration. If we

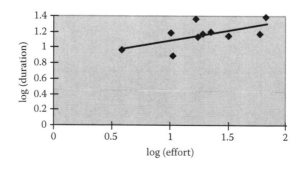

FIGURE 6.24   Logarithm of effort plotted against logarithm of duration.

use both size and effort in a duration prediction equation, we may predict a longer time period than is really required, because we are (in a sense) double counting; the duration suggested by the size measures may also be represented in our effort value. We must take care to ensure that the variables contributing to our equation are as independent of one another as possible.

In general, if we want to investigate the relationship among several attributes, we want to be sure that subsets of related attributes do not present a misleading picture. The *principal component analysis* generates a linear transformation of a set of correlated attributes, such that the transformed variables are independent. In the process, it simplifies and reduces the number of measures with which we must deal. Thus, we begin with n measures x_i, and our analysis allows us to obtain n new variables of the form

$$y_i = a_{i1}x_1 + a_{i2}x_2 + \cdots + a_{in}x_n$$

The technique identifies values for the a_{ij} and determines the contribution of y_i to the overall variability of the set of transformed variables.

Although the process creates the same number of transformed variables as original variables, we can reduce the number of variables by examining the *variability* in the relationship. We need fewer transformed variables to describe the variability of the dataset than we had originally, because the attributes that are virtually identical will contribute to a single variable on the new scale. The analysis indicates what proportion of the total variability is explained by each transformation. The transformation accounting for the most variability is called the *first principal component*, the transformation accounting for the next largest amount is called the second principal component, etc. The principal components that account for less than 5% of the variability are usually ignored.

EXAMPLE 6.21

During the first phase of the Software Certification Programme in Europe (SCOPE), 39 software metrics related to maintainability were collected from modules produced by five industrial software projects. A principal component analysis reduced the set to six, accounting for nearly 90% of the variation. Size alone explained almost 57% of the variation. In other words, more than half of the variation in the maintainability measure was explained by changes to the size measure (Neil 1992).

Principal component analysis is available in most statistical packages because it is useful for several purposes:

1. To identify the underlying dimensionality of a set of correlated variables.

2. To allow a set of correlated variables to be replaced by a set of non-correlated variables in multivariate regression analysis.

3. To assist in outlier detection. Each set of attribute values for a software item can be transformed into one or more new variables (i.e., the principal components). If the first principal component is plotted against the second, there will be no relationship; however, points that are very distant from the central mass of points are regarded as anomalies.

EXAMPLE 6.22

Table 6.7 shows the results of a principal component analysis performed on the four structure and size variables of dataset 2. The variables were normalized prior to performing the analysis (i.e., transformations were applied, so that each variable had mean 0 and standard deviation 1). Table 6.7 shows the correlation between each of the original variables and the transformed variables. It also indicates the percentage of the variability accounted for by each component. The analysis suggests that the four original variables can be represented adequately by the first three transformed variables (i.e., the first three principal components). It also shows that the first principal component is related to LOC, fan-out and paths, since the correlation coefficients are positive and high. Similarly, the second principal component is related to fan-in, and the third principal component is related to LOC.

TABLE 6.7 Correlations between Principal Components and the Normalized Variables for Dataset 2 Size and Structure Measures

Original	Principal Component 1	Principal Component 2	Principal Component 3	Principal Component 4
LOC	0.81	−0.13	0.57	0.02
Fan-out	0.92	0.06	−0.28	0.27
Fan-in	−0.10	0.99	0.12	0.03
Path	0.92	0.15	−0.21	−0.28
% Variation explained	59.1	24.4	11.5	3.9

6.4.3.2 Cluster Analysis

Cluster analysis can be used to assess the similarity of modules in terms of their measurable characteristics. It assumes that modules with similar attributes will evidence similar behavior. First, a principal component analysis is performed, producing a reduced set of principal components that explain most of the variation in the behavior being investigated. For example, we saw in Example 6.21 that the behavior, maintainability, was explained by six principal components. Next, cluster analysis specifies the behavior by separating it into two categories, usually exhibiting the behavior and not; in Example 6.21, we can think of the categories as "easy to maintain" or "not easy to maintain."

There are many cluster algorithms available with statistics packages. These algorithms can be used much as box plots were used for single variables: to cluster the data and identify outliers or unusual cases.

6.4.3.3 Discriminant Analysis

Discriminant analysis allows us to separate data into two groups, and to determine to which of the two groups a new data point should be assigned. The technique builds on the results of a cluster analysis to separate the two groups. The principal components are used as discriminating variables, helping to indicate to which group a particular module is most likely to belong.

The groups generated by a cluster analysis sometimes have overlap between them. The discriminant analysis reduces this overlap by maximizing the variation between clusters and minimizing the variation within each cluster.

6.5 MULTICRITERIA DECISION AIDS

Most of the techniques we have presented so far are based on classical statistical methods. In this section, we present some newer ways of addressing data analysis, especially when decision-makers must solve problems by taking into account several different points of view. In other words, we have assumed so far that the goal of measurement is clear, but there are times (as we saw in Chapter 3) when some people interpret goals differently from others. We need to be able to analyze data and draw conclusions that address many different interpretations of the same goal, even when the interpretations conflict. Thus, rather than "solving" a problem by finding a universal truth or law exhibited by the data, we need to find, instead, a subjective problem resolution that is consistent and satisfies all parties involved, even if it is not optimal in the mathematical sense.

To understand why this situation is common in software measurement, consider a software system that is being built to meet requirements about safety or reliability. Data are collected to help in answering the question, Is this software safe? But the interpretation of "safe" can differ from one person to another, and it is not always clear that we have sufficient evidence to answer the question. Moreover, we often have to balance several considerations. We may ask, which option provides the most reliability for the least cost? and we must make decisions that reflect our priorities when there is no clear-cut "best" answer.

Multicriteria decision aids can help us to address such questions. They draw heavily on methods and results in operations research, measurement theory, probability, fuzzy sets, the theory of social choice, and expert systems. In the 1980s and 1990s, the multicriteria decision aids made great strides, and several computer-based tools have been developed to implement the methods. The book by Vincke provides additional information (Vincke 1992).

In this section, we begin with the basic concepts of multicriteria decision-making. Then, we examine two classes of methods that have received a great deal of attention: (1) multiple attribute utility theory (including the analytical hierarchy process) and (2) outranking methods. The latter is far less stringent than the former, and it allows far more realistic assumptions. Our examples are presented at a very high level, so that you can see how the concepts and techniques are applied to software engineering problems. But these techniques are quite complex, and the details are beyond the scope of this book. The end of this chapter suggests other sources for learning the techniques in depth.

6.5.1 Basic Concepts of Multicriteria Decision-Making

We often have a general question to answer, and that question is really composed of a set of more specific decision problems. For example, it is not enough to ask, How do we build a system that we know is safe? We must ask an assortment of questions:

- Which combination of development methods is most appropriate to develop a safe system in the given environment?

- How much effort should be spent on each of a set of agreed-upon testing techniques in order to assure the safety of this system?

- Which compiler is most appropriate for building a safe system under these conditions?

- What is considered valid evidence of safety, and how do we combine the evidence from different subsets or sources to form a complete picture?

- Which of a set of possible actions should we take after system completion to assess the system safety?

In each case, we have a set of *actions*, A, to be explored during the decision procedure. A can include objects, decisions, candidates, and more. For instance, in the first question above, A consists of all combinations of mutually compatible methods selected from some original set. An enumeration of A might consist of combinations such as

(Alloy specification, correctness proofs)

(Alloy specification, UML design)

(OCL specification, Z formal verification, proof)

(OOD)

(Specification from Python rapid prototyping, agile development)

Once we have a set of actions, we define a *criterion*, g, to be a function from the set of actions to a totally ordered set, T. That T is *totally ordered* means that there is a relation R on pairs of elements of T that satisfies four properties:

1. R is *reflexive*: for each element x in T, the pair (x, x) is in R.

2. R is *transitive*: if (x, y) and (y, z) are in R, then (x, z) must be in R.

3. R is *antisymmetric*: if (x, y) and (y, x) are in R, then x must equal y.

4. R is *strongly complete*: for any x and y in T, either (x, y) is in R or (y, x) is in R.

These conditions guarantee that any action can be compared to any other action using a relation on T. We can use the total ordering to compare software engineering techniques, methods, and tools.

EXAMPLE 6.23

Suppose A is a set of verification and validation techniques used on a software project. We can define a criterion that maps A to the set of real numbers by defining g as follows: For each element a of A,

$g(a)$ = the total effort in person-months devoted to using technique a

This measure gives us information about experience with each element of A. Alternatively, we can define another criterion, g', to be a mapping to the set of nonnegative integers, so that we can compare the effectiveness of two techniques:

$g'(a)$ = the total number of faults discovered when using technique a

We can also have criteria that map to sets that are not numbers. For example, define the set T to be {poor, moderate, good, excellent}, representing categories of ease of use. Then we can define a criterion that maps each element of A to an element of T that represents the ease of use for a particular technique, as rated subjectively by an expert. In this example, T is totally ordered, since

poor < moderate < good < excellent

Suppose we consider a family of criteria defined on a set of actions, A. If the family is *consistent* (in a way that is described more fully by Roy 1990), then a multicriteria decision problem can be any one of the following:

- Determine a subset of A considered to be best with respect to the family of criteria (the *choice problem*).

- Divide A into subsets according to some norms (the *sorting problem*).

- Rank the actions of A from best to worst (the *ranking problem*).

To solve these problems, a decision-maker must compare each pair actions, a and b, in one of the three ways:

1. Strict preference for one of the actions

2. Indifference to either action

3. Refusal or inability to compare the actions

Each of these choices defines a relation between a and b. The set of all of these relations forms a *preference structure* on A. We can also define a *preference relation* S using only conditions 1 and 2, where a is related to b if and only if either a is strictly preferred to b, or there is indifference between the two. A classic problem of decision optimization is to define a numerical function that preserves S. In other words, we want to define a function, f, from the set A to some number system, N, which satisfies both of these conditions:

$$f(a) > f(b) \text{ if and only if } a \text{ is strictly preferred to } b$$

$$f(a) = f(b) \text{ if and only if there is indifference between } a \text{ and } b$$

This function may look familiar to you, as it is a mathematical way of describing the representation condition, introduced in Chapter 2. In multicriteria decision-making, a key problem is optimizing f. Here, "optimization" means that f must satisfy the stated conditions in a way that optimizes one or more attributes of the elements of A, such as cost or quality. When N is the set of real numbers, then no such function f exists if there are two actions that are incomparable. Unfortunately, there are many real-world situations where incomparabilities exist; we can include the incomparabilities by mapping to other types of number systems, such as vectors of real numbers, as we did in Chapter 2.

When we can preserve the preference structure by mapping to the real numbers, then we can rank the actions from best to worst, with possible ties when there is indifference between two actions; this mapping is called the *traditional model*. Such a relation is called a *complete preorder*. If there are no ties, then the preference structure is a *complete order*. Any criterion for which the underlying preference structure is a complete preorder is called a *true criterion*.

In the traditional model, indifference must be transitive; that is, if there is indifference between a and b, and indifference between b and c, then there must be indifference between a and c. However, sometimes this condition is not realistic, as when there are "sensibility thresholds" below which we are indifferent; multicriteria decision-making can be extended to cover this case, but here we focus on the traditional model and assume transitivity.

Suppose a and b are possible actions, and we have a set of n criteria $\{g_i\}$. We say that a *dominates* b if $g_i(a) \geq g_i(b)$ for each i from 1 to n. An action is said to be *efficient* if it is strictly dominated by no other action.

EXAMPLE 6.24

We are considering eight software packages, P1 through P8, to select one for use in a safety-critical application. All of the packages have the same functionality. There are four criteria that govern the selection: cost (measured in dollars), speed (measured in number of calculations per minute on a standardized set of test data), accuracy, and ease of use. The latter two criteria are measured on an ordinal scale of integers from 0 to 3, where 0 represents *poor*, 1 represents *fair*, 2 represents *good*, and 3 represents *very good*. Table 6.8 contains the results of the ratings for each criterion. To ensure that the dominance relation holds, we have changed the sign of the values for cost. In this example, we see that

P2 dominates P1
P5 dominates both P4 and P6
P3, P7, and P8 dominate no other package
P2, P5, P3, P7, and P8 are efficient

The first task in tackling a multicriteria decision problem is by reducing the set of actions to a (probably smaller) subset of efficient actions. We may find that there is only one efficient action, in which case that action is the simple solution to our problem. However, more often the dominance relation is weak, and multicriteria decision activities involve enriching the dominance relation by considering all relevant information.

To do this, we consider a vector formed by evaluating all of the criteria for a given action. That is, for each action, a, we form a vector $<g_1(a), g_2(a), \ldots, g_n(a)>$. Then the collection of all of these n-tuples is called the *image* of A. In Example 6.24, the image of A is the set of eight 4-tuples that correspond to the rows in Table 6.8. For each i from 1 to n, we can identify

TABLE 6.8   Ratings for Each Software Package Against Four Criteria

Software Package	g_1: Cost	g_2: Speed	g_3: Accuracy	g_4: Ease of Use
P1	−1300	3000	3	1
P2	−1200	3000	3	2
P3	−1150	3000	2	2
P4	−1000	2000	2	0
P5	−950	2000	2	1
P6	−950	1000	2	0
P7	−900	2000	1	0
P8	−900	1000	1	1

the (not necessarily unique) action a_i^* in A that is best according to the criterion g_i. The *ideal point* is the point $(z_1, z_2, ..., z_n)$, where $z_i = g_i(a_i^*)$. For instance, the ideal point in Example 6.24 is $(-900, 3000, 3, 2)$.

We can apply all of the criteria to each ideal point, so that we compute all possible $G_{ij} = g_j(a_i^*)$. The $n \times n$ matrix formed by the G_{ij} is called the *payoff matrix*; this matrix is unique only if each criterion achieves its maximum at only one action. The diagonal of a payoff matrix is the ideal point.

EXAMPLE 6.25

There are two different payoff matrices for the actions and criteria in Example 6.24:

$$
M = \begin{bmatrix}
a_1^* = P8 & -900 & 1000 & 1 & 1 \\
a_2^* = P2 & -1200 & 3000 & 3 & 2 \\
a_3^* = P2 & -1200 & 3000 & 3 & 2 \\
a_4^* = P3 & -1150 & 3000 & 2 & 2
\end{bmatrix}
$$

$$
M' = \begin{bmatrix}
a_1^* = P7 & -900 & 2000 & 1 & 0 \\
a_2^* = P3 & -1150 & 3000 & 2 & 2 \\
a_3^* = P1 & -1300 & 3000 & 3 & 1 \\
a_4^* = P2 & -1200 & 3000 & 3 & 2
\end{bmatrix}
$$

For a given payoff matrix, the *nadir* is the point whose ith coordinate is the minimum of the values in the ith column. In Example 6.25, the nadir for matrix M is $(-1200, 1000, 1, 1)$; for M', the nadir is $(-1300, 2000, 1, 0)$.

These concepts can be used to help solve the multicriteria decision problem. For instance, it can be shown that a positive linear combination of criteria always yields an efficient action, and that efficient actions can be characterized as those that minimized a certain distance (called the *Tchebychev distance*) to a point that slightly dominates the ideal point. The parametric optimization techniques can then be used to determine efficient actions.

However, there are often more constraints on the problem than these. For instance, in Table 6.8, we see that, although P2 is $50 more expensive, its accuracy is greater than that of P3. Our decision-makers may believe that it is worth paying the extra money to increase the accuracy. Such observations are called *substitution rates of criteria*; they allow us to add a particular amount to one criterion to compensate for a loss of one unit of another criterion. Although the assignment of substitution rates is subjective, the technique allows us to control the insertion of subjectivity, and to be consistent in applying our subjective judgments.

Suppose that, because of the safety-criticality of the application, the decision-makers prefer to pay a lot more for a package to get the highest degree of accuracy, no matter what the speed or ease of use. In this situation, two of the criteria (namely, speed and accuracy) are said to be *preferentially independent.*

6.5.2 Multiattribute Utility Theory

Multiattribute utility theory (MAUT) takes constraints such as a preference for accuracy over costs into account, providing another approach to organize subjective judgments for use in choosing between multiple alternatives. MAUT assumes that a decision-maker want to maximize a function of various criteria. Two types of problems present themselves:

1. *The representation problem*: What properties must be satisfied by the decision-maker's preferences so that they can be represented by a function with a prescribed analytical form?

2. *The construction problem*: How can the maximization function be constructed, and how can we estimate its parameters?

EXAMPLE 6.26

Suppose we are given four criteria for assessing which verification and validation techniques should be used on a software module; these criteria are described in Table 6.9.

A decision-maker rates four techniques, and the results are shown in Table 6.10. We have several choices for defining a utility function, U, which is a sum of functions U_i on each g_i. One possibility is to define each U_i to be a transformation onto the unit interval [0, 1]. For instance, we may define:

$U_1(\text{little}) = 0.8 \quad U_1(\text{moderate}) = 0.5 \quad U_1(\text{considerable}) = 0.2 \quad U_1(\text{excessive}) = 0$
$U_2(\text{bad}) = 0 \quad U_2(\text{reasonable}) = 0.1 \quad U_2(\text{good}) = 0.3 \quad U_2(\text{excellent}) = 0.6$
$$U_3(\text{no}) = 0.2 \quad U_3(\text{yes}) = 0.7$$
$$U_4(x) = 1/x$$

Thus, for any technique, we rate the availability of tool support as having greater utility than excellent coverage, and we rate the greatest utility to be ranked at the top by an expert. The final result can be evaluated by summing the functions for each technique:

U(inspections) = 0.2 + 0.6 + 0.2 + 1 = 2
U(proof) = 0 + 0.3 + 0.2 + 0.5 = 1
U(static analysis) = 0.5 + 0 + 0.7 + 0.25 = 1.45
U(black box) = 0.8 + 0.3 + 0.2 + 0.33

It is easy to see that the criteria in Table 6.9 are nonorthogonal, as the criteria are not independent. In an ideal situation, the criteria should be recast into an orthogonal set, so that the relative importance of each is clear and sensible. However, in many situations, it is not possible or practical to derive an orthogonal set. Nevertheless, MAUT helps us to impose order and consistency to the analysis process. Sometimes, the act of applying MAUT makes the decision-maker more aware of the need for orthogonality, and the criteria are changed. Otherwise, MAUT helps to generate a reasonable decision when the "best" is not possible.

The analytic hierarchy process (AHP) is a popular MAUT technique (Saaty and Vargas 2012). AHP begins by representing the decision problem as a hierarchy graph, where the top node is the main problem objective and the bottom nodes represent actions. At each level of the hierarchy, a decision-maker makes a pairwise comparison of the nodes from the viewpoint of their contribution to each of the higher-level nodes to which they are linked. The pairwise comparison involves *preference ratios* (for actions) or *importance ratios* (for criteria), evaluated on a particular numerical scale. Then, the "value" of each node is calculated, based on computing the eigenvalues of the matrix of pairwise comparisons. Each action's global contribution to the main objective is calculated by aggregating the weighted average type. A software package called Expert Choice supports the use of AHP.[*]

TABLE 6.9   Criteria for Assessing Verification and Validation Techniques

Criterion	Measurement Scale
g_1: Effort required	{Little, moderate, considerable, excessive}
g_2: Coverage	{Bad, reasonable, good, excellent}
g_3: Tool support	{No, yes}
g_4: Ranking of usefulness by expert	{1, 2, …, n} where n is number of techniques

---

[*] http://expertchoice.com/.

TABLE 6.10 Evaluation of Four Verification and Validation Techniques

Technique	g_1	g_2	g_3	g_4
Code inspection	Considerable	Excellent	No	1
Formal proof	Excessive	Good	No	2
Static analysis	Moderate	Bad	Yes	4
Black box testing	Little	Good	No	3

AHP has been used in dependability assessment (Auer 1994), and Vaisanene and colleagues have applied it to a safety assessment of programmable logic components (Vaisanene et al. 1994). AHP was also used to help NASA choose from among several strategies for improving a safety feature of the Space Shuttle (Frank 1995).

6.5.3 Outranking Methods

MAUT can be very valuable, resulting in a ranking of actions from best to worst. However, such rankings are based on assumptions that are unrealistic in many situations. For many problems, we do not have a complete ranking, so we can do without the unrealistic assumptions. However, the dominance relation alone is generally too weak to be useful. The outranking methods seek to enrich the dominance relation without having to make the assumptions of MAUT.

An *outranking relation* is a binary relation on a set of actions defined in the following way. Consider two actions *a* and *b*. Given what is known about the decision-maker's preferences, the quality of the valuations of the actions, and the nature of the problem, if there are enough arguments to decide that *a* is at least as good as *b* and no essential reason to refute that statement, then we say that *a* is related to *b*. This definition is informal, and published outranking methods differ in the way that they are formalized. No matter the definition, the outranking relation is neither necessarily complete nor transitive (Roy 1990).

Any outranking method has two steps:

1. Building the outranking relation

2. Exploiting the relation with regard to the chosen problem statement

To give you an idea of how outranking works, we present a brief example, using the Electre I method (Figueira et al. 2005).

EXAMPLE 6.27

Table 6.11 lists four criteria for assessing combinations of verification and validation techniques. For instance, action 1 might involve formal proof followed by code inspection, while action 2 might be formal proof plus static code analysis. Each criterion must map onto a totally ordered set, and each is assigned a weight, indicated by the number in parentheses. (For example, the weight of the first criterion is 5.)

For each ordered pair of actions (a, b), we compute a *concordance index* by adding the weights of all criteria g_i for which $g_i(a)$ is greater than $g_i(b)$. Thus, for the pair $(1, 2)$, action 1 is at least as good as action 2 with respect to all but the first criterion, so we sum the weights $(4 + 3 + 3)$ to yield a concordance index of 10. The full set of concordance indices is shown in Table 6.12. In a sense, this table is the "first draft" of the preference structure (i.e., the outranking relation). We say that a is preferable to b when the concordance index for (a, b) is larger than for (b, a).

Next, we restrict the structure by defining a *concordance threshold*, t, so that a is preferred to b only if the concordance index for (a, b) is at least as large as t. Suppose t is 12. Then, according to Table 6.12, action 2 is still preferable to action 1, but we no longer have preference of action 1 to action 7 because the concordance index for $(1, 7)$ is not large enough.

We further refine the preference structure to include other constraints. For instance, suppose that for criterion g_1 (effort required) we never allow action a to outrank action b if $g_1(a)$ is excessive and $g_1(b)$ is little. In other words, regardless of the values of the other criteria, b is so superior to a with respect to g_1 that we veto it being outranked by a. In general, we handle this situation by defining a *discordance set* for each criterion, containing the ordered pairs of values for the criterion for which the outranking is refused.

For example, let us define

$$D_1 = \{(\text{excessive, little}), (\text{considerable, little})\}$$
$$D_2 = \{(\text{moderate, excellent})\}$$
$$D_3 = D_4 = \{\}$$

Then we can define the outranking relation by saying that one action outranks another provided that its concordance index is at least as large as the threshold, and provided that for each criterion, the preference is not vetoed by the discordance set. The full outranking relation for the example is shown in Figure 6.25.

Finally, we exploit this relationship to find a subset of actions that is optimal with respect to the outranking. In graph theory, such a set (of nodes) is called a *kernel*. There are several graph-theoretic techniques for determining kernels; thus enabling us to find the best course of action, given our constraints. From Figure 6.25, we can see that there are three kernels, namely, $\{2,4,7\}$ and $\{2,5,7\}$; the actions 4 and 5 are considered to be tied.

TABLE 6.11 Criteria for Assessing Combined Verification and Validation Techniques

Action	g_1: Effort Required (Weight = 5)	g_2: Potential for Detecting Critical Faults (Weight = 4)	g_3: Coverage Achieved (Weight = 3)	g_4: Tool Support (Weight = 3)
1	Excessive	Excellent	Good	Yes
2	Considerable	Excellent	Average	Yes
3	Considerable	Good	Good	Yes
4	Moderate	Good	Good	No
5	Moderate	Good	Average	Yes
6	Moderate	Reasonable	Good	Yes
7	Little	Reasonable	Average	No

TABLE 6.12 Concordance Indices

	1	2	3	4	5	6	7
1	–	10	10	10	10	10	10
2	12	–	12	7	10	7	10
3	11	11	–	11	10	10	10
4	8	8	12	–	12	12	10
5	8	11	12	12	–	12	10
6	11	11	11	11	11	–	10
7	5	8	5	8	8	9	—

6.5.4 Bayesian Evaluation of Multiple Hypotheses

You may have noticed that AHP, MAUT, and outranking methods depend on operations on data that according to Chapter 2 represent inadmissible transformations. Many of the attributes and attribute weights appear to be rankings and thus are ordinal scale measures. Yet, the methods perform multiplication and/or addition on these values. It would be better if we could use an analysis using only well-defined and meaningful transformations.

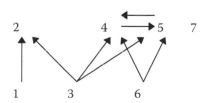

FIGURE 6.25 Graph of outranking relation. An arc from node x to node y indicates that y outranks x.

The Bayesian approach, based on a theory introduced by Thomas Bayes in 1763, analyzes multiple variables in terms of three key factors: the probability of the occurrence of a set of events, the outcome of each event should it occur, and the dependencies between events. The probabilities and outcomes are expressed in terms of ratio scales measures making the analyses meaningful. The Bayesian approach supports the use of qualitative beliefs concerning probabilities and outcomes, and is dynamic—it supports the updating beliefs in response to new evidence. In addition, unlike classical data analysis techniques described in Section 6.2, the Bayesian approach focuses on the magnitude of relations rather than their precision.

This elegant approach has been widely used in law, medicine, finance, and engineering. Tools that support Bayesian analyses are now readily available. This approach is well suited to analyzing software engineering problems. Thus, we devote Chapter 7 to an in-depth description of the Bayesian approach applied to software engineering problems.

6.6 OVERVIEW OF STATISTICAL TESTS

This chapter has mentioned several statistical tests that can be used to analyze your data. We have presented techniques in terms of the type of relationship we seek (linear, multivariate, etc.), but our choice of technique must also take into account the number of groups being compared, the size of the sample, and more. In this section, we describe a few statistical tests, so that you can see how each one is oriented to a particular experimental situation. The section is organized in terms of sample and experiment type, as summarized in Table 6.13.

6.6.1 One-Group Tests

In a one-group design, measurements from one group of subjects are compared with an expected population distribution of values. These tests are appropriate when an experimenter has a single set of data with an explicit null hypothesis concerning the value of the population mean. For example, you may have sampled the size of each of a group of modules, and you hypothesize that the average module size is 200 LOC.

For normal distributions, the parametric test is called the t-test (or, equivalently, the Student's t-test, single sample t-test, or one-group t-test). Since the mean of the data is involved, the test is appropriate only for data that are on the interval scale or above.

TABLE 6.13   Examples of Statistical Tests

Sample Type	Nominal	Ordinal	Interval	Ratio
One sample	Binomial	Kolmogorov–Smirnov		Single-sample *t*-test
	Chi-square	One-sample runs		
		Change point		
Two related samples	McNemar change	Sign test	Permutation	Matched groups *t*-test
		Wilcoxon signed ranks		
Two independent samples	Fisher exact	Median test	Permutation	Population *t*-test
	Chi-square	Wilcoxon–Mann–Whitney		
		Robust rank order		
		Kolmogorov–Smirnov two-sample		
		Siegel–Tukey		
K-related samples	Cochran Q-test	Friedman two-way ANOVA		Within-groups ANOVA
		Page test (ordered alternatives)		
K-independent samples	Chi-squared	Kruskal–Wallis one-way ANOVA		Between-groups ANOVA
		Jonckheere test		

There are several alternatives for nonparametric data.

6.6.1.1 Binomial Test
The binomial test should be used when

- The dependent variable can take only two distinct, mutually exclusive, and exhaustive values (such as "module has been inspected" and "module has not been inspected").

- The measurement trials in the experiment are independent.

6.6.1.2 Chi-Squared Test for Goodness of Fit
The chi-squared test is appropriate when

1. The dependent variable can take two or more distinct, mutually exclusive, and exhaustive values (such as "module has fewer than 100

LOC," "module has between 100 and 300 LOC," and "module has more than 300 LOC").

2. The measurement trials in the experiment are independent.

3. None of the categories has an observed or expected frequency of less than five occurrences.

6.6.1.3 Kolmogorov–Smirnov One-Sample Test

This test assumes that the dependent measure is a continuous, rather than discrete, variable. It assesses the similarity between an observed and an expected cumulative frequency distribution.

6.6.1.4 One-Sample Runs Test

This test determines if the results of a measurement process follow a consistent sequence, called a *run*. For example, it can be used in analyzing the reasons for system downtime, to determine if one particular type of hardware or software problem is responsible for consecutive system crashes. The test assumes that the successive measurement trials are independent.

6.6.1.5 Change-Point Test

This test determines whether the distribution of some sequence of events has changed in some way. It assumes that the observations form an ordered sequence. If, at some point in the distribution of observed measures, a shift in the median (or middle) score has occurred, the test identifies the change point.

6.6.2 Two-Group Tests

Single-sample designs and tests are often used to decide if the results of some process are straying from an already-known value. However, these designs are not appropriate when an experimentally established comparison is required.

Two-group designs allow you to compare two samples of dependent measures drawn from related or matched samples of subjects (as in a within-subjects design), or from two independent groups of subjects (as in a between-subjects design). The appropriate statistical test depends on whether the samples are independent or related in some way.

6.6.2.1 Tests to Compare Two Matched or Related Groups

The parametric *t*-test for matched groups applies when the dependent measurement is taken under two different conditions, and when one of the following additional conditions has been met:

1. The same subject is tested in both conditions (i.e., a within-subjects or repeated-measures design has been used).

2. Subjects are matched according to some criterion (i.e., a matched-groups design has been used).

3. Prescreening has been used to form randomized blocks of subjects (i.e., a randomized block design has been used).

This test assumes the following:

- Subjects have been randomly selected from the population. When different subjects have been used in each condition, assignment to the conditions should be random.

- If fewer than 30 pairs of dependent measures are available, the distribution of the differences between the two scores should be approximately normal.

Like the single-sample *t*-test, this test also requires measurement data to be at least on the interval scale.

Nonparametric alternatives to this test include the McNemar change test, the sign test, and the Wilcoxon signed ranks test. The *McNemar change test* is useful for assessing change on a dichotomous variable, after an experimental treatment has been administered to a subject. For example, it can be used to assess whether programmers switch preferences for language type after receiving training in procedural and object-oriented techniques. The test assumes that the data are frequencies that can be classified in dichotomous terms.

The *sign test* is applied to related samples when you want to establish that one condition is greater than another on the dependent measure. The test assumes that the dependent measure reflects a continuous variable (such as experience) rather than categories. For example, you may use this test to determine whether a particular type of programming construct is significantly easier to implement on a given system. The *Wilcoxon signed ranks test* is similar to the sign test, taking into account the magnitude as well as the direction of difference; it gives more weight to a pair that shows a large difference than to a pair that shows a small difference.

6.6.2.2 Tests to Compare Two Independent Groups

The parametric test appropriate for independent groups or between-subjects designs is the *t*-test for differences between population means. This test is applied when there are two groups of different subjects that are not paired or matched in any way. The test assumes the following:

1. Subjects have been randomly selected from the population and assigned to one of the two treatment conditions.

2. If fewer than 30 subjects are measured in each group, the distribution of the differences between the two scores should be approximately normal.

3. The variance, or spread, of scores in the two population groups should be equal or homogeneous.

As before, the data must be measured on at least an interval scale to be suitable for this test.

Several nonparametric alternatives to this test are available; they are described in most statistics books.

6.6.3 Comparisons Involving More than Two Groups

Suppose you have measured an attribute of k groups, where k is greater than two. For example, you are comparing the productivity rates of 10 projects, or you want to look at a structural attribute for each of 50 modules. Here, the appropriate statistical analysis technique is an ANOVA. This class of parametric techniques is suitable for data from between-subjects designs, within-subjects designs, and designs that involve a mixture of between- and within-subjects treatments. Complementary nonparametric tests are available but are beyond the scope of this book; some are listed in Table 6.13.

ANOVA results tell you whether at least one statistically significant difference exists somewhere within the comparisons drawn by the analysis. Statistical significance in an ANOVA is reflected in the F statistic that is calculated by the procedure. When an F statistic is significant, you then apply multiple comparisons tests to determine which levels of a factor differ significantly from the others.

As we noted in Chapter 4, when your experimental design is complex, it is best to consult a statistician to determine which analysis techniques are most appropriate. Since the techniques require different types of data and control, it is sensible to lay out your analysis plans and methods *before* you begin your investigation; otherwise, you risk collecting data that are insufficient or inappropriate for the analysis you need to do.

6.7 SUMMARY

Datasets of software attribute values must be analyzed with care because software measures are not usually normally distributed. We have presented several techniques here that address a wide variety of situations: differing data distributions, varying measurement scales, varying sample sizes, and differing goals. In general, it is advisable to

- Describe a set of attribute values using box plot statistics (based on median and quartiles) rather than on mean and variance

- Inspect a scatter plot visually when investigating the relationship between two variables

- Use robust correlation coefficients to confirm whether or not a relationship exists between two attributes

- Use robust regression in the presence of atypical values to identify a linear relationship between two attributes, or remove the atypical values before analysis

- Always check the residuals by plotting them against the dependent variable

- Use Tukey's ladder to assist in the selection of transformations when faced with nonlinear relationships

- Use principal component analysis to investigate the dimensionality of datasets with large numbers of correlated attributes

It is also helpful to consider using more advanced techniques, such as multiattribute utility theory, to help you choose a "good" solution, rather than the "best" solution, subject to the constraints on your problem. But most important, we have demonstrated that your choice of analysis technique must be governed by the goals of your investigation, so that you can support or refute the hypothesis you are testing.

EXERCISES

1. Why are statistics necessary? Why can experimenters not just look at the data and decide for themselves what is important? On the other hand, which of the following is more important: (a) the results of a statistical test or (b) the results of the intraocular significance test (i.e., when the effect leaps out and hits you between the eyes)?

2. Can a comparison that does not achieve statistical significance still be important?

3. Three years ago, the software development manager introduced changes to the development practices at your company. These changes were supposed to ensure that more time was spent up-front on projects, rather than coding. Suppose you have the following data giving actual effort by software development phase for the last five projects (in chronological order). Provide an appropriate graphical representation for the manager so that she can see whether the changes have had an effect.

	Project 1	Project 2	Project 3	Project 4	Project 5
Requirements	120	100	370	80	410
Specification	320	240	490	140	540
High-level design	30	40	90	40	60
Detailed design	170	190	420	120	340
Coding	1010	420	1130	250	1200
Testing	460	300	580	90	550

4. Construct a box plot for the paths measure in dataset 2 of Table 6.1.

5. Draw the scatter plot of the LOC and the path measures for the modules in dataset 2 of Table 6.1. Are there any unusual values?

6. Obtain the Spearman rank correlation coefficient and the Pearson correlation coefficient for paths and LOC for dataset 2 of Table 6.1. Why do you think the rank correlation is larger than the Pearson correlation? (*Hint*: Look at the scatter plot produced for Exercise 5.)

7. Using a statistical package or spreadsheet, calculate the least squares regression line for faults against LOC for all the modules in dataset 2 of Table 6.1. Calculate the residuals and plot the residual against the dependent variable (LOC). Can you identify any outlying points?

8. If you have access to a statistical package, do a principal component analysis of the normalized LOC, fan-out, fan-in, and paths measures from dataset 2 of Table 6.1. Produce a scatter plot of the first principal component against the second and identify the atypical values.

9. What result do you think you would get if you did a principal component analysis on the raw rather than the normalized size and structure measures in dataset 2 of Table 6.1? If you have a statistical package, perform a principal component analysis on the raw size and structure data and see if the results confirm your opinion.

10. Company X runs a large software system S that has been developed in-house over a number of years. The company collects information about software defects discovered by users of S. During the regular maintenance cycle, each defect is traced to one of the nine subsystems of S (labeled with letters A through I), each of which is the responsibility of a different team of programmers. The table below summarizes information about new defects discovered during the current year:

System	A	B	C	D	E	F	G	H	I
Defects	35	0	95	35	55	40	55	40	45
Size (KLOC)	40	100	5	50	120	70	60	100	40

Suppose you are the manager of system S.

a. Compute the defect density for each subsystem.

b. Use simple outlier analysis to identify unusual features of the system.

c. What conclusions can you draw from your outlier analysis?

d. What are the basic weaknesses of the metrics data as currently collected?

e. What simple additions or changes to the data collection strategy would significantly improve your diagnostic capability?

11. The table below contains three measures for each of 17 software modules. LOC is the number of LOC in the module, CFP is the number of control flow paths in the module, and Faults are the number of faults found in the module. Construct a box plot for each of the three measures, and identify any outliers. What conclusions can you draw? What recommendations can you make to address the problems you have discovered?

Module	LOC	CFP	Faults
A	15	4	0
B	28	6	15
C	40	2	10
D	60	26	1
E	60	14	0
F	95	18	15
G	110	12	9
H	140	12	6
I	170	54	6
J	180	36	20
K	185	28	14
L	190	32	20
M	210	54	15
N	210	46	18
P	270	128	58
Q	400	84	59
R	420	120	43

12. The table below contains metrics from 17 software systems under investigation: total number of thousands LOC for the system (KLOC), average module size (MOD) measured in LOC, and total number of faults found per thousand LOC (FD). Construct a box plot for each metric, and identify any outliers.

System	KLOC	MOD	FD
A	10	15	36
B	23	43	22
C	26	61	15
D	31	10	33
E	31	43	15
F	40	57	13
G	47	58	22
H	52	65	16
I	54	50	15
J	67	60	18
K	70	50	10
L	75	96	34
M	83	51	16
N	83	61	18
P	100	32	12
Q	110	78	20
R	200	48	21

13. *Test-first development* is one component of many agile software development processes such as Extreme Programming and SCRUM that is briefly described as follows:

    Test-first development: Unit tests are written for each class or component before the class or component body code is written. The tests are run continuously and the program must pass all tests. These tests act as executable specifications for the program units.

    The following table shows the results of a case study that examined the defect profile of a software system in which part of the system was developed by applying *test first* methods. The table shows the number of "defects" found at different phases, comparing the portion of the system with test first developed code to the rest of the system.

    a. What conclusions can you draw about the effectiveness of *test-first* development based on these results? Justify your conclusions and note limitations.

    b. List all assumptions that you used to arrive at your conclusions in (a).

   c. List any additional information/metrics data that you would like to have seen.

Test Phase	Defects Found in *Test-First* Developed Code (10,000 Lines Total)	Defects Found in Remainder of Code (8000 Lines Total)
Specification and design inspection	10	16
Unit test	89	32
System and integration test	15	80
12 months of operational use	16	16
Overall	130	144

REFERENCE

Figueira, J., Salvatore G., Matthias E., *Multiple Criteria Decision Analysis: State of the Art Surveys*, Springer Science + Business Media, Inc., New York, ISBN 0-387-23081-5, 2005.

FURTHER READING

There are several good books on statistics and their application.

Chatfield C., *Statistics for Technology: A Course in Applied Statistics Third Edition (Revised)*, Chapman and Hall, London, England, 1998.

Ott R.L. and Longnecker M.T., *An Introduction to Statistical Methods and Data Analysis 6th Edition*, Duxbury Press, Pacific Grove, California and United Kingdom, 2008.

The book by Siegel and Castellan is a classic text on nonparametric statistics, while the recent book by Kraska-Miller is geared for practical application by students and practitioners.

Kraska-Miller M., *Nonparametric Statistics for Social and Behavioral Science*, CRC Press, Boca Raton, 2014.

Siegel S. and N. Castellan N.J., Jr., *Nonparametric Statistics for the Behavioral Sciences*, 2nd edition, McGraw-Hill, New York, 1988.

For descriptions of box plots and other exploratory data analysis, consult Hoaglin, Mosteller and Tukey.

Hoaglin D. C., Mosteller F., and Tukey J.W., *Understanding Robust and Exploratory Data Analysis*, John Wiley and Sons, New York, 2000.

Sometimes, you have two datasets representing the same situation. Each is different from the other, in terms of distribution, mean or variance, but both were collected with reasonable care. This problem arises frequently but is rarely discussed in standard statistical texts; Barford explains how to tell which is the better dataset for answering your questions.

Barford, N. C., *Experimental Measurements: Precision, Error and Truth Second Edition*, Addison-Wesley, Reading, Massachusetts, 1985.

For an excellent overview of MCDA, you should consult:

Vincke P., *Multicriteria Decision Aid*, John Wiley and Sons, New York, 1992.

Additional references on multiattribute analysis methods are:

Roy B., Decision aid and decision making, *European Journal of Operational Research*, 45, 324–331, 1990.
Saaty T.L. and Vargas L.G., *Models, Methods, Concepts & Applications of the Analytic Hierarchy Process Second Edition*, Springer, New York, 2012.

Metrics for Decision Support

*The Need for Causal Models**

T HE ULTIMATE GOAL FOR software metrics is to help software professionals (be they developers, testers, managers, or maintainers) make decisions under uncertainty. As explained so far, we are ultimately interested in knowing things like how much effort and time will be required to produce a system to a particular set of requirements, how much more testing will be needed before the system has sufficiently few bugs, how much effort can we save by using some particular tool, etc. All of these decisions involve uncertainty, risk, and trade-offs. Whereas the other chapters in this book provide the tools for identifying and collecting the specific metrics to capture the underlying attributes involved in these problems, this chapter focuses on how to incorporate these metrics into a rational decision-making framework.

Several years ago at a leading international software metrics conference, which attracted nearly a thousand delegates, a keynote speaker recounted an interesting story about a company-wide metrics program that he had been instrumental in setting up. He said that one of the main objectives of the program was to achieve process improvement by learning from metrics what process activities worked and what ones did not.

---

* This chapter is based on collaborative work with Martin Neil.

To do this, the company looked at those projects that, in metrics terms, were considered most successful. These were the projects with especially low rates of customer-reported defects, measured by defects per thousand lines of code (KLOC). The idea was to learn what processes characterized such successful projects. A number of such "star" projects were identified, including some that achieved the magical perfect quality target of zero defects per KLOC in the first six months post-release. But, it turned out that what they learned from this was very different from what they had expected. Few of the star projects were, in fact, at all successful from any commercial or subjective perspective:

> The main explanation for the very low number of defects reported by customers was that they were generally so poor that they never got properly completed or used.

This story exposes the classic weakness of relying only on traditional static metrics: the omission of sometimes obvious and simple *causal* factors that can have a major explanatory effect on what is observed and learnt. If you are managing a software development project and you hear that very few defects were found during a critical testing phase, is that good news or bad news? Of course it depends on the testing effort, just as it does if you heard that a large number of defects were discovered. The danger is to assume that defects found in testing correlates with defects found in operation, and that it is thus possible to build a regression model (in the sense of Chapter 6) in which *defects in operation* (the uncertain quantity we are trying to predict) is a function of the variable *defects found in test* (the known quantity we can measure). Similarly, it is dangerous to assume that we can build regression models in which other metrics, such as size and complexity metrics, act as independent variables that can predict dependent variables like defects and maintenance effort. In Section 7.1, we will explain why such attempts at prediction based on correlation are not normally effective and cannot be used for risk assessment.

If you are interested in, say, predicting defects in operation from information gained during development and testing, then certainly metrics like defects found in test and other size and complexity metrics might be useful pieces of evidence. A rational way to use this evidence to update your beliefs about defects in operation is to use *Bayesian reasoning*. When there are multiple, dependent metrics then we use Bayesian reasoning together with causal models, that is, *Bayesian networks* (BNs). We explain the basics of Bayesian Reasoning and BNs in Section 7.2. In Section 7.3, we describe a range of BNs that have been effectively used in software projects, including BNs that enable risk trade-off analysis. We also discuss the practical issues in building and deploying BNs in software projects. In Section 7.4, we look at typical approaches to risk assessment used by software project managers and demonstrate how the BN approach can provide far more logic and insight.

There are numerous free and commercial tools available that make it very easy to build and run the BN models described in this chapter (the book (Fenton and Neil 2012) provides a comprehensive overview). In particular, the AgenaRisk BN tool (Agena 2014) can be downloaded free and it contains all of the models described in this chapter.

7.1 FROM CORRELATION AND REGRESSION TO CAUSAL MODELS

Standard statistical approaches to risk assessment and prediction in all scientific disciplines seek to establish hypotheses from relationships discovered in data. To take a nonsoftware example, suppose we are interested in the risk of fatal automobile crashes. Table 7.1 gives the number of crashes resulting in fatalities in the United States in 2008 broken down by month (*Source*: US National Highways Traffic Safety Administration). It also gives the average monthly temperature.

We plot the fatalities and temperature data in a scatterplot graph as shown in Figure 7.1.

There seems to be a clear relationship between temperature and fatalities—fatalities increase as the temperature increases. Indeed, using the

TABLE 7.1   Fatal Automobile Crashes per Month

Month	Total Fatal Crashes	Average Monthly Temperature (°F)
January	297	17.0
February	280	18.0
March	267	29.0
April	350	43.0
May	328	55.0
June	386	65.0
July	419	70.0
August	410	68.0
September	331	59.0
October	356	48.0
November	326	37.0
December	311	22.0

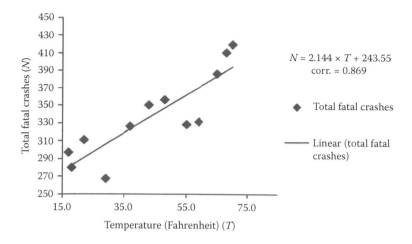

FIGURE 7.1 Scatterplot of temperature against road fatalities (each dot represents a month).

standard statistical tools of correlation and p-values, statisticians would accept the hypothesis of a relationship as "highly significant" (the correlation coefficient here is approximately 0.869 and it comfortably passes the criteria for a p-value of 0.01).

However, in addition to serious concerns about the use of p-values generally (as described comprehensively in Ziliak and McCloskey 2008), there is an inevitable temptation arising from such results to infer causal links such as, in this case, *higher temperatures cause more fatalities*. Even though any introductory statistics course teaches that correlation is not causation, the regression equation is typically used for prediction (e.g., in this case the equation relating N to T is used to predict that at 80°F we might expect to see 415 fatal crashes per month).

But there is a grave danger of confusing prediction with risk assessment. For risk assessment and risk management the regression model is useless, because it provides no explanatory power at all. In fact, from a risk perspective this model would provide irrational, and potentially dangerous, information: it would suggest that if you want to minimize your chances of dying in an automobile crash you should do your driving when the highways are at their most dangerous, in winter.

One obvious improvement to the model, if the data are available, is to factor in the number of miles traveled (i.e., journeys made). But there are other underlying causal and influential factors that might do much to explain the apparently strange statistical observations and provide better insights into risk. With some common sense and careful reflection, we can recognize the following:

- Temperature influences the highway conditions (which will be worse as temperature decreases).

- Temperature also influences the number of journeys made; people generally make more journeys in spring and summer and will generally drive less when weather conditions are bad.

- When the highway conditions are bad, people tend to reduce their speed and drive more slowly. So highway conditions influence speed.

- The actual number of crashes is influenced not just by the number of journeys, but also the speed. If relatively few people are driving, and taking more care, we might expect fewer fatal crashes than we would otherwise experience.

The influence of these factors is shown in Figure 7.2.

The crucial message here is that the model no longer involves a simple single causal explanation; instead, it combines the statistical information available in a database (the "objective" factors) with other causal "subjective" factors derived from careful reflection. These factors now interact in a nonlinear way that helps us to arrive at an explanation for the observed results. Behavior, such as our natural caution to drive slower when faced with poor road conditions, leads to lower accident rates (people are known to adapt to the perception of risk by tuning the risk to tolerable levels. This is formally referred to as *risk homeostasis*). Conversely, if we insist on driving fast in poor road conditions then, irrespective of the temperature, the risk of an accident increases and so the model is able to capture our intuitive beliefs that were contradicted by the counterintuitive results from the simple regression model.

The role played in the causal model by driving speed reflects human behavior. The fact that the data on the average speed of automobile drivers was not available in a database explains why this variable, despite its apparent obviousness, did not appear in the statistical regression model. The situation whereby a statistical model is based only on available data, rather than on reality, is called "conditioning on the data." This enhances convenience but at the cost of accuracy.

By accepting the statistical model we are asked to defy our senses and experience and actively ignore the role unobserved factors play. In fact, we

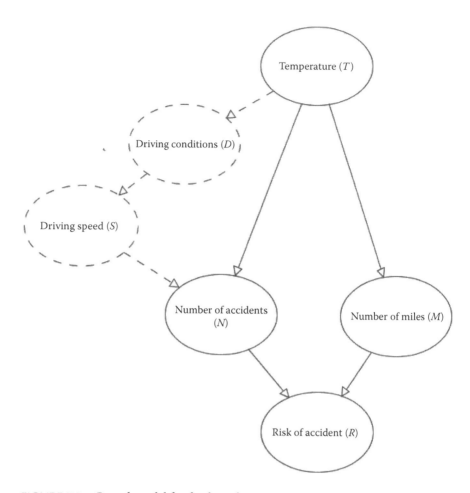

FIGURE 7.2 Causal model for fatal crashes.

cannot even explain the results without recourse to factors that do not appear in the database. This is a key point: with causal models we seek to dig deeper behind and underneath the data to explore richer relationships missing from over-simplistic statistical models. In doing so, we gain insights into how best to control risk and uncertainty. The regression model, based on the idea that we can predict automobile crash fatalities based on temperature, fails to answer the substantial question: how can we control or influence behavior to reduce fatalities. This at least is achievable; control of weather is not.

Statistical regression models have played a major role in software prediction with size (whether it be solution size, as in complexity metrics or LOC-based approaches, or problem size, as in function point-based approaches) being the key driver. Thus, it is assumed that we can fit a function F such that

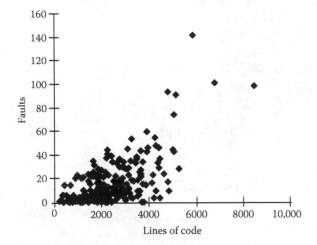

FIGURE 7.3   Scatterplot of LOC against all faults for major system (each dot represents a module).

$$\text{Defects} = F \text{ (size, complexity)}$$

Figures 7.3 through 7.6 show data from a major case study (Fenton and Ohlsson 2000) that highlighted some of the problems with these approaches. In each figure the dots represent modules (which are typically self-contained software components of approximately 2000 LOC) sampled at random from a very large software system. Figure 7.3 (in which "faults" refers to all known faults discovered pre- and post-delivery) confirms what many studies have shown: module size is correlated with number of faults, but is not a good predictor of it. Figure 7.4 shows that complexity metrics, such as cyclomatic complexity (see Section 9.2.2.1 for the details of this metric), are not significantly better (and in any case are

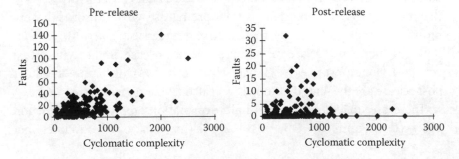

FIGURE 7.4   Scatterplots of cyclomatic complexity against number of pre- and post-release faults for release $n + 1$ (each dot represents a module).

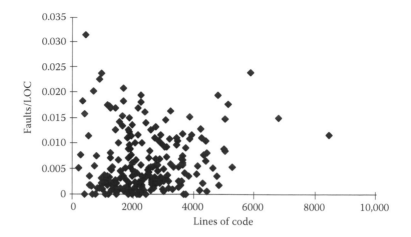

FIGURE 7.5   Scatter plot of module fault density against size (each dot represents a module).

very strongly correlated to LOC). This is true even when we separate out pre- and post-delivery faults.

Figure 7.5 shows that there is no obvious empirical support for the widely held software engineering theory that smaller modules are less "fault-prone" than large ones. The evidence here shows a random relationship.

But the danger of relying on statistical models is most evident when we consider the relationship between the number of pre-release faults (i.e., those found during pre-release testing) and the number of post-release faults (the latter being the number you really are most interested in predicting). The data alone, as shown in Figure 7.6 (an empirical result that has been repeated in many other systems) shows a relationship that is contrary to widely perceived assumptions. Instead of the expected strong (positive) correlation between pre-release and post-release faults (i.e., the expectation that modules which are most fault-prone pre-release will be the ones that are most fault-prone post-release), there is strong evidence of a *negative correlation*. The modules that are very fault-prone pre-release are likely to reveal very few faults post-release. Conversely, the truly "fault-prone" modules post-release are the ones that mostly revealed few or no faults pre-release.*

There are, of course, very simple explanations for the phenomenon observed in Figure 7.6. One possibility is that most of the modules that

---

* These results are potentially devastating for some regression-based fault prediction models, because many of those models were "validated" on the basis of using pre-release fault counts as a surrogate measure for operational quality.

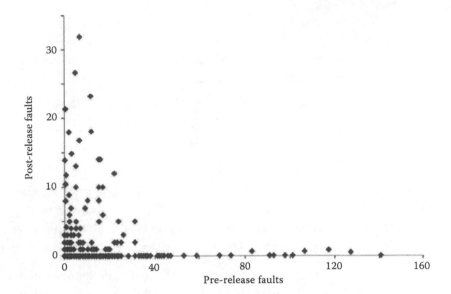

FIGURE 7.6   Scatter plot of pre-release faults against post-release faults for a major system (each dot represents a module).

had many pre-release and few post-release faults were very well tested. The *amount of testing* is therefore a very simple explanatory factor that must be incorporated into any predictive model of defects. Similarly, a module that is simply never executed in operation will reveal no faults no matter how many are latent. Hence, *operational usage* is another obvious explanatory factor that must be incorporated.

The absence of any causal factors that explain variation is a feature also of the classic approach to software metrics resource prediction, where again the tendency (as we will see in Chapter 8) has been to produce regression-based functions of the form:

$$\text{Effort} = F(\text{size, process quality, product quality})$$
$$\text{Time} = F(\text{size, process quality, product quality})$$

The basic problems with this approach are:

- It is inevitably based on limited historical data of projects that just happened to have been completed. Based on typical software projects these are likely to have produced products of variable quality, including many that are poor. It is therefore difficult to interpret what a figure for effort prediction based on such a model actually means.

- It fails to incorporate any true causal relationships, relying often on the fundamentally flawed assumption that somehow the solution size can influence the amount of resources required. This is contrary to the economic definition of a production model where

  Output = F(input)

  Rather than

  Input = F(output).

- There is a flawed assumption that projects do not have prior resource constraints. In practice all projects do, but these cannot be accommodated in the models. Hence, the "prediction" is premised on impossible assumptions and provides little more than a negotiating tool.

- The models are effectively "black boxes" that hide/ignore crucial assumptions explaining the relationships and trade-offs between the various inputs and outputs.

- The models provide no information about the inevitable inherent uncertainty of the predicted outcome; for a set of "inputs" the models will return a single point value with no indication of the range or size of uncertainty.

To provide genuine risk assessment and decision support for managers, we need to provide the following kinds of predictions:

- For a problem of this size, and given these limited resources, how likely am I to achieve a product of suitable quality?

- How much can I scale down the resources if I am prepared to put up with a product of specified lesser quality?

- The model predicts that I need 4 people over 2 years to build a system of this kind of size. But I only have funding for 3 people over one year. If I cannot sacrifice quality, how good must the staff be to build the systems with the limited resources? Alternatively, if my staff are no better than average and I cannot change them, how much required functionality needs to be cut in order to deliver at the required level of quality?

Our aim is to show that causal models, using Bayesian networks, can provide relevant predictions, as well as incorporating the inevitable uncertainty, reliance on expert judgment, and incomplete information that are pervasive in software engineering.

7.2 BAYES THEOREM AND BAYESIAN NETWORKS

While Section 7.1 provided the rationale for using causal models rather than purely statistically driven models, it provided no actual mechanism for doing so. The necessary mechanism is driven by Bayes theorem, which provides us with a rational means of updating our belief in some unknown hypothesis in light of new or additional evidence (i.e., observed metrics).

At their core, all of the decision and prediction problems identified so far incorporate the basic causal structure shown in Figure 7.7.

There is some unknown hypothesis H about which we wish to assess the uncertainty and make some decision. Does our system contain critical bugs? Does it contain sufficiently few bugs to release? Will it require more than 3 person months of effort to complete the necessary functionality? Will the system fail within a given period of time?

Consciously or unconsciously we start with some (unconditional) prior belief about H. Taking a nonsoftware example, suppose that we are in charge of a chest clinic and are interested in knowing whether a new patient has cancer; so H is the hypothesis that the patient has cancer. Suppose that 10% of previous patients who came to the clinic were ultimately diagnosed with cancer. Then a reasonable prior belief for the probability H is true, written $P(H)$, is 0.1. This also means that the prior $P(\text{not } H) = 0.9$.

One piece of evidence E we might discover that could change our prior belief is whether or not the person is a smoker. Suppose that 50% of the people coming to the clinic are smokers, so $P(E) = 0.5$. If we discover that the person is indeed a smoker, to what extent do we revise our prior judgment about the probability $P(H)$ that the person has cancer? In other words we want to calculate the probability of H given the evidence E. We write this as $P(H|E)$; it is called the *conditional probability* of H given E and because it represents a revised belief about H once we have seen the evidence E we also call it the *posterior probability* of H. To arrive at the correct answer for the posterior probability, we use a type of reasoning

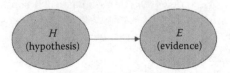

FIGURE 7.7   Causal view of evidence.

called *Bayesian inference*, named after Thomas Bayes who determined the necessary calculations for it in 1763.

What Bayes recognized was that (as in this and many other cases) we might not have direct information about $P(H|E)$ but we do have *prior* information about $P(E|H)$. The probability $P(E|H)$ is called the *likelihood* of the evidence—it is the chance of seeing the evidence E if H is true[*] (generally we will also need to know $P(\text{not } E|H)$). Indeed, we can find out $P(E|H)$ simply by checking the proportion of people with cancer who are also smokers. Suppose, for example, that we know that $P(E|H) = 0.8$.

Now Bayes theorem tells us how to compute $P(H|E)$ in terms of $P(E|H)$. It is simply:

$$P(H|E) = \frac{P(E|H) \times P(H)}{P(E)}$$

In this example, since $P(H) = 0.1$; $P(E|H) = 0.8$; $P(E) = 0.5$, it follows that $P(H|E) = 0.16$. Thus, if we get evidence that the person is a smoker we revise our probability of the person having cancer (it increases from 0.1 to 0.16).

In general, we may not know $P(E)$ directly, so the more general version of Bayes theorem is one in which $P(E)$ is also determined by the prior likelihoods:

$$P(H|E) = \frac{P(E|H) \times P(H)}{P(E)} = \frac{P(E|H) \times P(H)}{P(E|H) \times P(H) + (E|notH) \times P(notH)}$$

EXAMPLE 7.1

Assume one in a thousand people has a particular disease. Then our prior belief is

$$P(H) = 0.001, \text{ so } P(not\ H) = 0.999$$

Also, assume a test to detect the disease has 100% sensitivity (i.e., no false negatives) and 95% specificity (meaning 5% false positives). Then if E represents the Boolean variable "Test positive for the disease," we have

---

[*] For simplicity, we assume for the moment that the hypothesis and the evidence are simple Boolean propositions, that is, they are simply either true or false. Later on, we will see that the framework applies to arbitrary discrete and even numeric variables.

$$P(E|notH) = 0.05$$

$$P(E|H) = 1$$

Now suppose a randomly selected person tests positive. What is the probability that the person actually has the disease? By Bayes theorem this is

$$P(H|E) = \frac{P(E|H)P(H)}{P(E|H)P(H) + (E|notH)P(notH)} = \frac{1 \times 0.001}{1 \times 0.001 + 0.05 \times 0.999}$$

$$= 0.01963$$

So there is a less than 2% chance that a person testing positive actually has the disease. For every thousand people tested, there will be on average 50 false positives and only 1 true positive. Bayes theorem gives us a way to show the impact of the false positives.

While Bayes theorem is a rational way of revising beliefs in light of observing new evidence, it is not easily understood by people without a statistical/mathematical background. Moreover, as Example 7.1 shows, the results of Bayesian calculations can appear, at first sight, as counterintuitive. Indeed, in a classic study (Casscells et al. 1978) when Harvard Medical School staff and students were asked to calculate the probability of the patient having the disease (using the exact assumptions stated in Example 7.1) most gave the wildly incorrect answer of 95% instead of the correct answer of less than 2%.

If Bayes theorem is difficult for lay people to compute and understand in the case of a single hypothesis and piece of evidence (as in Figure 7.7), the difficulties are obviously compounded when there are multiple related hypotheses and evidence as in the example of Figure 7.8.

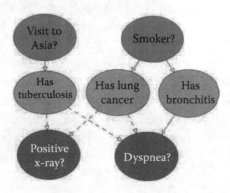

FIGURE 7.8   Bayesian network for diagnosing disease.

Yes	0.01
No	0.99

Probability table for "visit to Asia?"

Smoker?	Yes	No
Yes	0.6	0.3
No	0.4	0.7

Probability table for "bronchitis?"

FIGURE 7.9   Node Probability Table (NPT) examples.

As in Figure 7.7 the nodes in Figure 7.8 represent variables (which may be known or unknown) and the arcs* represent causal (or influential) relationships. In addition to the graphical structure we specify, for each node, a *node probability table* (NPT) (such as the examples shown in Figure 7.9). These tables capture the relationship between a node and it parents by specifying the probability of each outcome (state) given every combination of parent states. The resulting model is called a *Bayesian network (BN)*.

So, a BN is a directed graph together with a set of probability tables. The directed graph is referred to as the "qualitative" part of the BN, while the probability tables are referred to as the "quantitative" part.

The BN in Figure 7.8 is intended to model the problem of diagnosing diseases (TB, cancer, bronchitis) in patients attending a chest clinic. Patients may have symptoms (like dyspnea—shortness of breath) and can be sent for diagnostic tests (x-ray); there may be also underlying causal factors that influence certain diseases more than others (such as smoking, visit to Asia).

To use Bayesian inference properly in this type of network necessarily involves multiple applications of Bayes theorem in which evidence is "propagated" throughout. This process is complex and quickly becomes infeasible when there are many nodes and/or nodes with multiple states. This complexity is the reason why, despite its known benefits, there was for many years little appetite to use Bayesian inference to solve real-world decision and risk problems. Fortunately, due to breakthroughs in the late 1980s that produced efficient calculation algorithms, there are now widely available tools that enable anybody to do the Bayesian calculations without ever having to understand, or even look at, a mathematical formula. These developments were the catalyst for an explosion of interest in BNs. Using such a tool we can do the kind of powerful reasoning shown in Figure 7.10.

---

* Some of the arcs in Figure 7.8 are dotted. This simply means there are some "hidden" nodes through which the causal or influential relationship passes.

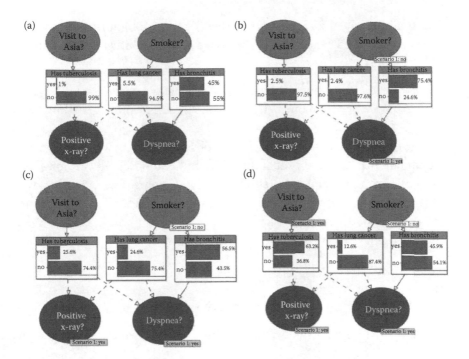

FIGURE 7.10 Reasoning within the Bayesian network. (a) Prior beliefs point to bronchitis as most likely, (b) patient is "nonsmoker" experiencing dyspnea (shortness of breath): strengthens belief in bronchitis, (c) positive x-ray result increases probability of TB and cancer but bronchitis still most likely, (d) visit to Asia makes TB most likely now.

Specifically,

- With the prior assumptions alone (Figure 7.10a), Bayes theorem computes what are called the prior marginal probabilities for the different disease nodes (note that we did not "specify" these probabilities—they are computed automatically; what we specified were the conditional probabilities of these diseases given the various states of their parent nodes). So, before any evidence is entered the most likely disease is bronchitis (45%).

- When we enter evidence about a particular patient, the probabilities for all of the unknown variables get updated by the Bayesian inference. So, in Figure 7.10b once we enter the evidence that the patient has dyspnea and is a nonsmoker, our belief in bronchitis being the most likely disease increases (75%).

- If a subsequent x-ray test is positive (Figure 7.10b) our belief in both TB (26%) and cancer (25%) are raised but bronchitis is still the most likely (57%).

- However, if we now discover that the patient visited Asia (Figure 7.10d) we overturn our belief in bronchitis in favor of TB (63%).

Note that we can enter any number of observations anywhere in the BN and update the marginal probabilities of all the unobserved variables. As the above example demonstrates, this can yield some exceptionally powerful analyses that are simply not possible using other types of reasoning and classical statistical analysis methods.

In particular, BNs offer the following benefits:

- Explicitly model causal factors.

- Reason from effect to cause and vice versa.

- Overturn previous beliefs in the light of new evidence (also called "explaining away").

- Make predictions with incomplete data.

- Combine diverse types of evidence including both subjective beliefs and objective data.

- Arrive at decisions based on visible auditable reasoning (unlike blackbox modeling techniques there are no "hidden" variables and the inference mechanism is based on a long-established theorem).

With the advent of the BN algorithms and associated tools, it is therefore no surprise that BNs have been used in a range of applications that were not previously possible with Bayes theorem alone.

7.3 APPLYING BAYESIAN NETWORKS TO THE PROBLEM OF SOFTWARE DEFECTS PREDICTION

Once we think in terms of causal models that relate unknown hypotheses and evidence, we are able to make much better use of software metrics and arrive at more rational predictions that not only avoid the kind of fallacies we introduced earlier, but are also able to explain apparent empirical anomalies. This is especially evident in the area of defect prediction (Fenton and Neil 1999).

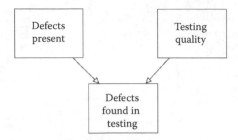

FIGURE 7.11 Simple causal model for defects found.

7.3.1 A Very Simple BN for Understanding Defect Prediction

When we count *defects found in testing* what we actually have is evidence about the unknown variable we are most interested in, namely the *number of defects present*. A very simple, but rational, causal model is shown in Figure 7.11.

Clearly, the number of defects present will influence the number of defects found, but the latter will also be influenced by the testing quality (as we will see later, this will turn out to be a fragment of a larger model, with for example the node representing defects present being a synthesis of a number of factors including process quality and problem complexity).

To keep things as simple as possible to start with, we will assume that the number of defects is classified into just three states (low, medium, high) and that testing quality is classified into just two (poor, good). Then the NPT for the node *defects found in testing* might reasonably be specified as shown in Table 7.2. For example, this specifies that if testing quality is "poor" and there are a "medium" number of defects present then there is a 0.9 probability (90% chance) that defects found is "low," 0.1 probability (10% chance) that defects found is "medium," and 0 probability that defects found is "high."

So, if testing quality is "poor," then defects found is likely to be "low" even when there are a high number of defects present. Conversely, if testing quality is "good" we assume that most (but not all) defects present will

TABLE 7.2 NPT for defects found node

Defects Present	Low		Medium		High	
Testing Quality	Poor	Good	Poor	Good	Poor	Good
Low	1.0	1.0	0.9	0.1	0.7	0.0
Medium	0.0	0.0	0.1	0.9	0.2	0.2
High	0.0	0.0	0.0	0.0	0.1	0.8

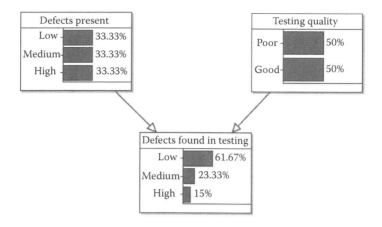

FIGURE 7.12   Model in its initial (marginal) state.

be found. Assuming that the prior probabilities for the states of the nodes *defect present* and *testing quality* are all equal, then in its marginal state the model is shown in Figure 7.12.

One of the major benefits of the causal models that we are building is that they enable us to reason in both a "forward" and a "reverse" direction. That is, we can identify the possible causes given the observation of some effect. In this case, if the number of defects found is observed to be "low" the model will tell us that low testing quality and a low number of defects present are both possible explanations (perhaps with an indication as to which one is the most likely explanation) as shown in Figure 7.13.

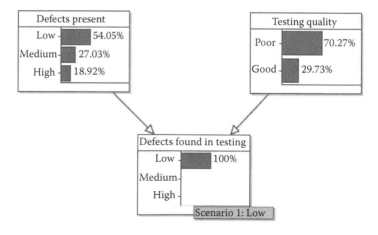

FIGURE 7.13   Low number of defects found.

The power of the model becomes clearer as we discover more "evidence." Suppose, for example, that we also have independent evidence that the testing quality is good. Then, as shown in Figure 7.14a, we can deduce with some confidence that there were indeed a low number of defects present. However, if it turns out that the testing quality is poor, as shown in Figure 7.14b we are little wiser than when we started; the evidence of low number of defects found in testing has been "explained away" by the poor testing quality; we now have no real knowledge of the actual number of defects present, despite the low number found in testing.

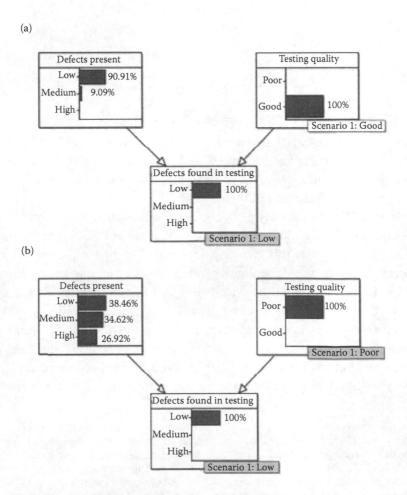

FIGURE 7.14 Effect of different evidence about testing quality. (a) Testing quality is good, (b) testing quality is poor.

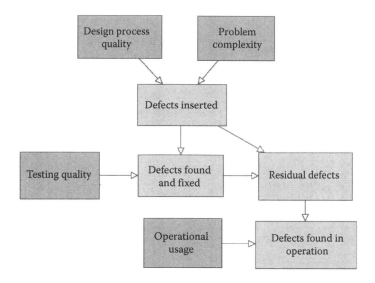

FIGURE 7.15   BN model for software defects and reliability prediction.

7.3.2 A Full Model for Software Defects and Reliability Prediction

A more complete causal model for software defects and reliability prediction is shown in Figure 7.15 (this is itself a simplified version of a model that has been extensively used in a number of real software development environments, Fenton et al. 2007).

In this case the *number of defects found in operation** (i.e., those found by customers) in a software module is what we are really interested in predicting. We know this is clearly dependent on the number of *residual defects*. But it is also critically dependent on the amount of *operational usage*. If you do not use the system you will find no defects irrespective of the number there. The *number of residual defects* is determined by the *number you introduce during development* minus the *number you successfully find and fix*. Obviously defects found and fixed are dependent on the *number introduced*. The number introduced is influenced by *problem complexity* and *design process quality*. The better the design the fewer the

---

* Handling numeric nodes such as *the number of defects found in operation* in a BN was, until relatively recently, a major problem because it was necessary to "discretize" such nodes using some predefined range and intervals. This is cumbersome, error prone, and highly inaccurate [see Fenton et al. (2008) for a comprehensive explanation]. Such inaccuracies, as well as the wasted effort over selecting and defining discretization intervals, can now be avoided by using *dynamic discretization* (in particular, the algorithm described in Neil et al. (2007) and implemented in AgenaRisk). Dynamic discretization allows users to simply define a numeric node by a single range (such as—infinity to infinity, or 0–100, 0 to infinity, etc.).

defects and the less complex the problem the fewer defects. Finally, how many defects you find is influenced not just by the number there to find but also by the amount of *testing effort*.

The task of defining the NPTs for each node in this model is clearly more challenging than the previous examples. In many situations we would need the NPT to be defined as a *function* rather than as an exhaustive table of all potential parent state combinations. Some of these functions are deterministic rather than probabilistic: for example, the "Residual defects" is simply the numerical difference between the "Defects inserted" and the "Defects found and fixed." In other cases, we can use standard statistical functions. For example, in this version of the model we assume that "Defects found and fixed" is a binomial $B(n,p)$ distribution where n is the number of defects inserted and p is the probability of finding and fixing a defect (which in this case is derived from the "testing quality"); in more sophisticated versions of the model the p variable is also conditioned on n to reflect the increasing relative difficulty of finding defects as n decreases. Table 7.3 lists the full set of conditional probability distributions for the nodes (that have parents) of the BN model of Figure 7.15.

The nodes "design quality," "complexity," "testing quality," and "operational usage" are all examples of what are called *ranked nodes* (Fenton et al. 2007); ranked nodes have labels like {very poor, poor, average, good, very good} but they have an underlying [0,1] scale that makes it very easy to define relevant probability tables for nodes which have them as parents (as in some of the functions as described in Table 7.3).* The nodes without parents are all assumed to have a prior uniform distribution, that is, one in which any state is equally as likely as any other state (in the "real" models the distributions for such nodes would normally not be defined as

TABLE 7.3   Probability Distributions for the Nodes of BN Model in Figure 7.15

Node Name	Probability Distribution
Defects found in operation	Binomial (n, p) where $n =$ "residual defects" and $p =$ "operational usage"
Residual defects	Defects inserted − Defects found (and fixed) in testing
Defects found in testing	Binomial (n, p) where $n =$ "defects inserted" and $p =$ "testing quality
Defects inserted	This is a distribution based on empirical data from a particular organization. For full details see Fenton et al. (2007)

---

* Again note that AgenaRisk provides comprehensive support for ranked nodes and their associated NPTs.

uniform but would reflect the historical distribution of the organization either from data or expert judgment).

We next illustrate the BN calculations (that are performed automatically by the BN tool), which show that the case for using BNs as causal models for software defects and reliability prediction is both simple and compelling.

Figure 7.16 shows the marginal distributions of the model before any evidence has been entered. So this represents our uncertainty before we enter any specific information about this module. Since we assumed uniform distributions for nodes without parents we see, for example, that the module is just as likely to have very high complexity as very low, and that the number of defects found and fixed in testing is in a wide range where

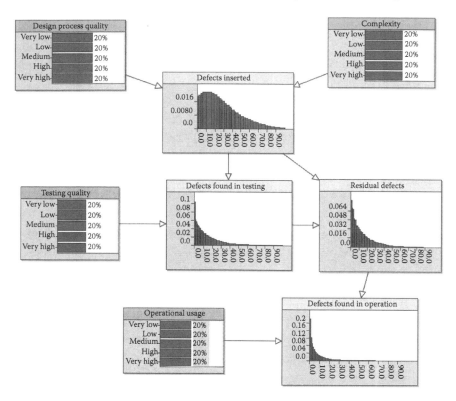

FIGURE 7.16 BN model with marginal distributions for variables superimposed on nodes. All of the graphs are probability distributions, but there is a standard convention to represent discrete distributions (such as the node *testing quality*) with horizontal bars (i.e., the probability values are on the *x*-axis), whereas continuous/numeric distributions (such as the node *defects found in testing*) have vertical bars (i.e., the probability values are on the *y*-axis).

the median value is about 18–20 (the prior distributions here were for a particular organization's modules).

Figure 7.17 shows the result of entering two observations about this module:

1. That it had zero defects found and fixed in testing; and

2. That the problem complexity is "High"

Note that all the other probability distributions updated. The model is doing both forward inference to predict defects in operation and backward inference about, say, design process quality. Although the fewer than expected defects found does indeed lead to a belief that the post-release faults will drop, the model shows that the most likely explanation is inadequate testing.

So far, we have made no observation about operational usage. If, in fact, the operational usage is "Very High" (Figure 7.18) then what we have done is replicate the apparently counter-intuitive empirical observations we

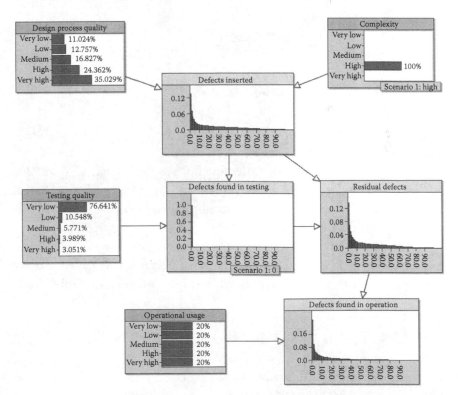

FIGURE 7.17   Zero defects in testing and high complexity observed.

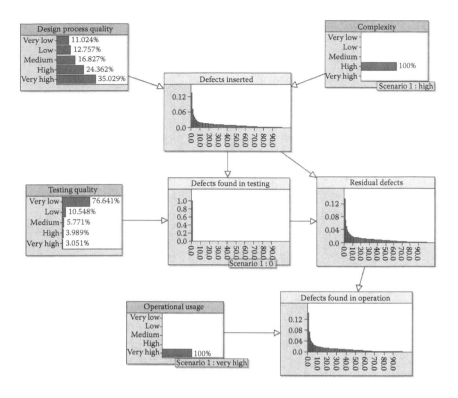

FIGURE 7.18 Very high operational usage.

discussed in Section 7.1 whereby a module with no defects found in testing has a high number of defects post-release.

But suppose we find out that the test quality was "Very High" (Figure 7.19).

Then we completely revise out beliefs. We are now fairly certain that the module will be fault free in operation. Note also that the "explanation" is that the design process is likely to be very high quality. This type of reasoning is unique to Bayesian networks. It provides a means for decision makers (such as quality assurance managers in this case) to make decisions and interventions dynamically as new information is observed.

7.3.3 Commercial Scale Versions of the Defect Prediction Models

The ability to do the kind of prediction and what-if analysis described in the model in Section 7.3.2 has proved to be very attractive to organizations that need to monitor and predict software defects and reliability, and that already collect defect-type metrics. Hence, organizations such as Motorola (Gras 2004), Siemens (Wang et al. 2006), and Philips (Fenton et al. 2007)

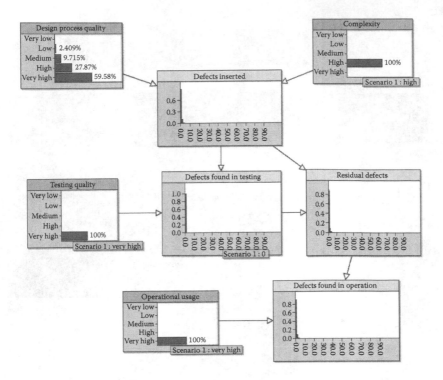

FIGURE 7.19   Testing quality is very high.

have exploited models and tools originally developed in Fenton et al. (2002) to build large-scale versions of the kind of model described in Section 7.3.2.

It is beyond the scope of this chapter to describe the details of these models and how they were constructed and validated, but what typifies the approaches is that they are based around a sequence of testing phases, by which we mean those testing activities such as system testing, integration testing, and acceptance testing that are defined as part of the companies' software processes (and hence for which relevant defect and effort data is formally recorded). In some cases a testing phase is one that does not involve code execution, such as design review. The final "testing" phase is generally assumed to be the software in operation. Corresponding to each phase is a "subnet" like that in Figure 7.20, where a subnet is a component of the BN with interface nodes to connect the component subnet to other parent and child subnets. For the final "operational" phase, there is, of course, no need to include the nodes associated with defect fixing and insertion.

The distributions for nodes such as "probability of finding defect" derive from other subnets such as that shown in Figure 7.21. The particular nodes and distributions will, of course, vary according to the type of testing phase.

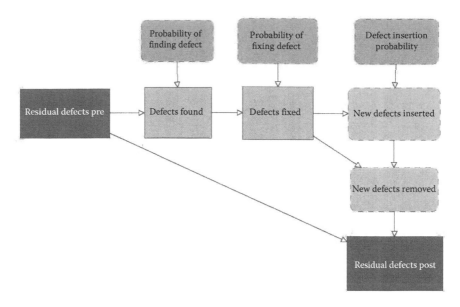

FIGURE 7.20 Defects phase subnet.

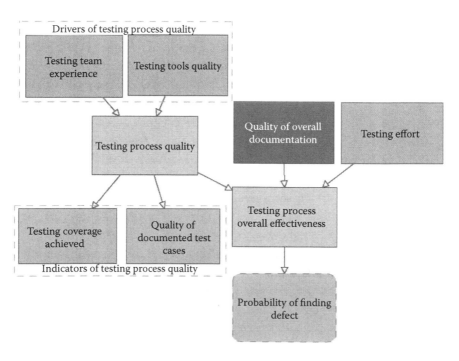

FIGURE 7.21 Typical subnet for testing quality.

TABLE 7.4   NPT for Node "Probability of Finding a Defect"

Parent (Overall Testing Process Effectiveness) State	Probability of Finding a Defect
Very low	TNormal (0.01, 0.001, 0, 1)
Low	TNormal (0.1, 0.001, 0, 1)
Average	TNormal (0.2, 0.001, 0, 1)
High	TNormal (0.35, 0.001, 0, 1)
Very high	TNormal (0.5, 0.001, 0, 1)

To give a feel for the kind of expert elicitation and data that was required to complete the NPTs in these kinds of models, we look at two examples, namely the nodes "probability of finding a defect" and "testing process overall effectiveness":

EXAMPLE 7.2

The NPT for the node "probability of finding a defect." This node is a continuous node in the range [0,1] that has a single parent "testing process overall effectiveness" that is a ranked node (in the sense of Fenton et al. (2007) on a 5-point scale from "very low" to "very high"). For a specific type of testing phase (such as integration testing), the organization had both data and expert judgment that enabled them to make the following kinds of assessment:

- "Typically (i.e., for our average level of test quality) this type of testing will find approximately 20% of the residual defects in the system."
- "At its best (i.e., when our level of testing is at its best) this type of testing will find 50% of the residual defects in the system; at its worst it will only find 1%."

On the basis of this kind of information, the NPT for the node "probability of finding a defect" is a partitioned expression like the one in Table 7.4. Thus, for example, when overall testing process effectiveness is *Average*, the probability of finding a defect is a truncated normal distribution over the range [0,1] with mean 0.2 and variance 0.001.

EXAMPLE 7.3

The NPT for the node "testing process overall effectiveness." This node is a ranked node on a 5-point ranked scale from "very low" to "very high." It has three parents "testing process quality," "testing effort," and "quality of overall documentation," which are all also ranked nodes on the same 5-point ranked scale from "very low" to "very high." Hence, the NPT in this case is a table of 625 entries. Such a table is essentially impossible to elicit manually, but the techniques described in Fenton et al. (2007) (in which ranked nodes are mapped on to an underlying [0,1] scale) enabled experts to construct a sensible table in seconds using an appropriate "weighted expression" for the child node in terms of the parents. For example, the expression elicited in one case was a truncated normal (on the range [0,1]) with mean equal to the weighted minimum of the parent values (where the weights were: 5.0 for "testing effort," 4.0 for testing quality; and 1.0 for "documentation quality") and the variance was 0.001. Informally this weighted minimum expression captured expert judgment like the following:

> "Documentation quality cannot compensate for lack of testing effort, although a good testing process is important."

As an illustration, Figure 7.22 shows the resulting distribution for overall testing process effectiveness when testing process quality is average, quality of documentation is very high, but testing effort is very low.

Using a BN tool such as AgenaRisk the various subnets are joined, according to the BN Object approach (Koller and Pfeffer 1997) as shown in Figure 7.23. Here, each box represents a BN where only the "input" and "output" nodes are shown. For example, for the BN representing the defects in phase 2 the "input" node *residual defects pre* is defined by the marginal distribution of the output node *residual defects post* of the BN representing the defects in phase 1.

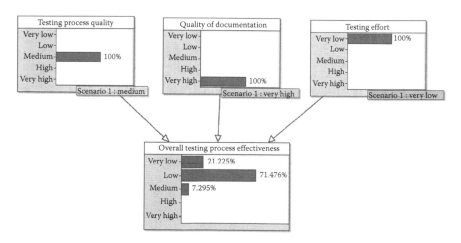

FIGURE 7.22   Scenario for "overall testing effectiveness."

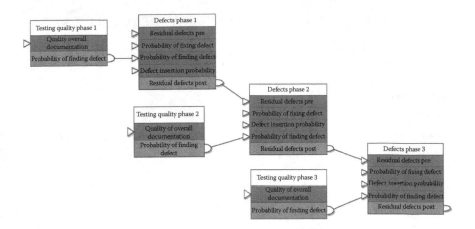

FIGURE 7.23 Sequence of software testing phases as linked BN objects.

The general structure of the BN model proposed here is relevant for any software development organization whose level of maturity includes defined testing phases in which defect and effort data are recorded. However, it is important to note that a number of the key probability distributions will inevitably be organization/project specific. In particular, there is no way of producing a *generic* distribution for the "probability of finding a defect" in any given phase (and this is especially true of the operational testing phase); indeed, even within a single organization this distribution will be conditioned on many factors (such as ones that are unique to a particular project) that may be beyond the scope of a workable BN model. At best, we assume that there is sufficient maturity and knowledge within an organization to produce a "benchmark" distribution in a given phase. Where necessary this distribution can then still be tailored to take account of specific factors that are not incorporated in the BN model. It is extremely unlikely that such tailoring will always be able to take account of extensive relevant empirical data; hence, as in most practically usable BN models, there will be a dependence on subjective judgments. But at least the subjective judgments and assumptions are made explicit and visible.

The assumptions about the "probability of finding a defect" are especially acute in the case of the operational testing phase because, for example, in this phase the various levels of "operational usage" will be much harder to standardize on. What we are doing here is effectively predicting reliability and to do this accurately may require the operational usage node to be conditioned on a formally defined operational profile such as described in the literature on statistical testing (see Chapter 11).

7.4 BAYESIAN NETWORKS FOR SOFTWARE PROJECT RISK ASSESSMENT AND PREDICTION

In Section 7.1, we suggested that effective metrics-driven software risk methods should be able to provide answers to the following types of questions:

- For a problem of this size, and given these limited resources, how likely am I to achieve a product of suitable quality?

- How much can I scale down the resources if I am prepared to put up with a product of specified lesser quality?

- The model predicts that I need 4 people over 2 years to build a system of this kind of size. But I only have funding for 3 people over one year. If I cannot sacrifice quality, how good does the staff have to be to build the systems with the limited resources? Alternatively, if my staff are no better than average and I cannot change them, how much required functionality needs to be cut in order to deliver at the required level of quality?

We now describe a "project level software risk" model that has been widely used and which provides support to help answer exactly these types of questions. The model is a very general-purpose quality and risk assessment model for large software projects. It was developed as part of a major international consortium, with key empirical and expert judgment provided by a range of senior software managers and developers. Although it is beyond the scope of this chapter to describe the full details of the model and its validation (these details are provided in Fenton et al. 2004), we can show how the model is used to predict different aspects of resources and quality while monitoring and mitigating different types of risks. The full model is shown in Figure 7.24, but is too complex to understand all at once.

Figure 7.25 provides an easier to understand schematic view of the model. We can think of the model as comprising six subnets (shown as square boxes).

The subnets are:

- *Distributed communications and management.* Contains variables that capture the nature and scale of the distributed aspects of the project and the extent to which these are well managed.

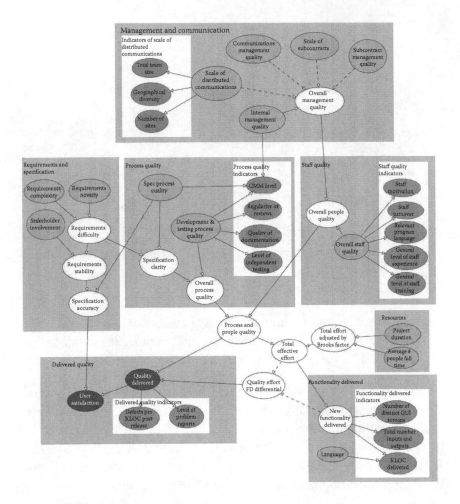

FIGURE 7.24   Full BN model for software project risk.

- *Requirements and specification.* Contains variables relating to the extent to which the project is likely to produce accurate and clear requirements and specifications.

- *Process quality.* Contains variables relating to the quality of the development processes used in the project.

- *People quality.* Contains variables relating to the quality of people working on the project.

- *Functionality delivered.* Contains all relevant variables relating to the amount of new functionality delivered on the project, including the effort assigned to the project.

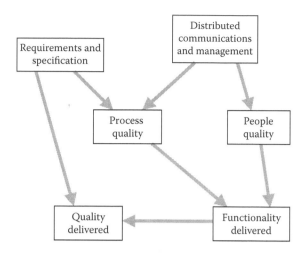

FIGURE 7.25   Schematic for the project level model.

- *Quality delivered.* Contains all relevant variables relating to both the final quality of the system delivered and the extent to which it provides user satisfaction (note the clear distinction between the two).

The full model enables us to cope with variables that cannot be observed directly. Instead of making direct observations of the process and people quality, the functionality delivered and the quality delivered, the states of these variables are inferred from their causes and consequences. For example, the process quality is a synthesis of the quality of the different software development processes—requirements analysis, design, and testing.

The quality of these processes can be inferred from "indicators." Here, the causal link is from the "quality" to directly observable values like the results of project audits and of process assessments, such as the CMM. Of course, only some organizations have been assessed to a CMM level, but this need not be a stumbling block since there are many alternative indicators. An important and novel aspect of our approach is to allow the model to be adapted to use whichever indicators are available.

At its heart the model captures the classic trade-offs between:

- *Quality* (where we distinguish and model both *user satisfaction*—this is the extent to which the system meets the user's true requirements—and *quality delivered*—this is the extent to which the final system works well).

- *Effort* (represented by the *average number of people full-time* who work on the project).

- *Time* (represented by the *project duration*).

- *Functionality* (meaning *functionality delivered*).

So, for example, if you want a lot of functionality delivered with little effort in a short time then you should not expect high quality. If you need high quality then you will have to be more flexible on at least one of the other factors (i.e., use more effort, use more time, or deliver less functionality).

What makes the model so powerful, when compared with traditional software cost models, is that we can enter observations anywhere in the model to perform not just predictions but also many types of trade-off analysis and risk assessment. So we can enter *requirements* for quality and functionality and let the model show us the distributions for effort and time. Alternatively, we can specify the effort and time we have available and let the model predict the distributions for quality and functionality delivered (measured in function points, which are described in Chapter 8). Thus, the model can be used like a spreadsheet—we can test the effects of different assumptions.

To explain how this works we consider two scenarios called "New" and "Baseline" (Figure 7.26 shows how, in AgenaRisk, you can enter

Risk Map	Risk Table	New	Baseline
Project resources			
Project duration			
Average # people full time			
Total effort adjusted by Brooks factor			
Total effective effort			
		New	Baseline
Product size			
New functionality delivered		4000	4000
---KLOC delivered			
---Language		No Answer	No Answer
---Total number Inputs and Outputs			
---Number of distinct GUI screens			
		New	Baseline
Product quality			
Quality delivered		Perfect	No Answer
User satisfaction		No Answer	No Answer
Quality effort FD differential dummy		No Answer	No Answer

FIGURE 7.26 Two scenarios in risk table view.

observations for the different scenarios into a table view of the model). Suppose the new project is to deliver a system of size 4000 function points (this is around 270 KLOC of Java, an estimate you can see for the node KLOC by entering the observation "java" for the question "language"). In the baseline scenario we enter no observations other than the one for functionality. We are going to compare the effect against this baseline of entering various observations into the new scenario.

We start with the observations shown in Figure 7.26, that is, the only change from the baseline in the new project is to assert that the quality delivered should be "perfect." Running the model produces the results shown in Figure 7.27 for the factors *process and people quality, project duration*, and *average number of people full time*. First, note that the distributions for the latter factors have high variances (not unexpected given the minimal data entered) and that generally the new scenario will require a bit more effort for a bit longer. However, the factor *process and people quality* (which combines all the process and people factors) shows a very big difference from the baseline. The prediction already suggests that it will be unlikely (a 14% chance) to deliver the system to the required level of quality unless the quality of staff is better than average.

Suppose, however, that we can only assume *process and people quality* is "medium." Then the predictions for project duration and effort increase significantly. For example, the median value for project duration is up from 31 months in the baseline case to around 54 months (Figure 7.28) with full time staff increasing to 33.

Now, we withdraw the observation of *process and people quality* and suppose, as is typical in software projects, that we have a hard schedule deadline of 18 months in which to complete (i.e., a target that is significantly lower than the one the model predicts). With this observation we get the distributions shown in Figure 7.29 for *process and people quality* and *average number of people full time*. Now, not only do we need much higher quality people, we also need a lot more of them compared with the baseline.

But typically, we will only have a fixed amount of effort. Suppose, for example that additionally we enter the observation that we have only 10 people full-time (so the project is really "under-resourced" compared with the predictions). Then the resulting distribution for *process and people quality* is shown in Figure 7.30.

What we see now is that the probability of the overall *process and people quality* being "very high" (compared to the industry average) is 0.9966. Put a different way, if there is even a tiny chance that your processes and

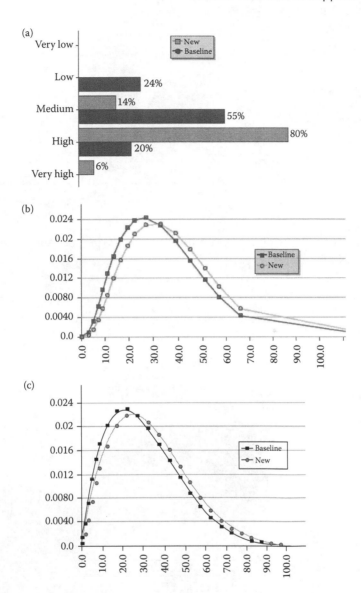

FIGURE 7.27 Distributions when functionality delivered is set as "perfect" for new project (compared with baseline). (a) Process and people quality, (b) project duration (median 31, 39), (c) average number of people full time (19, 23).

people are NOT among the best in the industry then this project will NOT meet its quality and resource constraints. In fact, if we know that the *process and people quality* is just "average" and now remove the observation "perfect" for quality delivered, then Figure 7.31 shows the likely quality to be delivered; it is very likely to be "abysmal" (with probability 0.69).

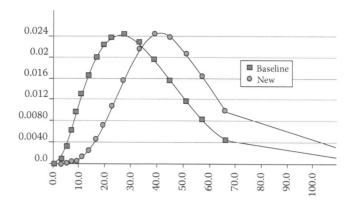

FIGURE 7.28   When staff quality is medium, project duration jumps to median value of 54 months.

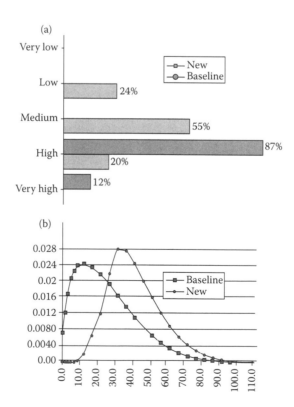

FIGURE 7.29   Project duration set to 18 months. (a) Process and people quality, (b) average number of people full time.

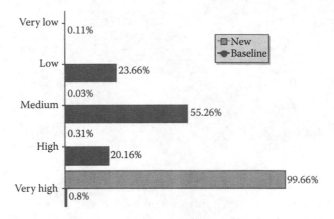

FIGURE 7.30   Project duration = 12 months, people = 10.

Nevertheless, suppose we insist on perfect quality and all the previous resource constraints. In this case the *only* thing left to "trade-off" is the functionality delivered. So we remove the observation 4000 in the new scenario. Figure 7.32 shows the result when we run the model with these assumptions: we are likely to deliver only a tenth of the functionality originally planned. Armed with this information, a project manager can make an informed decision about how much functionality needs to be dropped to meet the quality and resource constraints.

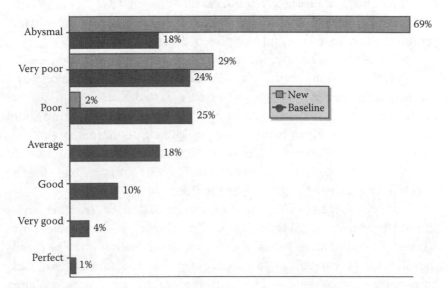

FIGURE 7.31   Quality delivered if process and people quality = medium with resource constraints set.

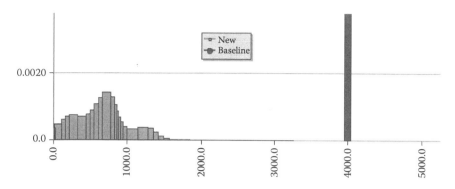

FIGURE 7.32 Functionality (function points) delivered if process and people quality = medium with resource constraints set.

7.5 SUMMARY

There have been many noncausal models for software defect prediction and software resource prediction. Some of these have achieved very good accuracy (see Fenton et al. 2008 (ESE) for a detailed overview) and they provide us with an excellent empirical basis. However, in general these models are typically data-driven statistical models; they provide us with little insight when it comes to effective risk management and assessment. What we have shown is that, by incorporating the empirical data with expert judgment, we are able to build causal Bayesian network models that enable us to address the kind of dynamic decision making that software professionals have to confront as a project develops.

The BN approach helps to identify, understand, and quantify the complex interrelationships (underlying even seemingly simple situations) and can help us make sense of how risks emerge, are connected and how we might represent our control and mitigation of them. By thinking about the causal relations between events we can investigate alternative explanations, weigh up the consequences of our actions and identify unintended or (un)desirable side effects. Above all else the BN approach quantifies the uncertainty associated with every prediction.

We are not suggesting that building a useful BN model from scratch is simple. It requires an analytical mindset to decompose the problem into "classes" of event and relationships that are granular enough to be meaningful, but not too detailed that they are overwhelming. The states of variables need to be carefully defined and probabilities need to be assigned that reflect our best knowledge. Fortunately, there are tools that help avoid much of the complexity of model building, and once built the tools provide dynamic and automated support for decision making. Also, so we have presented some pre-defined models that can be tailored for different organizations.

EXERCISES

1. It is known that a particular type of software test is certain to identify virus X if it has been inserted into a computer system. However, there is also a 5% probability of a false alarm (i.e., there is a 5% probability that the test will be positive when virus X has not been inserted). You run the test and the outcome is positive.

   a. What can you conclude about whether or not the system really is infected with the virus X?

   b. How would your answer to (i) change if, additionally, it was known that virus X had been inserted into approximately one in every thousand computer systems?

   c. Suppose the only known effective fix for virus X costs $250,000 when the full costs of system shutdown and repair are considered. From a risk assessment perspective what action would you recommend (you should state any assumptions about additional information needed)?

2. Look again at the story (on the first page of this chapter) recounted by the keynote speaker. Draw a BN model (with 5 nodes) that "explains" the phenomenon he observed.

3. Look again at Figure 7.6. Use your answer to Question 2 to explain what could be going on here.

4. A tool that computes the values of a set of code complexity metrics $C_1, C_2, \ldots, C_n$ is applied to a large number of systems and subsystems for which the number of defects D found in operations is known. Using Excel, a positive correlation is observed and a regression model of the form

$$D = f(C_1, C_2, \ldots, C_n)$$

is computed. The following quality assurance procedure is subsequently recommended:

Before any major system release the tool should be applied to extract the metrics $C_1, C_2, \ldots, C_n$ for each subsystem and the function f is then computed. Any subsystem for which f is above 25 must undergo additional testing.

What concerns do you have about this quality assurance procedure?

5. Explain how a Bayesian network approach can help answer the questions posed at the end of Question 7.1.

FURTHER READING

For an introduction and historical perspective of Bayes theorem and its applications we recommend:

McGrayne S.B., *The Theory That Would Not Die*, Yale University Press, CT, 2011.
Simpson E., Bayes at Bletchley Park, *Significance*, 7(2), 76–80, 2010.

For a comprehensive and not overtly mathematical overview of Bayesian networks and their applications and support, see:

Fenton N.E. and Neil M., *Risk Assessment and Decision Analysis with Bayesian Networks*, 2012, CRC Press, Boca Raton, FL, ISBN: 9781439809105, ISBN 10:1439809100, 2012.

There are also extensive resources available on the associated website: http://www.bayesianrisk.com/.

To understand the limitations of statistical modeling techniques and their tests of significance and p-values, see the following for a devastating critique of their widespread abuse across a range of empirical disciplines:

Ziliak S.T. and McCloskey D.N., *The Cult of Statistical Significance: How the Standard Error Costs Us Jobs, Justice, and Lives*, University of Michigan Press, Ann Arbor, USA, 2008.

Mathematically adept readers seeking more in depth understanding of the theoretical underpinnings of Bayesian networks and their associated algorithms should consider the following books:

Jensen F.V. and Nielsen T., *Bayesian Networks and Decision Graphs*, Springer-Verlag Inc, New York, 2007.
Madsen A.L., *Bayesian Networks and Influence Diagrams*, Springer-Verlag, New York, 2007.
Neapolitan R.E., *Learning Bayesian Networks*, Upper Saddle River Pearson Prentice Hall, 2004.

Pearl J., *Causality: Models Reasoning and Inference*, Cambridge University Press, Cambridge, UK, 2000.

Readers who are interested in building BN models and running the models in this chapter can do so by downloading the Agenarisk software and following the instructions contained on the webpage describing the models:

Agena 2014, http://www.agenarisk.com

The following papers and books were referenced in the chapter:

Casscells W., Schoenberger A. and Graboys T.B., Interpretation by physicians of clinical laboratory results, *New England Journal of Medicine*, 299, pp. 999–1001, 1978.

Fenton N.E., Marsh W., Cates P., Forey S., and Tailor M., Making resource decisions for software projects, *Proceedings of the 26th International Conference on Software Engineering (ICSE2004)*, Edinburgh International Conference Centre, Edinburgh, UK, IEEE Computer Society, pp. 397–406, 2004.

Fenton N.E. and Neil M., A critique of software defect prediction models, *IEEE Transactions on Software Engineering*, 25(5), 675–689, 1999.

Fenton N.E., Neil M., and Gallan J., Using ranked nodes to model qualitative judgments in Bayesian networks, *IEEE Transactions on Knowledge and Data Engineering*, 19(10), 1420–1432, 2007.

Fenton N.E., Neil M., and Krause P., Software measurement: Uncertainty and causal modelling, *IEEE Software*, 10(4), 116–122, 2002.

Fenton N.E., Neil M., Marsh W., Hearty P., Marquez D., Krause P., and Mishra R., Predicting software defects in varying development lifecycles using Bayesian nets, *Information & Software Technology*, 49, 32–43, 2007.

Fenton N.E., Neil M., Marsh W., Hearty P., Radlinski L., and Krause P., On the effectiveness of early life cycle defect prediction with Bayesian nets, *Empirical Software Engineering*, 13, 499–537, 2008.

Fenton N.E., Neil M., and Marquez D., Using Bayesian networks to predict software defects and reliability, *Proceedings of the Institution of Mechanical Engineers, Part O, Journal of Risk and Reliability*, 222(O4), pp. 701–712, 2008.

Fenton N.E. and Ohlsson N. Quantitative analysis of faults and failures in a complex software system. *IEEE Transactions on Software Engineering*, 26(8), 797–814, 2000.

Gras J.-J., End-to-end defect modeling. *IEEE Software*, 21(5), 98–100, 2004.

Koller D. and Pfeffer A. Object-oriented Bayesian networks, *Proceedings of the 13th Annual Conference on Uncertainty in AI (UAI)*, Providence, Rhode Island, pp. 302–313, 1997.

Neil M., M. Tailor M., and Marquez D., Inference in hybrid Bayesian networks using dynamic discretization, *Statistics and Computing*, 17(3), 219–233, 2007.

Wang H., Peng F., Zhang C., and Pietschker A., Software project level estimation model framework based on Bayesian belief networks, in *Sixth International Conference on Quality Software (QSIC'06)*, Beijing, China, pp. 209–218, 2006.

II

Software Engineering Measurement

CHAPTER 8

Measuring Internal Product Attributes

Size

T HE FIRST PART OF this book laid the groundwork for measuring and evaluating software products, processes, and resources. Now, we turn to the actual measurement, to see which metrics we can use and how they are applied to software engineering problems. In the remaining chapters, we look at the products themselves. Here and in Chapter 9, we focus on internal product attributes; Chapters 10 and 11 will turn to external attributes.

We saw in Chapter 3 that internal product attributes describe software products (including documents) in a way that is dependent only on the product itself. The most obvious and useful of such attributes is the *size* of a software system, which can be measured statically, without having to execute the system. In this chapter, we look at different ways to measure the size of any development product. We start the discussion by describing the properties that should be satisfied by any software size measure. Then we will show, in turn, how size can be measured on a variety of software entities, including code, designs, requirements, and problem statements. We also consider applications of size measures, including the use of size to normalize other measures, size measures used to quantify testing attributes, and those that indicate the amount of *reuse* (the extent to which the software is genuinely new).

8.1 PROPERTIES OF SOFTWARE SIZE

Each product of software development is expressed in a concrete form and can be treated in a manner similar to physical entities. Like physical entities, software products can be described in terms of their *size*. The size of physical objects is easily measured by length, volume, mass, or other standard measure, as well as their length (width, height, depth). Measuring the size of software is straightforward as long as we use relatively simple measures that are consistent with measurement theory principles.

Remember, size measures only indicate how much of an entity we have. Size alone cannot directly indicate external attributes such as *effort, productivity*, and *cost*. We have seen many complaints about the limitations of size measures. For example, Conte, Dunsmore, and Shen assert that

> … there is a major problem with the lines-of-code measure: it is not consistent because some lines are more difficult to code than others …. One solution to the problem is to give more weight to lines that have more 'stuff' in them.

CONTE ET AL. 1986

Although size measures do not indicate external attributes like "difficulty of coding," they are very useful. Clearly, when all other attributes are similar, the size of a software entity really matters. In general, a 100,000 line program will be more difficult to test and maintain than a 10,000 line program. A large program is more likely to contain faults than a small program. Problem size is a good attribute to use to predict software development time and resources. Size is commonly used as a component to compute indirect attributes such as productivity:

$$\text{Productivity} = \text{Size}/\text{Effort}$$

Another example, which was discussed earlier in the book, is defect density:

$$\text{Defect density} = \text{Defect count}/\text{Size}$$

Also, size is commonly used in many cost estimation models, which are often used for project planning.

Rejecting simple size measures because they do not indicate attributes like coding difficulty reflects a misunderstanding of the basic measurement principles discussed in Chapter 2. Those who reject a measure because it does not provide enough information may be expecting too much of a single measure. Consider the following analogy:

EXAMPLE 8.1

If human *size* is measured in terms of only a single attribute, say a person's weight or mass, then we can use it to determine how much additional fuel is needed for a satellite launch vehicle to carry someone into orbit. If human *size* is measured as a single attribute, say *height*, then we can use it to predict whether or not a person would bump his or her head when entering a doorway. These applications of the definitions make mass and height useful measures. However, we cannot use them effectively alone to determine whether a person can lift a heavy object.

Similarly, if we measure software code size as the number of bytes, for instance, the fact that it is not useful in measuring quality does not negate its value in predicting the amount of space it requires in a file.

EXAMPLE 8.2

If human size is measured in terms of two attributes, say height *and* weight or mass, then we can use it to determine (on the basis of empirical understanding) whether a person is obese. However, we cannot use it effectively to determine how fast such a person can run nor how intelligent the person might be.

Likewise, models that define an attribute in terms of several internal attributes may be useful, even if they are not complete. As we have seen in Chapter 3, the measures and models are derived from the goals set for them, and applying them to different goals does not invalidate them for their original purpose.

Ideally, we want to define general properties or empirical relations of software size and other properties that are analogs to properties of humans such as weight. We want any measure of size to satisfy the properties so that the measure actually quantifies the attribute and thus satisfies the representation condition of measurement and is valid in the narrow sense.

We examine the empirical relation systems for measures of size by applying a published set of software size properties (Briand et al. 1996). The properties are defined in terms of a system of modules. Modules contain elements (really graph nodes), and elements have links (graph edges) with other elements. An element may be in more than one module. The modules may be nested and may be disjoint. The properties are defined in terms of the effects of system changes to any valid measure of size. As we will see in Chapter 9, this model supports the definition of properties of software complexity, length, coupling, and cohesion.

This model of a system can represent most software entities at any level of abstraction. It can represent source code using nodes to represent expressions or statements and edges to represent control flow or data links. It can represent various kinds of UML diagrams with nodes and edges. It can also be used to model requirements documents. The model works as long as we can think of a system as a set of modules that contain elements that can be linked.

We use our intuition about the size of things in the physical world to develop measures of the size of software entities.

EXAMPLE 8.3

Consider boxes of marbles, where size is measured in terms of the number of marbles. A box cannot have negative size, but it may be empty and have no size. If we put several boxes of marbles inside another box, the size of the enclosing box will be the sum of the marbles in each box as long as we make sure not to count a marble more than once.

Following this intuition, Briand, Morasco, and Basili define the following three properties for any valid measure of software size:

1. *Nonnegativity*: All systems have nonnegative size.

2. *Null value*: The size of a system with no elements is zero.

3. *Additivity*: The size of the union of two modules is the sum of sizes of the two modules after subtracting the size of the intersection.

These three properties are, of course, empirical relations in the sense of the representational theory of measurement described in Chapter 2; they constitute the minimum empirical relation system for the notion of "size." Many of the size measures that we will use are calculated by counting elements. Such measures easily satisfy the size properties.

Knowing the size of most development products can be very useful. The size of a requirements specification can predict the size and complexity of a design, which can predict the code size. Size attributes of the design and code can determine the required effort for testing, as well as the effort required to add features. Size measures are commonly used to indicate the amount of reuse in a system.

8.2 CODE SIZE

Program code is an integral component of software. Such code includes source code, intermediate code, byte code, and even executable code. We look at approaches for directly measuring code size.

8.2.1 Counting Lines of Code to Measure Code Size

The most commonly used measure of source code program size is the *number of lines of code (LOCs)*, introduced in Chapter 2. But some LOCs are different from others. For example, many programmers use spacing and blank lines to make their programs easier to read. If LOCs are being used to estimate programming effort, then a blank line does not contribute the same amount of effort as a line implementing a difficult algorithm. Similarly, comment lines improve a program's understandability, and they certainly require some effort to write. But they may not require as much effort as the code itself. Many different schemes have been proposed for counting lines, each defined with a particular purpose in mind, so there are many ways to calculate LOCs for a given program. Without a careful model of a program, coupled with a clear definition of an LOC, confusion reigns. We must take great care to clarify what we are counting and how we are counting it. In particular, we must explain how each of the following is handled:

- Blank lines

- Comment lines

- Data declarations

- Lines that contain several separate instructions

Jones reports that one count can be as much as five times larger than another, simply because of the difference in counting technique (Jones 2008).

EXAMPLE 8.4

There is some general consensus that blank lines and comments should not be counted. Conte et al. define an LOC as any line of program text that is not a comment or blank line, regardless of the number of statements or fragments of statements on the line. This definition specifically includes all lines containing program headers, declarations, and executable and nonexecutable statements (Conte et al. 1986). Grady and Caswell report that Hewlett-Packard defines an LOC as a noncommented source statement: any statement in the program except for comments and blank lines (Grady and Caswell 1987).

This definition of an LOC is still the most widely accepted. To stress the fact that an LOC according to this definition is actually a *noncommented* line, we use the abbreviation NCLOC, sometimes also called *effective lines of code*. The model associated with this definition views a program as a simple file listing, with comments and blank lines removed, giving an indication of the extent to which it is self-documented.

NCLOC is useful for comparing subsystems, components, and implementation languages. For example, a study of the Debian 2.2 Linux release found that 71% of the NCLOC of Debian code was written in C. The remainder of the system was written in 10 other programming languages. The study also measured the sizes of the largest packages (González-Barahona et al. 2001).

Another use of NCLOC is to evaluate the growth of systems over time. Godfrey and Tu used NCLOC analyzed the growth of the Linux system and key subsystems over multiple releases. They found that Linux had been growing at a superlinear rate (Godfrey and Tu 2000). Another study of the growth of evolving systems using NCLOC included both Linux and FreeBSD, but found no evidence of superlinear growth; the systems grew at a linear rate (Izurieta and Bieman 2006).

In a sense, valuable size information is lost when comment lines are not counted. In many situations, program size is important in deciding how much computer storage is required for the source code, or how many pages are required for a print-out. Here, the program size must reflect the blank and commented lines. Thus, NCLOC is not a valid measure (in the sense of Chapter 3) of total program size; it *is* a measure of the uncommented size. Uncommented size is a reasonable and useful attribute to measure, but only when it addresses appropriate questions and goals. If we are relating size with effort from the point of view of productivity assessment, then uncommented size may be a valid input. But even here, there is room for doubt, as uncommented size carries with it an implicit assumption that comments do not entail real programming effort and so should not be considered.

As a compromise, we recommend that the *number of comment lines of program text (CLOC)* be measured and recorded separately. Then we can define

$$\text{Total size (LOC)} = \text{NCLOC} + \text{CLOC}$$

and some useful indirect measures follow. For example, the ratio

$$\frac{\text{CLOC}}{\text{LOC}}$$

measures of the *density of comments* in a program.

EXAMPLE 8.5

A large UK organization has written the code for a single application in two different kinds of COBOL. Although the code performs the same kinds of functions, the difference in comment density is striking, as shown in Table 8.1.

As a single measure of program size, LOC may be preferable to NCLOC, and it is certainly easier to measure automatically. In general, it may be useful to gather both measures.

Other code size measures try to take into account the way code is developed and run. Some researchers and practitioners acknowledge that sometimes programs are full of data declarations and header statements, and there is very little code that actually executes. For some purposes (such as testing), it is important to know how much *executable* code is being produced. Here, they prefer to measure the *number of executable statements* (ES). This measure counts separate statements on the same physical line as distinct. It ignores comment lines, data declarations, and headings.

Other developers recognize that the amount of code delivered can be significantly different from the amount of code actually written. The

TABLE 8.1   Comment Density by Language Type

	Number of Programs	Total Lines of Code	Comment Density (%)
Batch COBOL	335	670,000	16
CICS COBOL	273	507,000	26

programming team may write drivers, stubs, prototypes, and "scaffolding" for development and testing, but these programs are discarded when the final version is tested and turned over to the customer. Here, the developers want to distinguish the amount of *delivered* code from the amount of *developed* code. The *number of delivered source instructions* (DSI) captures this aspect of size; it counts separate statements on the same physical line as distinct, and it ignores comment lines. However, unlike ES, DSI includes data declarations and headings as source instructions.

Clearly, the definition of code size is influenced by the way in which it is to be used. Some organizations use size to compare one project with another, to answer questions such as

- What is our largest, smallest, and average-sized project?
- What is our productivity?
- What are the trends in project size over time?

Other organizations measure size only within a project team, asking

- What is the largest, smallest, and average module size?
- Does module size influence the number of faults?

The US Software Engineering Institute developed a set of guidelines to help you in deciding how to measure an LOC in your organization (Park 1992). It takes into account the use of automatic code generators, distinguishes physical LOCs from logical ones, and allows you to tailor the definition to your needs.

EXAMPLE 8.6

The Software Engineering Institute did a pilot study of the use of its guidelines at the Defense Information Systems Agency (Rozum and Florac 1995). The checklist was used to decide that a lines-of-code count would include the following:

- All executable code
- Nonexecutable declarations and compiler directives

Comment and blank lines are not counted. The checklist also requires the user to specify how the code was produced, and the pilot project classified code produced in the following ways:

- Programmed
- Generated with source code generators
- Converted with automatic translators
- Copied or reused without change
- Modified

But code that was removed from the final deliverable was not counted.

The checklist takes into account the origin of code, so that code that was new work or adapted from old work was counted, as was local library software and reuse library software. However, code that was furnished by the government or a vendor was not included in the count.

Figure 8.1 presents an example of a simple program. A large number of practitioners were asked to count independently (i.e., without conferring

```
with TEXT _ IO; use TEXT _ IO;
procedure Main is

        —This program copies characters from an input
        —file to an output file. Termination occurs
        —either when all characters are copied or
        —when a NULL character is input

        NullchAr, Eof: exception;
        Char: CHARACTER;
        Input _ file, Ouptu _ file, Console: FILE _ TYPE;
Begin
        loop
                Open (FILE => Input _ file, MODE => IN _ FILE,
                              NAME => "CharsIn");
                Open (FILE => Output _ file, MODE = > OUT _ FILE,
                              NAME => "CharOut");
                Ge (Input _ file, Char);
                if END _ OF _ FILE (Input _ file) then
                        raiseEof;
                elseif Char = ASCII.NUL then
                        raiseNullchar;
                else
                        Put(Output _ file, Char);
                endif;
        end loop;
exception
        whenEof => Put (Console, "no null characters");
        whenNullchar => Put (Console, "null terminator");
end Main
```

FIGURE 8.1   A simple program.

or seeking clarification) the "number of LOCs" in the program. Among the various valid answers were the following:

30   Count of the number of physical lines (including blank lines)

24   Count of all lines except blank lines and comments

20   Count of all statements except comments (statements taking more than one line count as only one line)

17   Count of all lines except blank lines, comments, declarations and headings

13   Count of all statements except blank lines, comments, declarations and headings

6   Count of only the ESs, not including exception conditions

Thus, even with a small program, the difference between least and most compact counting methods can be a factor of 5:1, as observed in practice by Jones. When other size measures are used, the same variations are encountered:

Number of ESs: Ranges from 6 to 20

Number of characters   Ranges from 236 to 611

You can obtain consistency by using a tool to count LOCs. For example, the UNIX™ (or Linux) command "wc –l code" will count the number of lines (including comments and blank lines) in file "code." You can improve consistency by formatting source code with a "pretty printer" program. The key is to be consistent in formatting and counting lines.

We now review the approach developed by Maurice Halstead, which is based on the programming language tokens in program code.

8.2.2 Halstead's Approach

Maurice Halstead made an early attempt to capture notions of size and complexity beyond the counting of LOCs (Halstead 1977). Although his work has had a lasting impact, Halstead's software science measures provide an example of confused and inadequate measurement, particularly when used to capture attributes other than size.

EXAMPLE 8.7

Halstead's software science attempted to capture attributes of a program that paralleled physical and psychological measurements in other disciplines. He began by defining a program P as a collection of tokens, classified as either operators or operands. The basic metrics for these tokens are the following:

μ_1 = Number of unique operators
μ_2 = Number of unique operands
N_1 = Total occurrences of operators
N_2 = Total occurrences of operands

For example, the FORTRAN statement

$$A(I) = A(J)$$

has one operator (=) and two operands ($A(I)$ and $A(J)$).

The *length* of P is defined to be $N = N_1 + N_2$, while the *vocabulary* of P is $\mu_1 = \mu_2 + \mu_3$. The *volume* of a program, akin to the number of mental comparisons needed to write a program of length N or the minimum number of bits to represent a program, is

$$V = N \times \log_2 \mu$$

Halstead derived measures for a number of other attributes including *program level*, *difficulty*, and *effort*. According to Halstead, one can use these metrics and an estimate of the number of mental discriminations per second to compute the time required to write a program.

The software science *program length* measure does satisfy the properties of a software size measure. However, the other metrics derived from the primitive counts of operators and operands are not true measures. The metrics are presented in the literature as a definitive collection, with no corresponding consensus on the meaning of attributes such as *volume*, *difficulty*, or *program level*. In other words, the relationship between the empirical relational system (i.e., the real world) and the mathematical model is unclear. Further, Halstead gives no real indication of the relationships among different components of his theory. And we cannot tell if he is defining measures or prediction systems.

Using the conceptual framework described in Chapter 3, we can discuss Halstead's metrics in more depth. The focus of Halstead's measurements is a product: the source code for an imperative language. We have seen that a model is prerequisite to any measurement; in this case, we require a model of source code sufficiently precise to identify unambiguously the operators and operands as objects, as well as occurrences of these objects. But no such model is provided, other than the formulae that interrelate the metrics. Four internal attributes are measured, and each is measured on an absolute scale: the number of distinct operators, μ_1 (corresponding to the source code attribute "having operators"), the number of distinct operands, μ_2, and total number of respective occurrences of these, N_1 and N_2.

The formula $N = N_1 + N_2$ is a proposed measure of the internal program attribute of *length*. From the perspective of measurement theory, N is a reasonable measure of the size of the actual code (without comments), since it does not contradict any intuitively understood relations among programs and their sizes. A similar argument applies to the formula for the internal program attribute of vocabulary. The program attribute *volume* is supposed to correspond to the amount of computer storage necessary for a uniform binary encoding; the assumption of a uniform encoding on an arbitrary machine suggests that this measure should be viewed as an internal attribute.

Thus, Halstead has proposed reasonable measures of three internal program attributes that reflect different views of size. Halstead's approach becomes problematic when we examine the remaining measures, which are generally unvalidated and, some would argue, intuitively implausible prediction systems in which the prediction procedures are not properly articulated. Abran provides a comprehensive discussion and evaluation of the Halstead suite of metrics (Abran 2010).

8.2.3 Alternative Code Size Measures

To define size differently, we have two other alternatives to explore, both of which are acceptable on measurement theory grounds as ratio measures:

1. We can measure size in terms of the *number of bytes* of computer storage required for the program text. This approach has the advantage of being on the same scale as the normal measure of size for object code. It is at least as well understood as LOC, and it is very easy to collect.

2. We can measure size in terms of the *number of characters (CHAR)* in the program text, which is another easily collected measure. For example, most modern word processors compute this count routinely for any text file. (Both the UNIX and Linux operating systems have the command *wc <filename>* to compute it.)

Because these size measures, as well as LOCs, are on the ratio scale, we can rescale any one of the size measures by multiplying by a suitable (empirical) constant.

EXAMPLE 8.8

If α is the average number of characters per line of program text, then we have the rescaling

$$CHAR = \alpha \, LOC$$

which expresses a stochastic relationship between LOC and CHAR. Similarly, we can use any constant multiple of the proposed measures as an alternative valid size measure. Thus, we can use KLOC (thousands of LOCs) or KDSI (thousands of delivered source instructions) to measure program size.

So far, we have assumed that one LOC is much like another. However, many people argue that an LOC is dependent on language; a line of APL or LISP is very different from a line of C, C++, or Java. Some practitioners use conversion factors, so that a module of k lines of APL code is considered to be equivalent to αk lines of C code, for some appropriate α. This conversion is important if you must use LOC to measure functionality and effort, rather than just size.

8.2.4 Dealing with Nontextual or External Code

There is another aspect of language dependence that causes a problem. Most code measures assume that software code consists purely of text. Up to 1990, this assumption was almost invariably true. However, the advent of visual programming and windowing environments (and to a lesser extent object orientation and fourth-generation languages) changed dramatically our notions of what a software program is.

EXAMPLE 8.9

In many program development environments, you can create a graphical user interface, complete with menus, icons and graphics, with almost no code in the traditional sense. For example, the executable code to produce a scroll bar is constructed automatically after you point at a scroll bar object in the programming environment. You only need to write code to perform the specific actions that result from, say, a click on a specific command button.

In an environment like the one described in Example 8.9, it is not at all clear how you would measure the size of your "program." The traditional code size may be negligible, compared to the "size" of the objects provided by the environment and other graphics objects. Thus, a program with just five hand-written source code statements can easily generate an executable program of 200 KB.

These approaches to programming raise two separate measurement issues:

1. How do we account in our size measures for objects that are not textual?

2. How do we account in our size measures for components that are constructed externally?

Issue 1 is relevant not just for code developed in new programming environments, but also for traditional specification and design documents, and we address this partially in Section 8.3. Issue 2 is an aspect of reuse, and we address it in Section 8.6.2.

8.3 DESIGN SIZE

We can measure the size of a design in a manner similar to that used to measure code size. We will count design elements rather than LOCs. The elements that we count depend on the abstractions used to express the design, and the design aspects of interest. Thus, the appropriate size measure depends on the design methodology, the artifacts developed, and the level of abstraction.

To measure the size of a procedural design, you can count the number of procedures and functions at the lowest level of abstraction. You can also measure the size of the procedure and function interfaces in terms of the number of arguments. Such measurements can be taken without code, for example, by analyzing the APIs of a system. At higher levels of abstraction, you can count the number of packages and subsystems. You can measure the size of a package or subsystem in terms of the number functions and procedures in the package.

Object-oriented designs add new abstraction mechanisms: objects, classes, interfaces, operations, methods, associations, inheritance, etc. Object-oriented design can also include realizations of design patterns (Gamma et al. 1994). When quantifying size, our focus is generally on the static entities rather than the links between entities, or runtime entities. Thus, we will measure size in terms of packages, design patterns, classes, interfaces, abstract classes, operations, and methods.

- *Packages*: Number of subpackages, number of classes, interfaces (Java), or abstract classes (C++)

- *Design patterns*:

  - Number of different design patterns used in a design

  - Number of design pattern realizations for each pattern type

  - Number of classes, interfaces, or abstract classes that play roles in each pattern realization

- *Classes, interfaces, or abstract classes*: Number of public methods or operations, number of attributes

- *Methods or operations*: Number of parameters, number of over-loaded versions of a method or operation

One of the earliest object-oriented design size measures was the *weighted methods per class (WMC)* measure (Chidamber and Kemerer 1994). As proposed, *WMC* is measured by summing the weights of the methods in a class, where weights are unspecified complexity factors for each method. The weight used in most studies is 1. With a weight of 1, *WMC* becomes a size measure—the number of methods in a class. For example, using a weight of 1, Zhou and Leung found that the *WMC* was one of a set of independent variables that can effectively predict the severity of faults in data from a NASA C++ system (Zhou and Leung 2006). Another study by Pai and Dugan applied a Bayesian analysis to develop a causal model-related object-oriented design measures including two size measures—*WMC* (again using a weight of 1) and the number LOCs, along with several object-oriented structure measures (to be discussed in the next chapter)—to predict fault proneness (Pai and Dugan 2007).

Both the number of methods and the number of attributes can serve as class size measures. One set of studies found that the number of methods is a better predictor of class change-proneness than the number of attributes (Bieman et al. 2001, 2003). The studies focused on the effects of design pattern use on change-proneness in three proprietary systems and two open sources systems. The studied systems included from 4 to 7 unique design pattern types, with a range of 6–35 pattern realizations—actual implementations of a pattern, as one pattern can be "realized" several times in a design. Here, the number of design pattern realizations served as a measure of design size. The studies found that the classes that played roles in design patterns were more, not less, change prone than the classes that were not part of a pattern.

If we are interested in run time properties of a design, we can measure the size of a running system in terms of the number of active objects over time.

8.4 REQUIREMENTS ANALYSIS AND SPECIFICATION SIZE

Requirements and specification documents generally combine text, graphs, and special mathematical diagrams and symbols. The nature of the presentation depends on the particular style, method, or notation used. When measuring code or design size, you can identify *atomic* entities to count (lines, statements, bytes, classes, and methods, for example). However, a requirements or specification document can consist of a mixture of text and diagrams. For example, a use case analysis may consist of a UML use case diagram along with a set of use case scenarios that may be expressed as either text or as UML activity diagrams. Because a requirements analysis often consists of a mix of document types, it is difficult to generate a single size measure.

EXAMPLE 8.10

There are obvious atomic elements in a variety of requirements and specification model types that can be counted:

  i. *Use case diagrams*: Number of use cases, actors, and relationships of various types
  ii. *Use case*: Number of scenarios, size of scenarios in terms of steps, or activity diagram model elements
  iii. *Domain model (expressed as a UML class diagram)*: Number of classes, abstract classes, interfaces, roles, operatons, and attributes
  iv. *UML OCL specifications*: Number of OCL expressions, OCL clauses
  v. *Alloy models*: Number of alloy statements—signatures, facts, predicates, functions, and assertions (Jackson 2002)
  vi. *Data-flow diagrams used in structured analysis and design*: Processes (bubbles nodes), external entities (box nodes), data-stores (line nodes) and data-flows (arcs)

vii. *Algebraic specifications*: Sorts, functions, operations, and axioms

viii. *Z specifications*: The various lines appearing in the specification, which form part of either a type declaration or a (nonconjunctive) predicate (Spivey 1993)

We can enforce comparability artificially by defining a *page* as an atomic entity, so that both text and diagrams are composed of a number of sequential pages. Thus, the *number of pages* measures length for arbitrary types of documentation and is frequently used in industry.

8.5 FUNCTIONAL SIZE MEASURES AND ESTIMATORS

Many software engineers argue that size is misleading, and that the amount of *functionality* inherent in a product paints a better picture of product size. In particular, those who generate effort and duration estimates from early development products often prefer to evaluate the amount of required functionality rather than product size (which is not available early). As a distinct attribute, required functionality captures an intuitive notion of the amount of function contained in a delivered product or in a description of how the product is supposed to be.

There have been several serious attempts to measure functionality of software products. We examine two approaches in this chapter: Albrecht's function points (FPs) and the COCOMO II approach. These approaches were developed as part of a larger effort to supply size information to a cost or productivity model, based on measurable early products, rather than as estimates of LOCs.

Both approaches measure the functionality of specification documents, but each can also be applied to later life-cycle products to refine the size estimate and therefore the cost or productivity estimate. Indeed, our intuitive notion of functionality tells us that if a program P is an implementation of specification S, then P and S should have the same functionality.

Neither approach adheres to the rigorous view of measurement that we have described. As we discuss each method, we indicate where the problems are and how they might be addressed rigorously. However, the current lack of rigor should not prevent us from using and refining the current approaches, as functionality is a very important product attribute. Moreover, these methods have been used to good effect in a number of industrial applications.

8.5.1 Function Points

Albrecht's effort estimation method was largely based on the notion of FPs. As their name suggests, FPs are intended to measure the amount of functionality in a system as described by a specification. We can compute FPs without forcing the specification to conform to the prescripts of a particular specification model or technique.

To compute the number of FPs we first compute an *unadjusted function point count* (UFC). To do this, we determine from some representation of the software the number of "items" of the following types:

- *External inputs:* Those items provided by the user that describe distinct application-oriented data (such as file names and menu selections). These items do not include inquiries, which are counted separately.

- *External outputs:* Those items provided to the user that generate distinct application-oriented data (such as reports and messages, rather than the individual components of these).

- *External inquiries:* Interactive inputs requiring a response.

- *External files:* Machine-readable interfaces to other systems.

- *Internal files:* Logical master files in the system.

EXAMPLE 8.11

Figure 8.2 describes a simple spelling checker. To compute the UFC from this description, we can identify the following items:

- The two external inputs are: document file-name, personal dictionary-name.
- The three external outputs are: misspelled word report, number-of-words-processed message, number-of-errors-so-far message.
- The two external inquiries are: words processed, errors so far.
- The two external files are: document file, personal dictionary.
- The one internal file is: dictionary.

Next, each item is assigned a subjective "complexity" rating on a three-point ordinal scale: *simple, average,* or *complex.* Then, a weight is assigned to the item, based on Table 8.2.

In theory, there are 15 different varieties of items (three levels of complexity for each of the five types), so we can compute the UFC by

Spell-checker spec: The checker accepts as input a document file and an optional personal dictionary file. The checker lists all words not contained in either of these files. The user can query the number of words processed and the number of spelling errors found at any stage during processing.

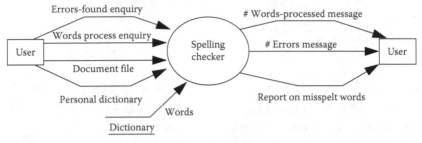

A = # external inputs = 2, B = # external outputs = 3, C = # inquiries = 2,
D = # external files = 2, and E = # internal files = 1

FIGURE 8.2   Computing basic function point components from specification.

multiplying the number of items in a variety by the weight of the variety and summing over all 15:

$$UFC = \sum_{i=1}^{15} (\text{Number of items of variety } i) \times (\text{weight}_i)$$

EXAMPLE 8.12

Consider the spelling checker introduced in Example 8.11. If we assume that the complexity for each item is *average*, then the UFC is

$$UFC = 4A + 5B + 4C + 10D + 7E = 58$$

If instead we learn that the dictionary file and the misspelled word report are considered *complex*, then

$$UFC = 4A + (5 \times 2 + 7 \times 1) + 4C + 10D + 10E = 63$$

TABLE 8.2   Function Point Complexity Weights

Item	Simple	Weighting Factor Average	Complex
External inputs	3	4	6
External outputs	4	5	7
External inquiries	3	4	6
External files	7	10	15
Internal files	5	7	10

TABLE 8.3   Components of the Technical Complexity Factor

F_1 Reliable backup and recovery	F_2 Data communications
F_3 Distributed functions	F_4 Performance
F_5 Heavily used configuration	F_6 Online data entry
F_7 Operational ease	F_8 Online update
F_9 Complex interface	F_{10} Complex processing
F_{11} Reusability	F_{12} Installation ease
F_{13} Multiple sites	F_{14} Facilitate change

To complete our computation of FPs, we calculate an *adjusted function-point count*, FP, by multiplying UFC by a *technical complexity factor*, TCF. This factor involves the 14 contributing factors listed in Table 8.3.

Each component or subfactor in Table 8.3 is rated from 0 to 5, where 0 means the subfactor is irrelevant, 3 means it is average, and 5 means it is essential to the system being built. Although these integer ratings form an ordinal scale, the values are used as if they are a ratio scale, contrary to the principles we introduced in Chapter 2. Also, we find it curious that the "average" value of 3 is not the median value.

The following formula combines the 14 ratings into a final technical complexity factor:

$$\mathrm{TCF} = 0.65 + 0.01 \sum_{i=1}^{14} F_i$$

This factor varies from 0.65 (if each F_i is set to 0) to 1.35 (if each F_i is set to 5). The final calculation of FPs multiplies the UFC by the technical complexity factor:

$$\mathrm{FP} = \mathrm{UFC} \times \mathrm{TCF}$$

EXAMPLE 8.13

To continue our FP computation for the spelling checker in Example 8.11, we evaluate the technical complexity factor. After having read the specification in Figure 8.2, it seems reasonable to assume that F_3, F_5, F_9, F_{11}, F_{12}, and F_{13} are 0, that F_1, F_2, F_6, F_7, F_8, and F_{14} are 3, and that F_4 and F_{10} are 5. Thus, we calculate the TCF as

$$\mathrm{TCF} = 0.65 + 0.01(18 + 10) = 0.93$$

Since UFC is 63, then

$$\mathrm{FP} = 63 \times 0.93 = 59$$

FPs can form the basis for an effort estimate.

EXAMPLE 8.14

Suppose our historical database of project measurements reveals that it takes a developer an average of two person-days of effort to implement an FP. Then we may estimate the effort needed to complete the spelling checker as 118 days (i.e., 59 FPs multiplied by 2 days each).

But FPs are also used in other ways as a size measure. For example, they can be used to express defect density in terms of defects per FP. They are also used in contracts, both to report progress and to define payment. For instance, Onvlee claimed that 50–60% of fixed priced software contracts in the Netherlands have had their costs tied to an FP specification (Onvlee 1995). In other words, many companies write software contracts to include a price per FP, and others track project completion by reporting the number of FPs specified, designed, coded, and tested.

8.5.1.1 Function Points for Object-Oriented Software

The notion of FPs has been adapted to be more suitable for object-oriented software. To apply FP analysis to object-oriented software and designs, we need to know how to identify the inputs, outputs, inquiries, and files in the early artifacts generated during object-oriented development (Lokan 2005). There are some obvious mappings: messages represent inputs and outputs, while classes with persistent state identified during domain analysis represent files. An analysis of system sequence diagrams developed during analysis can also identify FP items.

Rather than map object-oriented constructs to traditional FP items, *object points* can be computed directly from class diagrams using a weighting scheme similar to that used in FPs (Antoniol et al. 1999). To compute object points for classes with superclasses, inherited attributes are included as part of a subclass. An empirical study showed that object points can predict size in terms of LOC as well as other methods (Antoniol et al. 2003).

Since use case analysis is often performed early in each cycle of object-oriented development (Larman 2004), it makes sense to base FP analysis on use case documents. These documents may include UML use case models and associated scenarios or activity diagrams developed during use case analysis. *Use case points* are computed from counts of the actors, use cases, and the activities that are modeled along with associated complexity weights and technical complexity factors. The computation of use case points is very similar to the computation of FPs. Kusumoto et al. provide a simple example of the computation of use case points along with a description of an automated tool that computes use case points from use case models (Kusumoto et al. 2004). Empirical studies found that estimates produced by applying use case points can be more accurate than those produced by a group of professional developers (Anda et al. 2001; Anda 2002).

8.5.1.2 Function Point Limitations

Albrecht proposed FPs as a technology-independent measure of size. But there are several problems with the FPs measure, and users of the technique should be aware of its limitations.

1. *Problems with subjectivity in the technology factor.* Since the TCF may range from 0.65 to 1.35, the UFC can be changed by ±35%. Thus, the uncertainty inherent in the subjective subfactor ratings can have a significant effect on the final FP value.

2. *Problems with double counting.* It is possible to account for internal complexity twice: in weighting the inputs for the UFC, and again in the technology factor (Symons 1988). Even if the evaluator is careful in separating these two complexity-related inputs, double counting can occur in the use to which the final FP count is put. For example, the TCF includes subfactors such as performance requirements and processing complexity, characteristics often used in cost models. However, if FPs are used as a size input to a cost model that rates performance and complexity separately, these adjustment factors are counted twice.

3. *Problems with counterintuitive values.* In addition, the subfactor ratings lead to TCF values that are counterintuitive. For instance, when each *Fi* is *average* and rated 3, we would expect TCF to be 1; instead, the formula yields 1.07. Albrecht included the TCF as a means of improving resource predictions, and the TCF accentuates both internal system complexity and complexity associated with the user-view functionality. It may be preferable to measure these concepts separately.

4. *Problems with accuracy.* One study found that the TCF does not significantly improve resource estimates (Kitchenham and Känsälä

1993). Similarly, another study found that the UFC seems to be no worse a predictor of resources than the adjusted FP count (Jeffery et al. 1993). So the TCF does not seem useful in increasing the accuracy of prediction. Recognizing this, many FP users restrict themselves to the UFC.

5. *Problems with changing requirements.* FPs are an appealing size measure in part because they can be recalculated as development continues. They can also be used to track progress, in terms of number of FPs completed. However, if we compare an FP count generated from an initial specification with the count obtained from the resulting system, we sometimes find an increase of 400–2000% (Kemerer 1993). This difference may be not due to the FP calculation method but rather due to "creeping elegance," where new, nonspecified functionality is built into the system as development progresses. However, the difference occurs also because the level of detail in a specification is coarser than that of the actual implementation. That is, the number and complexity of inputs, outputs, enquiries, and other FP-related data will be underestimated in a specification because they are not well understood or articulated early in the project. Thus, calibrating FP relationships based on actual projects in your historical database will not always produce equations that are useful for predictive purposes (unless the FPs were derived from the original system specification, not the system itself).

6. *Problems with differentiating specified items.* Because evaluating the input items and technology factor components involves expert judgment, the calculation of FPs from a specification cannot be completely automated. To minimize subjectivity and ensure some consistency across different evaluators, organizations (such as the International Function Point User Group) often publish detailed counting rules that distinguish inputs and outputs from enquiries, to define what constitutes a "logical master file," etc. However, there is still great variety in results when different people compute FPs from the same specification.

7. *Problems with subjective weighting.* Symons notes that the choice of weights for calculating unadjusted FPs was determined subjectively from IBM experience. These values may not be appropriate in other development environments (Symons 1988).

8. *Problems with measurement theory.* Kitchenham et al. have proposed a framework for evaluating measurements, much like the framework presented in Chapters 2 and 3. They apply the framework to FPs, noting that the FP calculation combines measures from different scales in a manner that is inconsistent with measurement theory. In particular, the weights and TCF ratings are on an ordinal scale, while the counts are on a ratio scale, so the linear combinations in the formula are meaningless. They propose that FPs be viewed as a vector of several aspects of functionality, rather than as a single number (Kitchenham et al. 1995). A similar approach proposed by Abran and Robillard is to treat FPs as a multidimensional index rather than as a dimensionless number (Abran and Robillard 1994).

Several attempts have been made to address some of these problems. The International Function Points User Group meets regularly to discuss FPs and their applications, and they publish guidelines with counting rules. The organization also trains and tests estimators in the use of FPs, with the goal of increasing consistency of interpretation and application.

In general, the community of FP users is especially careful about data collection and analysis. If FPs are used with care, and if their limitations are understood and accounted for, they can be more useful than LOCs as a size or normalization measure. As use of FPs is modified, evaluated, and reported in the literature, we can monitor the success of the changes.

8.5.2 COCOMO II Approach

COCOMO is a model for predicting effort from a formula whose main independent variable is size. In investigating alternatives for LOCs as a size input to revise the COCOMO model, Boehm and his colleagues selected FPs for use when the system is completely specified. However, they sought a size measure that could be used even earlier in development, when feasibility is being explored and prototypes built. Although they are not specialized to object-oriented concerns, the term *object points* was unfortunately used for this early size measure in COCOMO II (Boehm et al. 2000). To distinguish from the more appropriate use of the "object point" as done in the prior section, we will use the term "C2 object point" here. The C2 object point approach is a synthesis of an approach for modeling estimation expertise (Kauffman and Kumar 1993) with productivity data (Banker et al. 1994a).

To compute C2 object points, counting the number of screens, reports, and third-generation language components that will be involved in the

TABLE 8.4　C2 Object Point Complexity Levels

For Screens				For Reports			
Number and Source of Data Tables				**Number and Source of Data Tables**			
Number of views contained	Total <4 (<2 server, <2 client)	Total <8 (2–3 server, 3–5 client)	Total 8+ (>3 server, >5 client)	Number of sections contained	Total <4 (<2 server, <2 client)	Total <8 (2–3 server, 3–5 client)	Total 8+ (>3 server, >5 client)
<3	Simple	Simple	Medium	0 or 1	Simple	Simple	Medium
3–7	Simple	Medium	Difficult	2 or 3	Simple	Medium	Difficult
8+	Medium	Difficult	Difficult	4+	Medium	Difficult	Difficult

Source: Boehm B.W. et al. *Software Cost Estimation with Cocomo II*, Prentice-Hall, Upper Saddle River, New Jersey, 2000.

application generates an initial size measure. It is assumed that these *entities* are defined in a standard way as part of an integrated software development environment. Next, each entity is classified as simple, medium, or difficult, much as are FPs. Table 8.4 contains guidelines for this classification.

The number in each cell is weighted according to Table 8.5. The weights reflect the relative effort required to implement an instance of that complexity level.

As with FPs, the weighted instances are summed to yield a single C2 object point number. Then, the procedure differs from FPs in that reuse is taken into account, since the C2 object points are intended for use in effort estimation. Assuming that $r\%$ of the objects will be reused from previous projects, the number of *new object points* is calculated to be

$$\text{New object points} = (\text{Object points}) \times (100 - r)/100$$

To use this number for effort estimation, COCOMO II determines a productivity rate (i.e., new object points per person-month) from a table based on developer experience and capability, coupled with ICASE maturity and capability.

TABLE 8.5　Complexity Weights for Object Points

Object Type	Simple	Medium	Difficult
Screen	1	2	3
Report	2	5	8
3GL component	–	–	10

8.6 APPLICATIONS OF SIZE MEASURES

Size measures are useful for quantifying many software attributes of interest. Here we look at three areas where size is useful: to normalize measures of other attributes, to quantify the amount of reuse, and to measure attributes related to software testing.

8.6.1 Using Size to Normalize Other Measurements

Besides indicating the size of software entities, the most common use of software size measures is to normalize measures of other attributes. For example, Chapters 1 through 6 introduced the commonly used software quality measure errors or faults per thousand lines of code (KLOC). This normalization can be done to compare the quality of functions or methods, classes, packages, subsystems, or systems.

Another attribute that is commonly normalized by size is cost. For example, the total project cost per delivered KLOC can be computed using accounting and code data. Normalized cost is often reported in terms of the cost per FP.

We can also normalize LOC in terms of other size measures, for example, number of classes or methods. Then we would report LOC per class and LOC per method. The average and distribution of method size is commonly reported.

EXAMPLE 8.15

A study of the change proneness of classes counted the number of changes to classes between releases. The number of changes was normalized by class size (measured by the number of operations) to create a new measure, *change density*. Change density is the number changes per operation—the total number of changes to a class over a specified period divided by the total number of operations defined for the class (Bieman et al. 2003).

EXAMPLE 8.16

Another study examined the frequency of the occurrence of coding concerns in software systems. A *coding concern* is an anomaly in the code that should be checked to confirm that it is not an error. Static analysis tools can identify dozens of coding concerns in Java code including having an empty catch block, overriding a base class with an empty method, unused private field or method, unused variable, hiding of names, etc. To get a better understanding of the frequency of coding concerns. Munger et al. reported their results as concern density: concerns per LOC, concerns per class, and concerns per package. It was surprising when they reported that average concerns per LOC in Netbeans 3.31 was 1.068, more than one concern per LOC (Munger et al. 2002).

8.6.2 Size-Based Reuse Measurement

We regularly use operating systems, compilers, database management systems, class libraries, statistical packages, and libraries of mathematical routines, rather than writing our own. This reuse of software (including requirements, designs, documentation, and test data and scripts as well as code) improves our productivity and quality, allowing us to concentrate on new problems, rather than continuing to solve old ones again. A review of empirical studies of software reuse reports that reuse led to lower problem density, decreased effort expended on fixing problems, and increased productivity (Mohagheghi and Conradi 2007). Table 8.6 compares the effects of reuse on subsystems of a large telecom system. You can see that the number of trouble reports (normalized by KLOC) is lower for the reused code. The code modification rate is the percentage of the total KLOC of code that were changed in response to trouble reports. The code modification rate is notably higher on the non-reused code.

Counting reused code is not as simple as it sounds. It is difficult to define formally what we mean by reused code. We sometimes reuse whole programs without modification, but more often we reuse some *unit* of code (a module, function, or procedure). And we often modify that unit to some extent. However, we account for reused code, we want to distinguish

TABLE 8.6   Trouble Reports (TRs) and Code Modification Rates for Subsystems in a Large Telecom System

	#Subsystems	# TRs	# TRs/ KLOC	# TRs/Modified KLOC	Code Modification Rate (%)
Release 2					
Reused	8	170	0.44	1.43	44
Nonreused	3	244	1.44	2.19	60
Release 3					
Reused	9	664	2.31	5.17	45
Nonreused	3	855	4.62	7.67	60

Source: Mohagheghi P. and Conradi R., An empirical investigation of software reuse benefits in a large telecom product, *ACM Transactions on Software Engineering and Methodology (TOSEM)*, 17(3), 1–31, June 2008.

a module with one modified line from a module with 100 modified lines. Thus, we consider the notion of *extent of reuse*, measured on an ordinal scale by NASA/Goddard's Software Engineering Laboratory in the following way (SPC 1995).

1. *Reused verbatim*: The code in the unit was reused without any changes.

2. *Slightly modified*: Fewer than 25% of the LOCs in the unit were modified.

3. *Extensively modified*: 25% or more of the LOCs were modified.

4. *New*: None of the code comes from a previously constructed unit.

EXAMPLE 8.17

Selby studied reuse in 25 large-scale NASA systems using the above ordinal scale for the extent of reuse. He found that the modules that were reused verbatim (level 1) exhibited fewer faults and required less effort for fault removal. The modules reused after extensive modifications (level 3) were the most fault prone (Selby 2005).

For a given program, we can define size in terms of total size and proportion of reuse at the different levels.

EXAMPLE 8.18

Hatton reports a high degree of reuse at Programming Research Ltd., where reuse was encouraged as part of a larger program to eliminate troublesome language constructs from company products (Hatton 1995). The reuse ratio is computed for each product as the proportion of reused lines out of the total number of lines. Table 8.7 illustrates the reuse ratio for Programming Research products.

Hewlett-Packard considers three levels of code: new code, reused code, and leveraged code. In this terminology, reused code is used as is, without modification, while leveraged code is existing code that is modified in some way. The Hewlett-Packard reuse ratio includes both reused and leveraged code as a percentage of total code delivered. The chart in Figure 8.3

TABLE 8.7   Reuse of Code at Programming Research Ltd.

Product	Reusable Lines of Code	Total Lines of Code	Reuse Ratio (%)
QAC	40,900	82,300	50
QA FORTRAN	34,000	73,000	47
QA Manager (X)	18,300	50,100	37
QA Manager (Motif)	18,300	52,700	35
QA C++	40,900	82,900	49
QA C Dynamic	11,500	30,400	38

illustrates the degree of reuse on six firmware products at Hewlett-Packard (Lim 1994).

8.6.3 Size-Based Software Testing Measurement

Various size attributes are related to software testing. The simplest is the size of the test suite in terms of SLOC of testing code or the number of test cases in a test suite. Measures related to the *test requirements* may be of greater interest.

A *test requirement* is a specific software element that must be executed or covered during testing to satisfy a particular testing criterion (Ammann and Offutt 2008). For example, each program statement (or node in a control flow graph) is a test requirement for the statement coverage criterion and each branch (or edge in a control flow graph) is a test requirement for

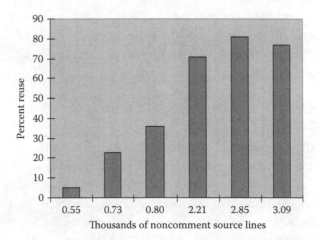

FIGURE 8.3   Reuse at Hewlett-Packard. The bars represent six different firmware projects.

the branch coverage criterion. Thus, the test requirements in a software system depend on the relevant testing criteria. The following are some testing criteria and associated measures of *the size of the test requirements*:

- *Statement/node coverage*: The number of statements or nodes in a control flow graph, or states in a state machine model

- *Branch/edge coverage*: The sum of the number of destinations for the program branches, the number of edges in a control flow graph, or the number of transitions in a state machine model

- *All use coverage*: The number of feasible definition–use pairs in a control flow graph that is annotated with variable definition and uses

- *All use cases*: The number of use cases generated by requirements analysis or in a UML use case diagram

- *Object-oriented requirements*: The number of object-oriented relationships (associations, realizations, dependencies, etc.) in a design or code

We can also use size-based measures to quantify the number of *test antipatterns* in a system, which are difficult to test constructs in an object-oriented program or design (Baudry and Sunye 2004; Baudry et al. 2001). Such measures would be the inverse of testability—a system with a greater number of antipatterns would be more difficult to test.

One study examined changes in the number of test requirements and test antipatterns between releases. It found that the number of object-oriented test requirements and test antipatterns increased as three open source software systems evolved, thus reducing testability (Izurieta and Bieman 2008, 2013).

8.7 PROBLEM, SOLUTION SIZE, COMPUTATIONAL COMPLEXITY

Some of the measures, particularly the functional size measures, use an adjustment based on the *size of the problem* being addressed by the requirements specification. For example, Albrecht's TCF is supposed to be a measure of the underlying problem. But a problem can have more than one solution, and the solutions can vary in their approaches and therefore in their attributes. Thus, as we noted at the beginning of this chapter, the *complexity or size of a solution* can be considered as a separate component of size, and it can be expressed in a more objective way.

Ideally, we would like the complexity of the solution to be no greater than the complexity of the problem, but that is not always the case. Let us informally define the *complexity of a problem* as the amount of resources required for an optimal solution to the problem. Then *complexity of a solution* can be regarded in terms of the resources needed to implement a particular solution.

We can view solution complexity as having at least two aspects (Garey and Johnson 1979):

1. *Time complexity*, where the resource is computer time.

2. *Space complexity*, where the resource is computer memory.

Both time and space are really size attributes—time and space measures must satisfy the three properties of size measures described in Section 8.1.

8.8 SUMMARY

Internal product attributes are important, and measuring them directly allows us to assess early products and predict likely attributes of later ones. In this chapter, we have seen the properties that any size measure should satisfy. We have also seen several ways to measure the size of code, designs, requirements, specifications, and functionality. In many ways, code size is the easiest to measure; it can be expressed in terms of LOCs, number of characters, and more. But we must take into account the dependence of code size measures on language, degree of reuse, and other factors. We can measure the size of designs, specifications, requirements, software models, etc. Object-oriented products can be measured in terms of numbers of classes, objects, methods, and links between objects and classes through associations, inheritance, and use dependencies.

Functional size can be derived from problem statements or specifications by using FPs, or COCOMO II object points; development products available later in the life cycle can be measured, too, and compared with earlier estimates. Measures of both functionality and length can be used to normalize other measures, for example, to express defect density in terms of defects per LOC or per FP, and progress and productivity can be tracked in terms of size or functionality of product.

EXERCISES

1. Recall several programs you have written in a particular programming language. Can you rank them according to length, functionality, and problem complexity? Is the ranking harder to perform if the programs are written in different languages?

2. How is the stochastic relationship

$$CHAR = \alpha\, LOC$$

   similar to this formula?

$$Time = Distance/Speed$$

   Manipulating the formula allows speed to be expressed as an indirect measure as a ratio of distance and time. Is a similar manipulation meaningful for the LOC relationship?

3. It is important to remember that "prediction" is not "prescription." How might a development policy that requires all programming staff to produce 50 lines of documentation per thousand LOC prove inappropriate?

4. Consider the following informal statement of requirements for a system:

   The system processes various commands from the operator of a chemical plant. The most important commands are

   a. Calculate and display average temperature for a specified reactor for the day period.

   b. Calculate and display average pressure for a specified reactor for the day period.

   c. Calculate and display a summary of the temperature and pressure averages.

   The operator can also choose to send the results to an urgent electronic bulletin board if necessary.

Compute both the unadjusted and adjusted function point count for this system, stating carefully any assumptions you are making.

5. For the specification in Figure 8.2, experiment with different possible values of TCF and different complexity weightings ranging from the lowest to highest possible values to see how FP may vary from the value 59 in Example 8.13. How does this variation scale up for systems involving hundreds or thousands of inputs, outputs, and interfaces?

6. Consider the following informal requirements for a safety-critical system:

The system will control a radiotherapy machine to treat cancer patients. This system is embedded in the machine and must run on a special purpose processor with limited memory (16 MB). The operator enters the type of treatment and radiation dose levels. In addition, the machine communicates with a patient database system to get patient information and, after treatment, record the radiation dose and other treatment details in a database. The system also displays this information to the operator.

a. Compute both the unadjusted and adjusted function-point count for this system, stating carefully any assumptions you are making.

b. Compare the function point values that you computed for the radiotherapy machine to those computed in Examples 8.12 and 8.13 for the system described in Figure 8.2. Do you believe that the computed values for these two systems accurately quantify the amount of functionality in these systems and the effort required to build them? Justify your answer.

7. Explain very briefly (five lines at most) the idea behind Albrecht's function points measure. List the main applications of function points, and compare function points with the lines of code measure.

8. Exam question

a. What do function points measure and what are the major drawbacks?

b. Give three examples of how you might use the function point measure in quality control or assurance (stating in each case why it might be preferable to some simpler alternative measure).

c. Consider the following specification of a simple system for collecting student course marks or grades.

The course marks system enables lecturers to enter student marks for a predefined set of courses and students on those courses. Thus, marks can be updated, but the lecturers cannot change the basic course information, as the course lists are the responsibility of the system administrator. The system is menu-driven, with the lecturer selecting from a choice of courses and then a choice of operations. Compute the number of function points in this system, stating carefully the assumptions you are making. The operations include

i. Enter coursework marks

ii. Enter exam marks

iii. Compute averages

iv. Produce letter grades

v. Display information (to screen or printer)

The information displayed is always a list of the students together with all the known marks, grades, and averages.

FURTHER READING

The book by Alain Abran describes and evaluates the entire suite of Halstead measures and both function points and object points. It also describes the development of the COSMIC measurement method to determine the functional size of real-time and embedded software. COSMIC was adopted as the ISO 19761 standard. The book also describes the relationship between function points and other ISO standards, including ISO 20926, 20968, 24570, and 29881.

Abran, A., *Software Metrics and Software Metrology*, Wiley, Hoboken, New Jersey, 2010.

For further information on COSMIC and ISO 19761, including tutorials and case studies, visit the following website:

The Common Software Measurement International Consortium (the "COSMIC" organization) http://www.cosmicon.com/

The latest definition of function point measurement is available from the International Function Point User Group (IFPUG). Information is available from http://www.ifpug.org/. For a comprehensive review of the background and evolution of function point analysis, see the review paper by Christopher Lokan.

Lokan C.J., Function Points, *Advances in Computers*, 65, 297–344, 2005.

A good source for information about the use of COCOMO II object points is the COCOMO II website:

http://sunset.usc.edu/csse/research/COCOMOII/cocomo_main.html

For details of measuring algorithmic efficiency and computational complexity, see Harel and Feldman's book.

Harel D. and Feldman Y., *Algorithmics*, 3rd Edition. Addison Wesley, Reading, Massachusetts, 2004.

Garey and Johnson have written the definitive text and reference source for computational complexity.

Garey M.R. and Johnson D.S., *Computers and Intractability: A Guide to the Theory of NP-Completeness*, W.H. Freeman, San Francisco, 1979.

CHAPTER 9

Measuring Internal Product Attributes

Structure

W̲E HAVE SEEN HOW SIZE MEASURES can be useful in many ways: as input to prediction models, as a normalizing factor, as a way to express progress during development, and more. But there are other useful internal product attributes. There is clearly a link between the structure of software products and their quality. The structure of requirements, design, and code can help us to understand the difficulty we sometimes have in converting one product into another (as, e.g., implementing a design as code), in testing a product (as, e.g., in testing code or validating requirements), or in predicting external software attributes from early internal product measures. Although structural measures vary in what they measure and how they measure it, they are often (and misleadingly) called "complexity" metrics, when they are actually quantifying other attributes.

We begin by describing aspects of structural measures based on the software entity to be measured, its level of abstraction, and the attribute that we want to measure. Then, we look in more detail at software control flow structure of individual program code-level functions, procedures, and methods. In the following sections, we turn to measures that are based on the information and data flow structure, metrics for object-oriented designs, and finally to data structure. We end this chapter with a discussion on how to compare general complexity measures.

9.1 ASPECTS OF STRUCTURAL MEASURES

The size of a development product tells us a lot about the effort that went into creating it. All other things being equal, we would like to assume that a large module takes longer to specify, design, code, and test than a small one. But experience shows us that such an assumption is not valid; the structure of the product plays a part, not only in requiring development effort but also in how the product is maintained. Thus, we must investigate the characteristics of product structure, and determine how they affect the outcomes we seek. In this chapter, we focus primarily on code and design measures, but many of the concepts and measures introduced here can be applied to other development products also.

A software module or design can be viewed from several perspectives. The perspective that we use depends on

1. The level of abstraction—program unit (function, method, class), package, subsystem, and system

2. The way the module or design is described—syntax and semantics

3. The specific attribute to be measured

To be sure that we are measuring the desired attribute, we generally represent the relevant aspects of a module or design using a model containing only the information relevant to the attribute. We can think of structure from at least two perspectives:

1. Control flow structure

2. Data flow structure

The *control flow* addresses the sequence in which instructions are executed in a program. This aspect of structure reflects the iterative and looping nature of programs. Whereas size counts an instruction or program element just once, control flow makes more visible the fact that an instruction or program element may be executed many times as the program is actually run.

Data flow follows the trail of a data item as it is created or handled by a program. Many times, the transactions applied to data are more complex than the instructions that implement them; data flow measures depict the behavior of the data as it interacts with the program.

In Chapter 8, we evaluated size measures in terms of three properties of the size attribute: nonnegativity, null value, and additivity (Briand et al. 1996). In this chapter, we will evaluate structure measures by applying the properties that Briand et al. also developed for the following structural attributes: *complexity*, *length*, *coupling*, and *cohesion*. As introduced in Chapter 8, the properties are defined in terms of a graph model of system of modules, where modules contain elements (nodes in the graph), and elements have links (graph edges) with other elements, and an element can be in more than one module. Also, modules may be nested or disjoint. The properties are defined in terms of the effect of system changes to a measure of the particular attribute. For each attribute, the properties are an empirical relation system and represent the required but not necessarily sufficient constraints on a valid measure. A measure may need to satisfy additional constraints to be intuitively valid.

9.1.1 Structural Complexity Properties

The relevant notion of *complexity* captures the complicatedness of the connections between elements in a system model. Here, complexity refers to the complexity of a system. The complexity of a system depends on the number of links between elements, and should, at a minimum, satisfy the following properties (Briand et al. 1996) (which are, of course, a set of empirical relations in the sense of Chapter 2):

1. *Nonnegativity*: System complexity cannot be negative.

2. *Null value*: The complexity of a system with no links is zero.

3. *Symmetry*: The complexity of a system does not depend on how links are represented.

4. *Module monotonicity*: System complexity "is no less than the sum of the complexities of any two of its modules with no relationships in common."

5. *Disjoint module additivity*: The complexity of a system of disjoint modules is the sum of the complexities of the modules.

9.1.2 Length Properties

Sometimes, we are only interested in the size of an entity in terms of one dimension of a model. For example, we are only interested in a person's *height* to predict whether or not a person would bump his or her head when entering a doorway. For software entities, we might be interested in the distance in terms of links from one element to another. The first two length properties are the same as those for size (in Chapter 8). However, the effect on system length from adding links depends on whether or not modules are disjoint (Briand et al. 1996):

1. *Nonnegativity*: System length cannot be negative.

2. *Null value*: A system with no links has zero length.

3. *Nonincreasing monotonicity for connected components*: Length does not increase when adding links between connected elements.

4. *Nondecreasing monotonicity for nonconnected components*: Length does not decrease when adding links between nonconnected elements.

5. *Disjoint modules*: The length of a system of disjoint modules is equal to the length of the module with the greatest length.

Property 5 reflects a difference between the notions of *length* from that of *size*. For example, height is a length attribute of a person, while *mass* or *weight* is a size attribute. The distribution of human heights in a population is more useful for determining the necessary height of a doorway than distribution of weights. The maximum depth of a tree and the maximum nesting depth in a program are examples of length measures.

9.1.3 Coupling Properties

Coupling is an attribute of an individual module and depends on a module's links to and from elements that are external to the module (Briand et al. 1996):

1. *Nonnegativity*: Module coupling cannot be negative.

2. *Null value*: A module without links to elements that are external to the module has zero coupling.

3. *Monotonicity*: Adding intermodule relationships does not decrease coupling.

4. *Merging modules*: Merging two modules creates a new module that has coupling that is at most the sum of the coupling of the two modules.

5. *Disjoint module additivity*: Merging disjoint modules without links between them creates a new module with coupling that is the sum of the coupling of the original modules.

9.1.4 Cohesion Properties

Cohesion is an attribute of an individual module and depends on the extent that related elements are contained within a module. Thus, a module with many connections between its internal elements will have greater cohesion than a module that contains unrelated elements (Briand et al. 1996):

1. *Nonnegativity and normalization*: Module cohesion is normalized so that it is between zero and one.

2. *Null value*: A module whose elements have no links between them has zero cohesion.

3. *Monotonicity*: Adding links between elements in a module cannot decrease the cohesion of the module.

4. *Merging modules*: Merging two unrelated modules creates a new module with a maximum cohesion no greater than that of the original module with the greatest cohesion.

Property 1 is pragmatic. It allows one to compare the cohesion of different sized modules. Zero cohesion is defined in property 2. Although Briand et al. did not define the notion of maximum cohesion—corresponding to a score of 1 on the normalized 0–1 scale, it is clear that a module with maximum cohesion is one in which every one of its elements is connected.

9.1.5 Properties of Custom Attributes

We can also define structural attributes that are defined for special purposes. For example, we may want a design to be consistent with a particular structure or architecture, or use specified design patterns. A system might have a design that requires a layered architecture. Then we might want to define measures based on the consistency with the architecture, or count violations of constraints implied by the architecture. Later in this chapter, we will see the notion of *tree impurity* defined as an attribute along with the properties of an associated tree impurity measure. Such a measure would indicate the difference between (1) a specific graph representation modeling some aspects of software and (2) a tree structure. We will also see that one can define measures based on the occurrences of realizations of design structures such as design patterns.

We now examine the measurement of the internal structure of software starting from the perspective of the control flow in individual program units.

9.2 CONTROL FLOW STRUCTURE OF PROGRAM UNITS

A great deal of early software metrics work was devoted to measuring the control flow structure of individual functions, procedures, or methods implemented as imperative language programs or algorithms. This work is still relevant, especially when applied to problems in software testing. The control flow measures are usually modeled with *directed graphs*, where each node (or point) corresponds to a program statement or basic block (code that always executes sequentially), and each arc (or directed edge) indicates the flow of control from one statement or basic block to another. We call these directed graphs *control flowgraphs* or *flowgraphs*.

Figure 9.1 presents a simple example of a program, *A*, and a reasonable interpretation of its corresponding flowgraph, *F(A)*. We say "reasonable interpretation" because it is not always obvious how to map a program *A* to a flowgraph *F(A)*. The flowgraph is a good model for defining measures of control flow structure, because it makes explicit many of the structural properties of the program. The nodes enumerate the program statements, and the arcs make visible the control patterns.

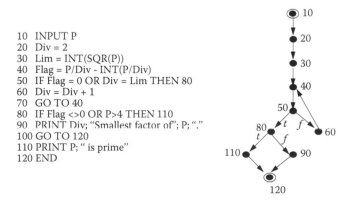

```
10   INPUT P
20   Div = 2
30   Lim = INT(SQR(P))
40   Flag = P/Div - INT(P/Div)
50   IF Flag = 0 OR Div = Lim THEN 80
60   Div = Div + 1
70   GO TO 40
80   IF Flag <>0 OR P>4 THEN 110
90   PRINT Div; "Smallest factor of"; P; "."
100  GO TO 120
110  PRINT P; " is prime"
120  END
```

FIGURE 9.1   A program and its corresponding flowgraph.

There are many flowgraph-based measures, some of which are incorporated into static analysis and program development tools. Many measures reflect an attempt to characterize a desirable (but possibly elusive) notion of structural complexity and preserve our intuitive feelings about complexity: if *m* is a structural measure defined in terms of the flowgraph model, and if program *A* is structurally "more complex" than program *B*, then the measure $m(A)$ should be greater than $m(B)$. In Chapter 2, we noted some theoretical problems with this simplistic approach. There are practical problems as well; since the measures are often based on very specific views of what constitutes good structure, there may be differing opinions about what constitutes a well-structured program. The structural complexity properties in Section 9.1.1 represent one perspective on complexity. The other structural attributes in Section 9.1 have different properties.

To support differing perspectives and multiple attributes, we introduce an approach for analyzing control flow structure that is independent of any particular view of desirable program structure. We can also use this technique to describe and differentiate views of desirable structure. The technique enables us to show that any program has a uniquely defined structural decomposition into primitive components. In the following sections, we show that this decomposition (which can be generated automatically) may be used to define a wide range of structural measures. These unifying principles objectively make clear the assumptions about structuredness inherent in each measure's definition. Most importantly, we show how the decomposition can be used to generate measures of specific attributes, such as those relating to test coverage. The decomposition is at the heart of many static analysis tools and reverse engineering tools.

9.2.1 Flowgraph Model and the Notion of Structured Programs

We begin our discussion by reviewing graph-related terminology. Recall that a *graph* consists of a set of points (or nodes) and line segments (or edges). In a *directed graph*, each edge is assigned a direction, indicated by an arrowhead on the edge. This directed edge is called an *arc*.

Thus, directed graphs are depicted with a set of nodes, and each arc connects a pair of nodes. We write an arc as an ordered pair, <*x*, *y*>, where *x* and *y* are the nodes forming the endpoints of the arc, and the arrow indicates that the arc direction is from *x* to *y*. The arrowhead indicates that something flows from one node to another node. Therefore, each node has some arcs flowing into it and some arcs flowing out (or possibly no arcs of each type). The *in-degree* of a node is the number of arcs arriving at the node, and the *out-degree* is the number of arcs that leave the node. We can move from one node to another along the arcs, as long as we move in the direction of the arrows. A *path** is a sequence of consecutive (directed) edges, some of which may be traversed more than once during the sequence. A *simple path†* is one in which there are no repeated edges.

EXAMPLE 9.1

In Figure 9.1, the node labeled "50" has in-degree 1 and out-degree 2. The following sequence S of edges is a path:

<30,40> <40,50> <50,60> <60,40> <40,50> <50,80 >

However, S is not a simple path, since it repeats edge <40,50>.

A *flowgraph* is a directed graph in which two nodes, the *start node* and the *stop node*, obey special properties: the stop node has out-degree zero, and every node lies on some path from the start node to the stop node. In drawing flowgraphs, we distinguish the start and stop nodes by encircling them.‡ Flowgraph nodes with out-degree equal to 1 are called *procedure nodes*; all other nodes (except the stop node) are termed *predicate nodes*.

EXAMPLE 9.2

In the flowgraph of Figure 9.1, the nodes labeled 50 and 80 are predicate nodes. All the other nodes (except the stop node) are procedure nodes.

When we model program control structure, certain flowgraphs occur often enough to merit special names. Figure 9.2 depicts those flowgraphs

---

* In graph theory, this is normally referred to as a *walk*. For historical reasons, computer scientists refer to walks as paths.
† In graph theory, this is normally referred to as a *path*. Ammann and Offutt define a *simple path* as a sequence of nodes connected by edges without any repeated nodes, except for the first and last node (Ammann and Offutt 2008).
‡ Flowgraphs are examples of *finite-state machines*. In general, a finite-state machine may have many start and stop nodes (states); the convention is to draw a circle around such nodes.

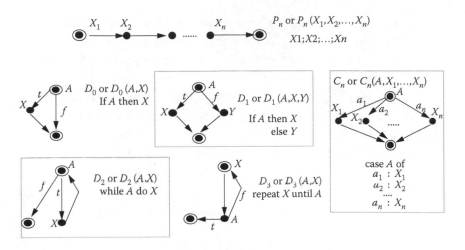

FIGURE 9.2   Common flowgraphs from program structure models.

that correspond to the basic control constructs in imperative language programming.

For example, the P_2 flowgraph* represents the construct *sequence*, where a program consists of a sequence of two statements. Beneath the name of each construct in Figure 9.2, we have written an example of program code for that construct. For instance, the example code for the D_0 construct is the *if–then* statement. Not all imperative languages have built-in constructs for each of the control constructs shown here, nor is the particular code for the constructs unique.

EXAMPLE 9.3

Figure 9.2 contains an example of the *repeat–until* statement. An alternate expression for that construct is

```
10 X
If A then goto 20 else goto 10
20 end
```

The flowgraph model for this program is identical to D_3 in Figure 9.2.

Strictly speaking, a flowgraph is "parameterized" when we associate with it the actual code that it represents. For example, the notation

---

* The sequence statement P_2 is a special case of a sequence of n statements P_n. Another special instance of P_n is the case $n = 1$, which represents a program consisting of a single statement.

$D_2(A,X)$ (meaning D_2 with parameters A and X) is an explicit denotation of the construct *while A do X*. Sometimes, for convenience, we refer only to the unparameterized names like D_2, meaning the generic *while–do* control construct.

Most imperative programs have built-in control constructs for the flowgraphs in Figure 9.2, but the same is not true for the control constructs shown in Figure 9.3. (We shall soon see that the reasons for this difference are more dogmatic than rational.) In theory, each of these additional constructs can be implemented using *goto* statements (which are available in C or C++, but not in many languages such as Java or Python). For example, the *two-exit loop* construct L_2 is equivalent to the following C code:

```
loop:
X;
if (A) goto end;
Y;
if (B) goto end;
else goto loop;
end: return;
```

Although we can construct the flowgraphs in Figures 9.2 and 9.3 using the goto construct, most developers avoid the goto construct because it allows you to create very complex control flow that is difficult to understand, debug, and modify. Experience suggests that it is best to build the control flow of procedures, functions, and methods using only the control flow constructs that implement those modeled in Figure 9.2 and possibly

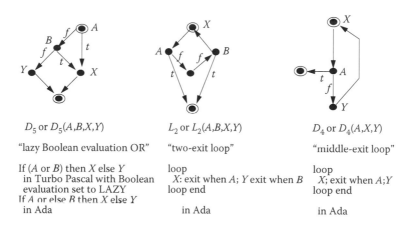

D_5 or $D_5(A,B,X,Y)$	L_2 or $L_2(A,B,X,Y)$	D_4 or $D_4(A,X,Y)$
"lazy Boolean evaluation OR"	"two-exit loop"	"middle-exit loop"

If $(A$ or $B)$ then X else Y in Turbo Pascal with Boolean evaluation set to LAZY
If A or else B then X else Y in Ada

loop
  X: exit when A; Y exit when B
loop end

in Ada

loop
  X; exit when A; Y
loop end

in Ada

FIGURE 9.3   Flowgraphs for less common control constructs.

Figure 9.3, as provided in an implementation language. There are many possible variations of the constructs in Figure 9.3. For example, a variation in D_5 is the *if–and–then–else* construct, as well as more complex conditions like

$$\text{if } (A \text{ or } B \text{ or } C) \text{ then } X \text{ else } Y, \quad \text{if } ((A \text{ or } B) \text{ and } (B \text{ or } C))$$

However, *all* of the flowgraphs in Figures 9.2 and 9.3 have an important common property that makes them suitable as "building blocks" for structured program units. To understand it, we must define formally the ways that we can build flowgraphs.

9.2.1.1 Sequencing and Nesting

There are just two legitimate operations we can use to build new flowgraphs from old: *sequencing* and *nesting*. Both have a natural interpretation in terms of program structure.

Let F_1 and F_2 be two flowgraphs. Then the *sequence of* F_1 *and* F_2 is the flowgraph formed by merging the stop node of F_1 with the start node of F_2. The resulting flowgraph is written as

$$F_1; F_2 \text{ or } Seq \ (F_1, F_2) \text{ or } P_2 \ (F_1, F_2)$$

EXAMPLE 9.4

Figure 9.4 shows the result of forming the sequence of the D_1 and D_3 flowgraphs.

The *flowgraph sequence* operation corresponds to the *sequence* operation (also called *concatenation*) in imperative language programming. In fact, the flowgraph operation preserves the program operation in the following

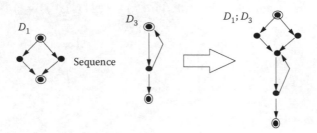

FIGURE 9.4   Applying the sequence operation.

way. Suppose A and A' are two blocks of program code, and recall that, in general, the flowgraph model of a program A is denoted by $F(A)$. Then

$$F(A; A') = F(A); F(A')$$

Thus, the flowgraph of the program sequence is equal to the sequence of the flowgraphs.

Let F_1 and F_2 be two flowgraphs. Suppose F_1 has a procedure node x. Then the *nesting of F_2 onto F_1 at x* is the flowgraph formed from F_1 by replacing the arc from x with the whole of F_2. The resulting flowgraph is written as

$$F_1 \, (F_2 \text{ on } x)$$

Alternatively, we write $F_1(F_2)$ when there is no ambiguity about the node onto which the graph is nested.

EXAMPLE 9.5

Figure 9.5 shows the result of nesting the D_3 flowgraph onto the D_1 flowgraph.

EXAMPLE 9.6

Figure 9.6 shows the construction of a flowgraph from a number of nesting and sequence operations.

The *flowgraph nesting operation* corresponds to the operation of *procedure substitution* in imperative language programming. Specifically, consider a program A in which the procedure A' is called by a parameter x. Then

$$F(A \text{ with } A' \text{ substituted for } x) = F(A) \, (F(A') \text{ on } x)$$

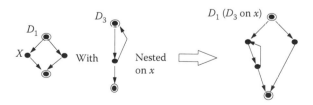

FIGURE 9.5   Applying the nesting operation.

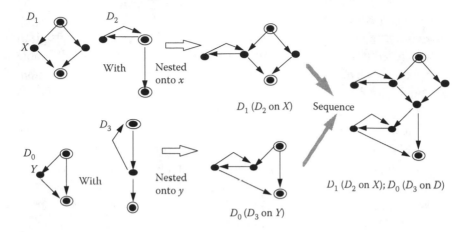

FIGURE 9.6   Combining the sequence and nesting operations.

Thus, the flowgraph of the program substitution is equal to the nesting of the flowgraphs.

In general, we may wish to nest n flowgraphs $F_1, \ldots, F_n$ onto n respective procedure nodes $x_1, \ldots, x_2$. The resulting flowgraph is written as

$$F(F_1 \text{ on } x_1, F_2 \text{ on } x_2, \ldots, F_n \text{ on } x_n.)$$

In many of our examples, the actual nodes onto which we nest are of no importance, and so we write: $F(F_1, F_2, \ldots, F_n)$.

Our examples describe the graphs that are constructed by using sequencing and nesting constructs. We can sometimes reverse the process, decomposing a flowgraph into smaller pieces that are related by sequencing and nesting. However, there are some graphs that cannot be decomposed that way. Formally, we say that *prime flowgraphs* are flowgraphs that cannot be decomposed nontrivially by sequencing or nesting. The building blocks of our approach to control structure analysis and measurement are these *prime flowgraphs*:

EXAMPLE 9.7

Each of the flowgraphs in Figures 9.2 and 9.3 is prime. The flowgraph on the right-hand side of Figure 9.5 is not prime, because it decomposes as: $D_1(D_3)$. In other words, it is constructed by nesting one flowgraph, D_3, onto another, D_1. Similarly, the flowgraph on the right-hand side of Figure 9.4 is not prime, because it decomposes as a sequence of a D_1 followed by D_3.

9.2.1.2 Generalized Notion of Structuredness

The common definitions of *structured programming* assert that a program is structured if it can be composed using only a small number of allowable constructs. Normally, these constructs must be only *sequence*, *selection*, and *iteration*. The rationale can be traced back to a classic result of Böhm and Jacopini, demonstrating that every algorithm may be implemented using just these constructs (Böhm and Jacopini 1966). Some authors have argued (wrongly, in light of Example 9.3) that this statement is equivalent to asserting that a program is structured only if it is written without any use of the *goto* statement.

We have to be able to assess an individual program and decide whether or not it is structured; the informal definition of structured programming offers us no help in answering this question. Ideally, we should be able to answer it by considering the flowgraph alone. However, to do so, we need a definition of *structuredness* that supports many different views, and a mechanism for determining the level of structuredness in an arbitrary flowgraph.

To form this definition, we must introduce more terminology involving a family S of prime flowgraphs. We say that a family of graphs is *S-structured* (or, more simply, that the family members are *S-graphs*) if it satisfies the following recursive rules:

1. Each member of S is S-structured

2. If F and F' are S-structured flowgraphs, then so are

    a. $F;F'$

    b. $F(F')$ (whenever nesting of F' onto F is defined)

3. No flowgraph is an S-structured graph unless it can be shown to be generated by a finite number of applications of the above steps.

Notice that, by definition, the members of S are themselves S-structured. We call these the *basic S-graphs*, and they correspond to the building blocks we seek. Since this definition is quite general, it permits you to choose which building blocks define your notion of structuredness.

EXAMPLE 9.8

For $S = \{P_1\}$, the set of S-structured graphs is the set $\{P_1, P_2, ..., P_n, ...\}$.

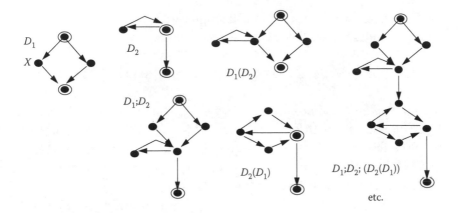

FIGURE 9.7   Examples of S-structured graphs when $S = \{D_1, D_2\}$.

EXAMPLE 9.9

For $S = \{D_1, D_2\}$, Figure 9.7 shows the examples of various S-structured graphs.

The definition allows us to nominate, for any particular development environment, a set of legal control structures (represented by the basic S-graphs) suited for particular applications. Then, by definition, any control structure composed from this nominated set will be "structured" in terms of this local standard; in other words, the set derived from the basic S-graphs will be S-structured.

EXAMPLE 9.10

Let $S^D = \{P_1, D_0, D_2\}$. Then the class of S^D-graphs is the class of flowgraphs commonly called (in the literature of structured programming) the D-structured (or sometimes just structured) graphs. Stated formally, the Böhm and Jacopini result asserts that every algorithm can be encoded as an S^D-graph. Although S^D is sufficient in this respect, it is normally extended to include the structures D_1 (if–then–else) and D_3 (repeat–until).

For reasons that have been discussed extensively elsewhere, it is now common to accept a larger set than S^D as the basis for structured programming. For example, there are very powerful arguments for including all of the primes in Figure 9.3. Many modern languages have constructs that support these primes. Thus, in these languages, the set of structured programs includes the set of S-graphs where

$$S = \{P_1, D_0, D_1, D_2, D_3, D_4, C_n \text{ (for all } n), L_2\}$$

9.2.1.3 Prime Decomposition

We can associate with any flowgraph a *decomposition tree* to describe how the flowgraph is built by sequencing and nesting primes (Fenton and Whitty 1986). Figure 9.8 illustrates how a decomposition tree can be determined from a given flowgraph.

Figure 9.9 presents another example, where a flowgraph is shown with its prime decomposition tree.

To understand this prime decomposition, consider the programming constructs that correspond to the named primes. By expanding

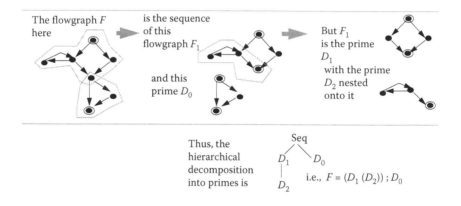

FIGURE 9.8 Deriving the decomposition tree of a flowgraph.

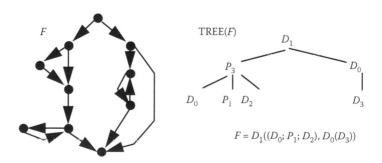

FIGURE 9.9 A flowgraph and its decomposition tree.

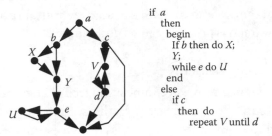

```
if a
  then
    begin
      If b then do X;
      Y;
      while e do U
    end
  else
    if c
      then do
        repeat V until d
```

FIGURE 9.10   The flowgraph of Figure 9.9 in terms of its program structure.

TREE(F) in Figure 9.9, we can recover the program text as shown in Figure 9.10.

Not only can we always decompose a flowgraph into primes, but we can be assured that the decomposition is always unique:

Prime decomposition theorem: Every flowgraph has a unique decomposition into a hierarchy of primes.

The proof of the theorem (see (Fenton and Whitty 1986)) provides a constructive means of determining the unique decomposition tree. For large flowgraphs, it is impractical to perform this computation by hand, but there are tools available commercially that do it automatically.

The prime decomposition theorem provides us with a simple means of determining whether an arbitrary flowgraph is S-structured or not for some family of primes S. We just compute the decomposition tree and look at the node labels; if every node is either a member of S or a P_n, then the flowgraph is an S-graph.

EXAMPLE 9.11

If $S = \{D_1, D_2\}$, then the flowgraph F in Figure 9.9 is *not* an S-graph, because one of the nodes in the decomposition tree is D_3. However, F is an S^D-graph, where $S^D = \{D_0, D_1, D_2, D_3\}$.

Every flowgraph must be S-structured for some family S (namely, where S is the set of distinct primes found in the decomposition tree); the question is whether any members of S are considered to be "nonstructured." The decomposition theorem shows that every program has a quantifiable degree of structuredness characterized by its decomposition tree. The only structures that cannot be decomposed in any way at all are primes. So, unless the whole flowgraph is prime, it can be decomposed to a certain extent.

9.2.2 Hierarchical Measures

The uniquely defined prime decomposition tree is a definitive description of the control structure of a program. In this section, we show that this tree is all we need to define a very large class of interesting measures. We are able to define many measures because they can be described completely in terms of their effect on primes and the operations of sequence and nesting.

EXAMPLE 9.12

Suppose we wish to measure formally the intuitive notion of *depth of nesting* within a program. Let the program be modeled by a flowgraph, F; we want to compute α, the depth of nesting of F. We can express α completely in terms of primes, sequence, and nesting:

Primes: The depth of nesting of the prime P_1 is zero, and the depth of nesting of any other prime F is equal to one. Thus, $\alpha(P_1) = 0$, and if F is a prime $\neq P_1$, then $\alpha((F) = 1$.

Sequence: The depth of nesting of the sequence $F_1, ..., F_n$ is equal to the maximum of the depth of nesting of the F_is. Thus, $\alpha(F_1; ...; F_n) = \max (\alpha(F_1); ...; \alpha(F_n))$.

Nesting: The depth of nesting of the flowgraph $F(F_1, ..., F_n)$ is equal to the maximum of the depth of nesting of the F_i *plus one* because of the extra nesting level in F. Thus, $\alpha(F(F_1, ..., F_n)) = 1 + \max(\alpha(F_1), ..., \alpha(F_n))$.

Next, consider the flowgraph F in Figure 9.9. We know that $F = D_1 ((D_1;P_1;D_2), D_0 (D_3))$, so we compute:

$\alpha(F) = \alpha (D_1 ((D_1;P_1;D_2), D_0 (D_3)))$

$\quad = 1 + \max (\alpha(D_1;P_1;D_2), \alpha(D_0 (D_3)))$ (Nesting rule)

$\quad = 1 + \max(\max(\alpha(D_1), \alpha(P_1), \alpha(D_2)),$ (Sequence rule and nesting
$\quad\quad 1 + \alpha(D_3))$ rule)

$\quad = 1 + \max(\max(1,0,1),2)$

$\quad = 1 + \max(1,2)$

$\quad = 3$

Intuitively, the depth of nesting measure α indicates the length of the path to the most deeply nested prime. Indeed, α satisfies the properties of a length measure described in Section 9.1.2.

Next, assume that S is an arbitrary set of primes. We say a measure m is a *hierarchical measure* if it can be defined on the set of S-graphs by specifying:

M_1: $m(F)$ for each $F \in S$

M_2: The sequencing function(s)

M_3: The nesting functions h_F for each $F \in S$

As we saw in Example 9.12, we may automatically compute a hierarchical measure for a program once we know M_1, M_2, M_3, and the decomposition tree.

The *uniqueness* of the prime decomposition implies that an S-graph can be constructed in only one way. Thus, we can construct new hierarchical measures simply by assigning a value $m(F)$ to each prime and a value to the sequence and nesting functions; in other words, we construct our own conditions M_1, M_2, and M_3. However, rather than generating arbitrary and artificial hierarchical measures in this way, we wish instead to show that many existing measures, plus many measures of specific intuitive structural attributes, are indeed hierarchical.

EXAMPLE 9.13

In Chapter 8, we discussed some of the problems involved in defining the lines of code measure unambiguously. Given the theory, which we have presented so far in this chapter, we define a formal size measure, v, which captures unambiguously the number of statements in a program when the latter is modeled by a flowgraph.

> M_1: $v(P_1) = 1$, and for each prime $F \neq P_1$, $v(F) = n + 1$, where n is the number of procedure nodes in F
>
> M_2: $v(F_1; \ldots; F_n) = \Sigma v(F_i)$
>
> M_3: $v(F(F_1, \ldots, F_m)) = 1 + \Sigma v(F_i)$ for each prime $F \neq P_1$

Condition M_1 asserts that the size of a procedure node (which will generally correspond to a statement having no control flow) will be 1. The size of a prime with n procedure nodes (which will generally be a control statement involving n noncontrol statements) is $n + 1$. This mapping corresponds to our intuitive notion of size, and satisfies the properties of the size attribute given in Section 8.1. The sequence and nesting functions given in M_2 and M_3, respectively, are equally noncontroversial.

To see how this measure works, we apply it to a flowgraph F in Figure 9.9:

$$
\begin{aligned}
v(F) &= v(D_1\,((D_0;P_1;D_2),\,D_0\,(D_3))) \\
&= 1 + (v(D_0;P_1;D_2) + v(D_0\,(D_3))) \quad \text{(Nesting rule)} \\
&= 1 + (v(D_0) + v(P_1) + v(D_2)) \quad \text{(Sequence rule} \\
&\quad + (1 + v(D_3)) \quad\quad\quad\quad\quad \text{and nesting rule)} \\
&= 1 + (2 + 1 + 2) + (1 + 2) \\
&= 9
\end{aligned}
$$

Once a hierarchical measure has been characterized in terms of the conditions M_1, M_2, and M_3, then we have all the information we need to calculate the measure for all S-graphs. We also have a constructive procedure for calculating measures using this information together with the prime decomposition tree. Some other simple but important hierarchical measures that capture very specific properties are shown in Figure 9.11.

Number of Nodes Measure n

M_1: $n(F)$ = number of nodes in F for each prime F

M_2: $n(F_1; \ldots; F_m) = \Sigma n(F_i) - k + 1$

M_3: $n(F(F_1, \ldots, F_p)) = n(F) + \Sigma n(F_i) - 2k$ for each prime F

Number of Edges Measure e

M_1: $e(F)$ = number of edges in F for each prime F

M_2: $e(F_1; \ldots; F_m) = \Sigma e(F_i)$

M_3: $e(F(F_1, \ldots, F_m)) = e(F) + \Sigma e(F_i) - n$ for each prime F

The "Largest Prime" Measure κ: (First Defined in Fenton (1985))

M_1: $\kappa(F)$ = number of predicates in F for each prime F

M_2: $\kappa(F_1; \ldots; F_n) = \max(\kappa(F_1), \ldots, \kappa(F_n))$

M_3: $\kappa(F(F_1, \ldots, F_n)) = \max(\kappa(F), \kappa(F_1), \ldots, \kappa(F_n))$ for each prime F

Number of Occurrences of Named Primes Measure p

M_1: $p(F) = 1$ if F is named prime, else 0

M_2: $p(F_1; \ldots; F_n) = \Sigma p(F_i)$

M_3: $p(F(F_1, \ldots, F_m)) = p(F) + \Sigma p(F_i)$

D-Structured Measure d

(This nominal scale measure yields the value 1 if the flowgraph is D-structured and 0 if it is not.)

M_1: $d(F) = 1$ for $F = P_1, D_0, D_1, D_2, D_3$ and 0 otherwise

M_2: $d(F_1; \ldots; F_n) = \min(d(F_1), \ldots, d(F_n))$

M_3: $d(F(F_1, \ldots, F_n)) = \min(d(F), d(F_1), \ldots, d(F_n))$

FIGURE 9.11 Some hierarchical measures.

Many flowgraph-based measures have been proposed as measures of structural complexity. Many of these measures are hierarchical, so they can be defined easily within our framework. By viewing each of them in the context of the three necessary and sufficient conditions, M_1, M_2, and M_3, we can see how such measures may be computed automatically. Furthermore, we can see easily what level of subjectivity has crept into the definition of the measure, and we can compare it with the notion of complexity in the real, empirical world that the measure is trying to capture.

In particular, we can isolate the three aspects of subjectivity in the measure's definition, namely

1. What is the "complexity" of the distinguished primes?

2. What is the "complexity" of sequencing?

3. What is the "complexity" of nesting onto a given prime?

The remainder of this section applies this analysis to several popular measures of the "complexity" of the control flow in program units.

9.2.2.1 McCabe's Cyclomatic Complexity Measure

As we noted in Chapter 2, McCabe proposed the cyclomatic number of a program's flowgraph as a measure of program complexity (McCabe 1976). For a program with flowgraph F, the *cyclomatic number* is calculated as

$$v(F) = e - n + 2$$

where F has e arcs and n nodes. The cyclomatic number actually measures the number of *linearly independent* paths through F, and an explanation of this term and its implications can be found in Appendix A.1.2. This measure is objective and useful when counting linearly independent paths, and it does satisfy the properties of a structural complexity measure as described in Section 9.1.1. See Briand et al. (1996) for details. However, it is not at all clear that $v(F)$ paints a complete or accurate picture of program complexity. To see why, let us examine v as a hierarchical measure.

For reasons detailed in Appendix A.1, we know that, for any flowgraph F,

$$v(F) = 1 + d$$

where d is the number of predicate nodes in F. Thus, v can be defined as a hierarchical measure in the following way:

M_1: $v(F) = 1 + d$ for each prime F, where d is the number of predicates in F

M_2: $v(F_1, \ldots, F_n) = \sum_{i=1}^{n} v(F_i) - n + 1$ for each n

M_3: $v(F(F_1, \ldots, F_n)) = v(F) + \sum_{i=1}^{n} v(F_i) - n$ for each prime F

Thus, if v is a measure of "complexity," it follows that

1. The "complexity" of primes is dependent only on the number of predicates contained in them.

2. The "complexity" of sequence is equal to the sum of the complexities of the components minus the number of components plus one.

3. The "complexity" of nesting components on a prime F is equal to the complexity of F plus the sum of the complexities of the components minus the number of components.

From a measurement theory perspective, it is extremely doubtful that any of these assumptions corresponds to intuitive relations about complexity. Thus, v cannot be used as a general complexity measure. However, the cyclomatic number is a useful indicator of how difficult a program or module will be to test and maintain. In this context, v could be used for quality assurance. In particular, McCabe has suggested that, on the basis of empirical evidence, when v is greater than 10 in any one module, the module may be problematic.

EXAMPLE 9.14

Grady reported a study at Hewlett-Packard, where cyclomatic number was computed for each module of 850,000 lines of FORTRAN code. The investigators discovered a close relationship between a module's cyclomatic number and the number of updates required. After examining the effects of cost and schedule on modules with more than three updates, the study team concluded that 15 should be the maximum cyclomatic number allowed in a module (Grady 1994).

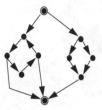

FIGURE 9.12   Flowgraph with essential complexity 4.

EXAMPLE 9.15

The quality assurance procedure for the software in the Channel Tunnel rail system requires that a module be rejected if its cyclomatic number exceeds 20 or if it has more than 50 statements, as determined by the LOGISCOPE tool (Bennett 1994).

9.2.2.2 McCabe's Essential Complexity Measure

McCabe also proposed a measure to capture the overall level of structuredness in a program. For a program with flowgraph F, the *essential complexity ev* is defined as

$$ev(F) = v(F) - m$$

where m is the number of subflowgraphs of F that are D-structured primes (i.e., either D_0, D_1, D_2, or $D_3)^*$ (McCabe 1976).

The flowgraph F in Figure 9.12 has a single D structured prime subflowgraph, namely, the D_2 construct on the right. Thus, $m = 1$, and $v(F) = 5$, yielding $ev(F) = 4$.

McCabe asserts that the essential complexity indicates the extent to which the flowgraph can be "reduced" by decomposing all those subflowgraphs that are D-primes. In this way, the essential complexity is supposed to measure the cyclomatic number of what remains after you decompose all the structured subflowgraphs. The essential complexity of a D-structured program is one, since only a P_n is left after decomposing all the structured primes.

EXAMPLE 9.16

The flowgraph on the left of Figure 9.13 is a truly unstructured "spaghetti" prime. Its essential complexity is 6, the same as its cyclomatic number. On

---

* McCabe (1976) actually refers to these curiously as "proper one-entry one-exit subflowgraphs."

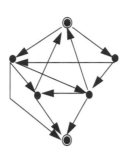

FIGURE 9.13   An unstructured prime.

the other hand, the flowgraph on the right is the sequence of three L_2 constructs. Since none is a D-prime, then according to McCabe's definition this flowgraph is nonreducible; therefore, its essential complexity should be the same as its cyclomatic number, namely, 7.

When a flowgraph is not D-structured, it is not at all clear that the definition of essential complexity corresponds to any natural intuition about structural complexity. For example, in Figure 9.13, the flowgraph on the right has a higher essential complexity value than the spaghetti structure on the left. This unusual circumstance happens because essential complexity is additive on sequences, except when the sequential components are D-structured. Another counterintuitive feature of essential complexity occurs when a nonstructured prime is nested *onto* a structured prime; the cyclomatic number of the structured prime is added to the overall essential complexity (because the structured prime is not a proper subflowgraph). This phenomenon is present in the flowgraph of Figure 9.12, where there is a D_0 at the highest level.

A more intuitive notion of essential complexity may simply be the cyclomatic number of the largest prime in the decomposition tree. This approach is similar to the largest prime metric described in Figure 9.11.

9.2.3 Code Structure and Test Coverage Measures

Many researchers approach the problem of measuring structure from the view of needing to understand how the module works. But there are other aspects of structure that reflect a different perspective. In particular, the structure of a module is related to the difficulty we find in testing it. We now turn to the application of control structure and decomposition for calculating the minimum number of test cases required to implement a test strategy.

We begin with the definitions and assumptions. Suppose that program P has been produced for a known specification S. To test P, we run P with an input i and check to see that the output satisfies the specification. We can formalize this procedure by defining a *test case* to be a pair, $(i, S(i))$. We are interested in whether $S(i)$, the expected output, is equal to $P(i)$, the actual output.

EXAMPLE 9.17

Suppose P is a program produced to meet the following specification, S:

> The inputs to the program are exam scores expressed as percentages. The outputs are comments, generated according to the following rules:
> For scores under 45, the program outputs "fail"
> For scores between 45 and 80 inclusive, the program outputs "pass"
> For scores above 80, the program outputs "pass with distinction"
> Any input other than a numeric value between 0 and 100 should produce the error message "invalid input"

A set of five test cases for P is

(40, "fail")
(60, "pass")
(90, "pass with distinction")
(110, "invalid input")
(fifty, "invalid input")

Suppose that on input "40," program P outputs "fail." Then we say that P is correct on that input. If, for the input "110," the program P outputs "pass with distinction," then P fails on this input; the correct output should have been the error message "invalid input."

As discussed in Section 8.6.3, the objective of a software testing strategy is to devise and run a set of test cases that satisfy or *cover* a set of test requirements. The test requirements are defined in terms of a software artifact or model of a software artifact, where an artifact may be the requirements, specification, design, program code, etc. (Ammann and Offutt 2008). These models are often graphs, for example, flowgraphs.

One testing strategy is to select test cases so that every program statement is executed at least once. This approach is called *statement coverage*— each statement is a test requirement. For deterministic software, each test case causes one *test path* through the program to execute. In terms of the program flowgraph, statement coverage is achieved by finding a set of test cases that execute test paths such that every flowgraph node lies on at least one test path.

EXAMPLE 9.18

Figure 9.14 contains a program written to meet the specification *S* of Example 9.17. We can achieve 100% statement coverage for this program by choosing just two of the test cases: one with input "90," and one with input "40." For "90," the program executes the path <ABDEFG>. For "40," the program executes the path <ABCEG>. Thus, C is covered in the second case, and all the other nodes are covered in the first case, so we have a complete statement coverage.

Another testing strategy is to select test cases so that every program branch is executed at least once. This approach, called *branch* (or *edge*) *coverage*, means finding test cases that cause the execution of a set of test paths such that every *edge* lies on at least one test path.

EXAMPLE 9.19

The set of test cases selected in Example 9.18 satisfies branch as well as statement coverage. However, this will not always be true. In the same program, the two paths ABCEFG and ABDEFG satisfy 100% statement coverage, but fail to cover the edge EG.

```
A  Input(score);
B  If score < 45
C     then print ("fail")
D     else print ("pass");
E  If score > 80
F     then print ("with distinction");
G  End
```

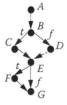

FIGURE 9.14   A simple program and corresponding flowgraph.

The most exhaustive structural testing strategy is to select test cases such that every possible test path is executed at least once. This approach is called *complete path coverage*. In terms of the program flowgraph, complete path coverage requires running every single test path through the flowgraph. This strategy, although appealing in theory, is normally impossible in practice. If the program has even a single loop, then complete path coverage can never be achieved because a loop gives rise to an infinite number of test paths.

There is another fundamental impediment to any structural testing strategy: *infeasible paths*. An *infeasible path* is a program path that cannot be executed for any input. In practice, infeasible paths can appear even in the simplest of programs.

EXAMPLE 9.20

In the program of Figure 9.14, the path ABCEFG is infeasible. To see why, notice that node C is executed only when the score is less than 45, while node F is executed only when the score is more than 80. Thus, no input score can cause the program to execute both nodes C and F.

The existence of infeasible paths can mean that, in practice, we may not be able to achieve 100% coverage for any structural testing strategy (even statement coverage).

Numerous other structural testing strategies have been proposed (see (Ammann and Offutt 2008) for a thorough account). In particular, researchers have tried to define strategies so that

1. They are more comprehensive than statement and branch coverage (in the sense of requiring more test cases).

2. The number of test cases required is finite even when there are loops.

Ammann and Offutt described numerous testing criteria for such strategies. Several of these strategies make use of the notion of simple paths and prime paths. They define a *simple path* as a path that does not contain the same node more than once except for the first and last node. A *prime path* is a simple path that "does not appear as a proper subpath of any other simple path." The following examples are from (Ammann and Offutt 2008):

- Edge-pair coverage: Testing "each reachable path of length up to 2" in the flowgraph.

- Simple-round-trip coverage: Testing "at least one round-trip path for each reachable node in" the flowgraph "that begins and ends a round-trip path," where a *round-trip path* is a prime path that starts and ends at the same node. This requires every cycle through a loop to be tested, but does require branch coverage or all cycles through loops that contain decisions.

- Prime path coverage: Testing all prime paths in the flowgraph. This requires testing all cycles that start at reachable nodes, and, depending on the branches inside a loop, requires testing execution paths that cause a loop to cycle two or more times. Prime path coverage satisfies branch coverage, and requires a finite number of tests.

We have already noted one problem with structural testing strategies (namely, infeasible paths). In practice, there are two further important obstacles:

1. *No structural strategy on its own (even 100% complete path coverage when feasible) can guarantee adequate software testing.* Consider, for example, the program in Figure 9.14, which implements the specification in Example 9.17. The path ABDEFG correctly outputs "distinction" when the input is "90." However, the correct output for *one* execution does not mean that *every* execution of the path ABDEFG is correct. For instance, the input score "110" also executes the path ABDEFG, but the program *wrongly* outputs "pass with distinction" when it should be producing "invalid input."

2. *Knowing the set of paths that satisfies a particular strategy does not tell you how to create test cases to match the paths.* This problem is undecidable in general—no automated method is guaranteed to either find the appropriate test cases or determine that such data does not exist; however, some commercial testing tools provide assistance.

Associated with every testing strategy are two important measures: the *minimum number of test cases* and the *test effectiveness ratio.*

9.2.3.1 Minimum Number of Test Cases

It is not enough for a test team to choose a test strategy. Before testing starts, the team members must also know the minimum number of test cases needed to satisfy the strategy for a given program. This information helps them to plan their testing, both in terms of generating data for each test case but also in understanding how much time testing is likely to take. Figure 9.15 illustrates, for some of the structural testing strategies described above, the test requirements as well as a minimal set of paths that satisfies the strategy. The flowgraph has a node corresponding to each statement. If a branch has no node on it, then the branch models a program in which no statement appears on that branch. For instance, the portion of the graph labeled with edges <2, 4>, <4, 6>, and <2, 6> represents an IF–THEN statement, rather than an IF–THEN–ELSE statement. When statement coverage is the testing strategy, the branch

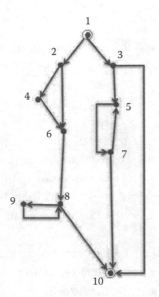

All test paths coverage (infinite number)
<1, 3, 10>, <1, 3, 5, 7, 10>, <1, 3, 5,(7, 5, 7)^n, 10>,
<1, 2, 6, 10>, <1, 2, 4, 6, 8, 10>, <1, 2, 6, 8, (9,8)^{n,10}>,
<1, 2, 4, 6, 8, (9,8)^n, 10> (any $n > 0$)

Prime path coverage (6)
Prime paths—test requirements: <1, 2, 3, 10>,
<1, 3, 5, 7, 10>, <5, 7, 5>, <1, 2, 6, 8, 10>, <1
2, 4, 6, 8, 10>, <8, 9, 8>, <9, 8, 9>, <9, 8, 10>.
Covered with 5 test path: traverse the 4 acyclic test
paths plus a test path that cycles through each loop
twice.

Branch coverage (4)
Branches—test requirements: <1, 2>, <2, 4>, <2, 6>,
<3, 5>, <3, 10>, <7, 10>, <7, 5>, <8, 9>, <8, 10>. Covered
with 4 test path: <1, 3, 10>, <1, 3, 5, 7, 5, 7, 10>,
<1, 2, 6, 8, 10>, <1, 2, 4, 6, 8, 9, 8, 10>.

Statement coverage (2)
Statements—test requirements: 1, 2, 3, 4, 5, 6, 7, 8,
9, 10. Covered with 2 test paths: <1, 3, 5, 7, 10>,
<1, 2, 4, 6, 8, 9, 8, 10>.

FIGURE 9.15 Test paths required to satisfy various structural testing criteria. (Numbers in parentheses represent the minimal number of test paths required for that criterion.)

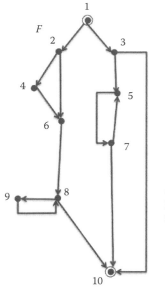

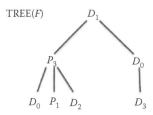

$$F = D_1((D_0; P_1; P_2), D_0(D_3))$$

Computing the minimal number of test paths for the branch coverage criterion:

$$m(F) = m(D_1(D_0((D_0; P_1; D_2), D_0(D_3)))$$
$$= m((D_0; P_1; P_2) + D_0(D_3)) \text{ formula for nesting on } D_1$$
$$= \max((m(D_0), m(P_1)), m(D_2)) + m(D_3) + 1$$
$$\quad\quad\quad \text{sequence formula and D0 nesting formula}$$
$$= \max(2, 1, 1) + 1 + 1 = 4$$

FIGURE 9.16 Computing the minimal number of test cases required using tables in Appendix A.1.2.

labeled <2, 4> does not have to be covered. Thus, the minimum number of test cases required for the branch coverage strategy in Figure 9.16 is four.

The decomposition theorem can help us to compute the minimum number of test cases for some criteria, including branch coverage and statement coverage. A test case corresponds to a test path through the flowgraph F. So to calculate the minimum number of test cases, we must compute the minimum number of test paths, $m(F)$, required to satisfy the strategy. We can compute $m(F)$ from the decomposition tree as long as we know how to compute $m(F)$ for the primes as well as for sequence and nesting.

Appendix A.2 supplies tables of relevant values for the various testing strategies. We have restricted the definition to a reasonable set of primes, which will be sufficient for most programs. If other primes are discovered in the decomposition tree, then their values must be added. Figure 9.16 illustrates the computation of the minimum number of test paths (or test cases) required to satisfy the "branch coverage" criterion.

The tables in Appendix A.2 help us to compute the various test coverage measures for arbitrary flowgraphs:

EXAMPLE 9.21

For the flowgraph in Figure 9.16, we compute the minimum number of test cases m for statement coverage using the tables in Appendix A.2:

$$m(F) = m(D_1(D_0;P_1;D_2),(m(D_0(D_3))$$

$$= m(D_0;P_1;D_2) + m(D_0(D_3))$$

$$= \max(m(D_0),m(P_1),m(D_2)) + m(D_3)$$

$$= \max(1,1,1) + 1 = 2$$

EXAMPLE 9.22

Bertolino and Marre noted several problems with using some of the strategies and their lower bound to drive test case selection (Bertolino and Marre 1995). For example, if a complex program is enclosed in a loop, then theoretically it is possible to satisfy branch coverage with a single path. In this case, the lower bound of one test case is not very meaningful, because it is unlikely that the single path required will be both feasible and obvious to construct. Thus, they propose a more practical strategy, based on finding a set of paths that provide reasonable coverage (in a very intuitive sense) which are much more likely to be feasible. The authors provide a hierarchical definition of the minimum number of paths needed to satisfy their testing strategy.

9.2.3.2 Test Effectiveness Ratio

For a given program and a set of cases to test it, we would like to know the extent to which the test cases satisfy a particular testing strategy.

EXAMPLE 9.23

Suppose we are testing the program in Figure 9.14, and we have run two test cases with input scores "60" and "90." The test cases traverse two paths, ABDEG and ABDEFG, covering 6 of the 7 statements, 7 of the 9 edges, and 2 of the 4 paths in the program. Thus, we say that statement coverage is 86%, branch coverage is 78%, and path coverage is 50%.

We can define these percentages formally. Let C be a testing criterion that requires us to cover a class of test requirements (such as paths, simple paths, linearly independent paths, LCSAJs, edges, or statements). For a given program and set of test cases, the test effectiveness ratio (TER$_C$) is defined as

$$\text{TER}_C = \frac{\text{Number of } C \text{ requirements executed at least once}}{\text{Total number of test requirements for criterion } C}$$

Some commercial tools compute various test effectiveness ratios when presented with a program and a set of test cases. The standard UNIX utility *lint* computes the TERs for statement and branch coverage testing of C programs.

EXAMPLE 9.24

Using the LDRA Testbed® commercial testing tool on actual test data used on commercial systems, Mike Hennel has made alarming observations about test effectiveness. For example, for the modest statement coverage strategy, most managers assume that a TER of 100% can be achieved routinely. However, the actual TER is typically no better than 40%. The typical TER for branch coverage is lower still, providing objective evidence that most software is not tested as rigorously as we like to think.

9.3 DESIGN-LEVEL ATTRIBUTES

So far, we have examined the attributes of individual modules. Measures of these are sometimes called *intramodular measures*. Next, we look at collections of modules, either at the design stage or when the program is implemented in code. It is the intermodule dependencies that interest us, and measures of these attributes are called *intermodular measures*. We restrict our discussion to design (unless we note otherwise), as the design structure is usually carried through to the code.

9.3.1 Models of Modularity and Information Flow

Although there is no standard definition of a module, we can rely on the one suggested by Yourdon: A *module* is a contiguous sequence of program statements, bounded by boundary elements, having an aggregate identifier (Yourdon and Constantine 1979). This deliberately vague definition permits a liberal interpretation. Also, a module should be (at least theoretically) separately compilable. Thus, a module can be any object that, at a given level of abstraction, you wish to view as a single construct. Thus,

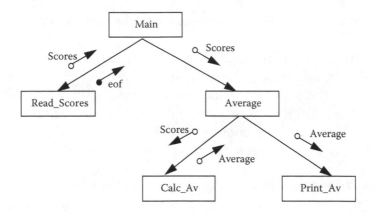

FIGURE 9.17   Design charts.

for example, in C, a procedure or function may be considered a module. In Java, a class or interface is considered a module. An individual Java or C++ method is considered a program unit, but not a module.

To describe intermodular attributes, we build models to capture the necessary information about the relationships between modules. Figure 9.17 contains an example of a diagrammatic notation capturing the necessary details about designs (or code).

This type of model describes the information flow between modules; that is, it explains which variables are passed between modules.

When measuring some attributes, we need not know the fine details of a design, so our models suppress some of them. For example, instead of examining variables, we may need to know only whether or not one module calls (or depends on) another module. In this case, we use a more abstract model of the design, a directed graph known as the module *call-graph*; a call-graph is not a flowgraph, as it has no circled start or stop node. An example is shown in Figure 9.18.

Usually, we assume that the call-graph has a distinguished root node, corresponding to the highest-level module and representing an abstraction

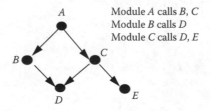

Module *A* calls *B, C*
Module *B* calls *D*
Module *C* calls *D, E*

FIGURE 9.18   Module call-graph.

of the whole system. This would correspond to a system with a centralized control structure.

For intramodular attributes, we consider models that capture the relevant details about information flow *inside* a module. Specifically, we look at the dependencies among data. A data dependency graph (DDG) is a model of information flow supporting this kind of measurement (Bieman and Debnath 1985). For instance, Figure 9.19 shows a simple program fragments and their corresponding DDG models.

9.3.2 Global Modularity

"Global modularity" is difficult to define because there are many different views of what modularity means. For example, consider *average module length* as an intuitive measure of global modularity. As defined by any of the measures in Chapter 8, module length is on a ratio scale, so we can meaningfully consider average module length for a software system in terms of the mean length of all modules. Boehm cautions us to distinguish this type of measure from "complexity" or "structuredness":

"A metric was developed to calculate the average size of program modules as a measure of structuredness. However, suppose one has a software product with n 100-statement control routines and a library of m 5-statement computational routines, which would be considered well structured for any reasonable values of m and n. Then, if $n = 2$ and $m = 98$, the average module size is 6.9, while if $m = 10$ and $n = 10$, the average module size is 52.5 statements" (Boehm et al. 1978).

We can describe global modularity in terms of several specific views of modularity (Hausen 1989), such as the following:

M_1 = modules/procedures

M_2 = modules/variables

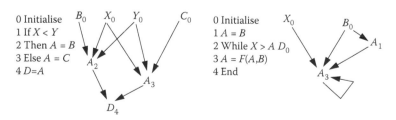

FIGURE 9.19   A data dependency graph model of information flow.

Both Hausen's and Boehm's observations suggest that we focus first on specific aspects of modularity, and then construct more general models from them.

9.3.3 Morphology

Yourdon and Constantine have analyzed what contributes to good design, and they suggest several ways to view design components and structure. They use the notion of *morphology* to refer to the "shape" of the overall system structure when expressed pictorially. Morphological characteristics such as *width* and *depth* can then be used to describe good and bad designs (Yourdon and Constantine 1979). Let us consider designs expressed in terms of *dependency graphs*, and determine which morphological characteristics of the graphs present important information about design quality. Here, we use an edge to connect two nodes if there is a defined dependence between the nodes. For instance, if a node represents a module and one module calls the other, then we connect the two corresponding nodes with an edge (similar to a call-graph, but without directed edges). Similarly, if one module passes data to another, then we can connect the corresponding nodes with an edge.

Many morphological characteristics are measurable directly, including the following:

- *Size*: Measured as number of nodes, number of edges, or a combination of these.

- *Depth*: Measured as the length of the longest path from the root node to a leaf node.

- *Width*: Measured as the maximum number of nodes at any one level.

- *Edge-to-node ratio*: Can be considered a connectivity density measure, since it increases as we add more connections among nodes.

Examples of these measures are illustrated in Figure 9.20.

If the likely length of each code module can be predicted within reason during design, then a number of quality assurance guidelines can be derived from the design morphology measures.

These measures are not restricted to dependency graphs. They are generally applicable to most models.

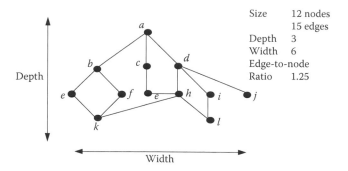

FIGURE 9.20 A design and corresponding morphology measures.

9.3.4 Tree Impurity

The graphs in Figure 9.21 represent different system structures that might be found in a typical design. They exhibit properties that may help us to judge good designs. We say that a graph is *connected* if, for each pair of nodes in the graph, there is a path between the two. All of the graphs in Figure 9.21 are connected. The *complete graph*, K_n, is a graph with n nodes, where every node is connected to every other node, so there are $n(n-1)/2$ edges. Graphs G_4, G_5, and G_6 in Figure 9.21 are complete graphs with three, four, and five nodes, respectively. The graph G_1 in Figure 9.21 is called a *tree*, because it is a connected graph having no *cycles* (i.e., no path that starts and ends at the same node). None of the other graphs is a tree, because each contains at least one cycle.

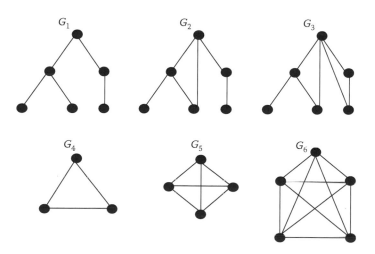

FIGURE 9.21 Dependency graphs with varying degrees of tree impurity.

Examining the trees in a graph tells us much about the design. Ince and Hekmatpour suggest that

> The more a system deviates from being a pure tree structure towards being a graph structure, the worse the design is ... it is one of the few system design metrics* to have been validated on a real project.

> INCE AND HEKMATPOUR 1988

Thus, we seek to create a measure, called *tree impurity,* to tell us how far a given graph deviates from being a tree. In what follows, we restrict our attention to undirected graphs.

To define tree impurity, we first describe several properties of graphs and trees. A tree with n nodes always has $n - 1$ edges. For every connected graph G, we can find at least one subgraph that is a tree built on exactly the same nodes as G; such a tree is called a *spanning subtree.* A *spanning subgraph G'* of a graph G is built on the same nodes of G, but with a minimum subset of edges so that any two nodes of G' are connected by a path. (Thus, a graph may have more than one spanning subgraph.) Intuitively, the tree impurity of G increases as the difference between G and G' increases. We want to make our definition formal, ensuring that it is consistent with the principles of measurement theory.

Any measure m of tree impurity must satisfy at least four properties:

Property 1: $m(G) = 0$ if and only if G is a tree.
> In other words, a graph that is actually a tree has no tree impurity. Thus, for example, any measure m of tree impurity must record $m(G_1) = 0$ for G_1 in Figure 9.21.

Property 2: $m(G_1) > m(G_2)$ if G_1 differs from G_2 only by the insertion of an extra edge (representing a call to an existing procedure).
> This property states that if we add an edge to a graph, then the resulting graph has greater tree impurity than the original. Thus, we have $m(G_3) > m(G_2)$ in Figure 9.21.

Property 3: For $i = 1$ and 2, let A_i denote the number of edges in G_i and N_i the number of nodes in G_i. Then if $N_1 > N_2$ and $A_1 - N_1 + 1 = A_2 - N_2 + 1$ (i.e., the spanning subtree of G_1 has

---

* Here, *metric* is synonymous with our term attribute.

more edges than the spanning subtree of G_2, but in both cases the number of edges additional to the spanning tree is the same), then $m(G_1) < m(G_2)$.

This property formalizes our intuitive understanding that we must take account of size in measuring deviation from a tree. Consider the two graphs G_2 and G_4 in Figure 9.21. Each has a single edge additional to its spanning subtree. However, since the spanning subtree of G_4 is smaller, we have an intuitive feeling that its tree impurity should be greater—a single deviation represents a greater proportional increase in impurity.

Property 4: For all graphs G, $m(G) \leq m(K_N) = 1$, where N = number of nodes of G and K_N is the complete graph of N nodes.

This property says that, of all the graphs with n nodes, the complete graph has maximal tree impurity. Since it is reasonable to assume that tree impurity can be measured on a ratio scale, we can consider our measure to map to some number between 0 and 1, with the complete graph measuring 1, the worst impurity.

EXAMPLE 9.25

We can define a measure of tree impurity that satisfies all four properties:

$$m(G) = \frac{\text{Number of edges more than the spanning tree}}{\text{Maximal number of edges more than the spanning tree}}$$

In a complete graph, the number of edges is computed as

$$e = \frac{n(n-1)}{2}$$

where e is the number of edges in G and n is the number of nodes. Since the number of edges in a spanning tree is always $n - 1$, the maximum number of edges more than the spanning tree is

$$\frac{n(n-1)}{2} - (n-1) = \frac{(n-1)(n-2)}{2}$$

The actual number of edges more than the spanning subtree must be $e - n + 1$, so our formal equation for the tree impurity measure is

$$m(G) = \frac{2(e - n + 1)}{(n - 1)(n - 2)}$$

We leave it as an exercise to check that m satisfies the four properties.

Applying the measure m to the graphs in Figure 9.21, we find that $m(G_2) = 1/10$, $m(G_3) = 1/5$, and $m(G_4) = 1$.

The measure m of Example 9.25 has been applied to graphs of a system design, to characterize its tree impurity. The results suggest a relationship between tree impurity and poor design. Thus, m may be useful for quality assurance of designs. System designs should strive for a value of m near zero, but not at the expense of unnecessary duplication of modules. As with all measures of internal attributes, m should be viewed in context with other quality measures, rather than judged by itself.

It is important to note that the measure in Example 9.25 is not the only proposed measure of tree impurity; others appear in the Exercises at the end of this chapter.

9.3.5 Internal Reuse

In Chapter 8, we mentioned the need to consider and measure reuse when looking at system size. In that context, we viewed reuse as the proportion of a system that had been constructed outside of or external to the project. In this chapter, we consider reuse in a different sense. We call *internal reuse* the extent to which modules within a product are used multiple times within the same product.

This informal definition leads to a more formally defined measure. Consider the graph in Figure 9.22, where each node represents a module,

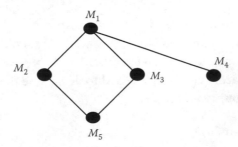

FIGURE 9.22   Example graph.

and two nodes are connected by an edge if one module calls the other. The graph shows that module M_5 is reused, in the sense that it is called by both M_2 and M_3. However, the graph model does not give us information about how many times a particular module calls another module. Thus, M_3 is not shown to be reused at all, but it may in fact be called several times by module M_1. If we think of module reuse solely as whether one module is called by another, then this graph model is sufficient.

Suppose we want to define a measure of this type of reuse. The first two properties for a tree density measure should also apply to a reuse measure. Similarly, property 4 seems reasonable, but we must drop the provision that $m(K_N) = 1$. However, property 3 is not applicable, but its converse may be a desirable property.

Yin and Winchester have proposed a simple measure of internal reuse, r, which satisfies properties 1, 2, and 4 plus the converse of 3. Calling the *system design measure*, it is defined by

$$r(G) = e - n + 1$$

where G has e edges and n nodes (Yin and Winchester 1978). Thus, the design measure is equal to the number of edges additional to the spanning subtree of G.

EXAMPLE 9.26

Applying this reuse measure to the graphs in Figure 9.21, we find that

$$r(G_1) = 0, \ r(G_2) = 1, \ r(G_3) = 2, \ r(G_4) = 1, \ r(G_5) = 3, \ r(G_6) = 6$$

The reuse measure is crude; not only does it take no account of possible different calls from the same module, it also takes no account of the size of the reused components. However, it gives an idea of the general level of internal reuse in a system design. Along with morphology measures and tree impurity, it helps to paint an overall quantifiable picture of the system structure. Empirical evidence is needed to determine the optimal balance between tree impurity and reuse.

9.3.6 Information Flow

Much of our discussion so far has been concerned with the notion of *information flow* between modules. For example, each type of coupling corresponds to a particular type of information flowing through the module. Researchers have attempted to quantify other aspects of information flow, including

- The total level of information flow *through a system*, where the modules are viewed as the atomic components (an intermodular attribute)

- The total level of information flow *between individual modules and the rest of the system* (an intramodular attribute)

Let us examine Henry and Kafura's information flow measure, a well-known approach to measuring the total level of information flow between individual modules and the rest of a system (Henry and Kafura 1981). To understand the measurement, consider the way in which data move through a system. We say a *local direct flow* exists if either

1. A module invokes a second module and passes information to it, or

2. The invoked module returns a result to the caller

Similarly, we say that a *local indirect flow* exists if the invoked module returns information that is subsequently passed to a second invoked module. A *global flow* exists if information flows from one module to another via a global data structure.

Using these notions, we can describe two particular attributes of the information flow. The *fan-in* of a module M is the number of local flows that terminate at M, plus the number of data structures from which information is retrieved by M. Similarly, the *fan-out* of a module M is the number of local flows that emanate from M, plus the number of data structures that are updated by M.

Based on these concepts, Henry and Kafura measure information flow "complexity" as

$$\text{Information flow complexity}(M) = \text{length}(M) \times ((\text{fan-in}(M) \times (\text{fan-out}(M))^2$$

Figure 9.23 shows an example of how this measure is calculated from a design's modular structure.

Briand, Morasca, and Basili evaluated the Henry–Kafura measure to see if it satisfies the properties of a structural complexity measure that are described in Section 9.1.1 (Briand et al. 1996). They found that properties

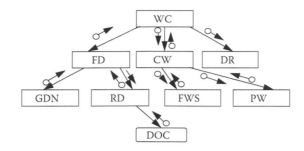

Module	Fan-in	Fan-out	$[(\text{Fan-in})(\text{Fan-out})]^2$	Length	"Complexity"
WC	2	2	16	30	480
FD	2	2	16	11	176
CW	3	3	27	40	1080
DR	1	0	0	23	0
GDN	0	1	0	14	0
RD	2	1	4	28	112
FWS	1	1	1	46	46
PW	1	1	1	29	29

FIGURE 9.23 Calculating the Henry and Kafura measure of information flow complexity.

1 through 4 are satisfied, but property 5, additivity, is violated due to the exponent in the equation and hybrid nature of the measure—it uses the product of fan-in and fan-out. Measured individually and without the exponent, fan-in and fan-out satisfy all of the complexity properties, as well as the coupling properties described in Section 9.1.3. Thus, fan-in and fan-out are often used as coupling measures.

9.3.7 Information Flow: Test Coverage Measures

We have discussed the use of control flow in structural testing, noting the utility of measuring the minimum number of test cases required to satisfy a given criterion. Now, we look at the role of data flow, or of a combination of data and control flow.

Most data-flow testing strategies focus on the program paths that link the *definition* and *use* of variables. Such a path is called a *du-path*. We distinguish between

- Variable uses within computations (*c-uses*)

- Variable uses within predicates or decisions (*p-uses*)

Figure 9.24 illustrates du-paths and describes the *c*-uses, *p*-uses, and definitions for each of the basic blocks of code.

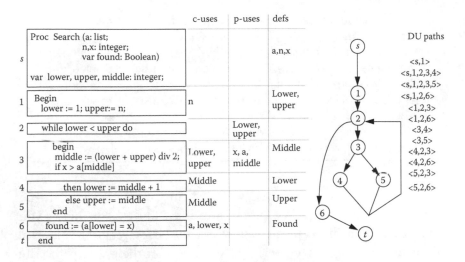

FIGURE 9.24 Example program and its du-paths.

The most stringent and theoretically effective data-flow strategy is to find enough test cases so that every du-path lies on at least one program path executed by the test cases; this is the *all du-paths* criterion. Weaker testing criteria involve *all c-uses, all p-uses, all defs*, and *all uses*.

Rapps and Weyuker note that, although the *all du-paths* criterion is the most discriminating, it requires a potentially exponential number of test cases (Rapps and Weyuker 1985). Specifically, if t is the number of conditional transfers in a program, then in the worst case there are 2^t du-paths. Despite this characteristic, empirical evidence suggests that du-paths testing is feasible in practice, and that the minimal number of paths P required for *all du-path* testing is a very useful measure for quality assurance purposes.

EXAMPLE 9.27

The worst-case scenario for the *all du-paths* criterion in Figure 9.24 is eight tests. Yet, in this case, P is 3, because just three paths can cover the set of du-paths:

<s,1,2,3,4,2,3,4,2,6,t>, <s,1,2,3,5,2,3,5,2,6,t>, <s,1,2,6,t>

EXAMPLE 9.28

An empirical study evaluated the the practicability of the *all du-paths* testing criterion (Bieman and Schultz 1992). The study looked at a commercial system consisting of 143 modules. For each module, they computed P, the minimum number of test paths required to satisfy *all du-paths*. Bieman and Schultz found that, for 81% of all modules, *P* was less than 11, and in only one module was *P* prohibitively large. Thus, the strategy seems to be practical for almost all modules. Moreover, Bieman and Schultz noted that the module with excessively high *P* was known to be the "rogue" module of the system. Thus, *P* may be a useful quality assurance measure, especially when used with outlier analysis as discussed in Chapter 6. To assure quality, we may want to review (and, if necessary, rewrite) those few modules with infeasible values of *P*, since it is likely that they are overly complex.

The *prime path coverage* criteria described in Section 9.2.3 actually subsumes the all du-paths criteria (Ammann and Offutt 2008). Thus, if your set of test cases achieves prime path coverage, the test set also achieves all of the data-flow testing criteria. The advantage of the prime path criteria is that you can identify the test requirements needed to achieve prime path coverage without any dataflow analysis—you only need to analyze the control flow graph. However, prime path coverage may require testing many additional requirements, including the testing of paths that do not involve related definitions and uses.

9.4 OBJECT-ORIENTED STRUCTURAL ATTRIBUTES AND MEASURES

Object-orientation has become a predominant way to structure software implementations, designs, and requirements. Object-oriented designs and implementations are built using classes, where each class defines a set of objects that may be instantiated. Classes contain methods that correspond to functions in procedural languages like C. Classes may be abstract because all methods may not be implemented. A system may be designed using the Java interface construct, which does not include implementations. Classes are generally part of a hierarchy of packages or namespaces, depending on the language. There are inheritance relations between classes and between interfaces. The use of dynamic binding, generics (Java), and templates (C++) allows names in programs to be bound to a wide variety of objects at run time or compile time. All of these constructs enhance the expressability of developers but can make designs and implementations more complicated.

Various tools and techniques have emerged to help manage the intricacies involved in developing and describing object-oriented systems. For example, the object management group (OMG) defined the unified modeling language (UML), which includes a set of diagram types for modeling object-oriented systems at various levels of abstraction to describe the structure, behavior, and interactions of a system. The commonly used UML diagram types include the following:

- *Class diagrams*: Model each class and its connections to other classes.

- *Object diagrams*: Model a configuration of run-time objects.

- *Activity diagrams*: Model the steps taken for a system to complete a task.

- *State machine diagrams*: Model of finite-state machine representations.

- *Use case diagrams*: Model external actors, the "use cases" that they take part in, and the dependencies between use cases.

- *Sequence diagrams*: Model the sequences of messages passed between objects in order to complete a task.

All of these diagram types can be treated as graphs with nodes and edges. Thus, we can apply the graph analysis described earlier in this chapter. For example, activity diagrams closely resemble control flow graphs. The UML diagrams have labels on their edges, and the nodes have content of various types. A class diagram node represents a class and contains a name, a set of attributes or instance variables, and a set of methods. Each class diagram component uses a different syntax. Class diagram edges represent inheritance or implements relationships, associations, or use dependencies. Associations may be labeled with names and multiplicity values. The UML diagrams are thus information-rich.

Different object-oriented languages have different constructs with different semantics for implementing similar structures. For example, Java generic classes support parameterized types in a manner that is different from C++ templates. While C++ supports multiple inheritance, Java does not. Java includes interfaces with inheritance hierarchies, while C++ does not. While C++ uses the *friend* construct to allow access to encapsulated entities, Java does not. The multitude of ways that object-oriented entities are defined and connected affect structural attributes can and should be measured. In the following sections, we will examine the various options for measuring key attributes.

9.4.1 Measuring Coupling in Object-Oriented Systems

Briand, Daly, and Wüst developed a framework for the measurement and analysis of coupling in object-oriented systems (Briand et al. 1999b). The framework supports a variety of perspectives on the notion of coupling, as well as mechanisms that couple object-oriented entities. Many object-oriented connections tend to be persistent—the connected entities can remain connected between method invocations, and may persist over the lifetime of an object. Persistent connections include objects coupled through associations that are implemented using instance or state variables. Generalization/specialization relations that are implemented using inheritance are persistent. There are also persistent connections between language-defined types, user-defined types, and classes.

Other connections may be transient. For example, a method call connects the calling entity (the client) with the called entity (server), but only while the server is active. The connection between a method body and the method's actual parameters lasts only as long as the method is active. When the method is called again, it is likely to be connected to different objects of the same type.

In addition to the coupling properties in Section 9.1.3, several orthogonal coupling properties can help to evaluate coupling measures (Briand et al. 1999b):

- Type: What kinds of entities are coupled?

- Strength: How many connections of a particular kind?

- Import or export: Are the connections import and/or export?

- Indirect: Is indirect coupling measured?

- Inheritance: Are connections to or from inherited entities counted?

- Domain: Are the measures used to indicate the coupling of individual attributes, methods, classes, sets of classes (e.g., packages), or the system as a whole?

Briand, Daly, and Wüst evaluated 30 different class-coupling measures in terms of these properties (Briand et al. 1999b). Most of the measures satisfied the coupling properties in Section 9.1.3. However, eight of the measures violated one or more coupling property. There is clearly a wide variety of ways to measure object-oriented coupling. However, most of the published coupling measures quantify coupling between classes, not objects. Thus, they are based on static connections—connections that can be identified at compile time, and do not account for run-time dependencies.

EXAMPLE 9.29

The *coupling between object classes (CBO)* is metric 4 in a commonly referenced suite of object-oriented metrics (Chidamber and Kemerer 1994). CBO is defined to be the number of other classes to which the class of interest is coupled. Briand et al. identified several problems with CBO (Briand et al. 1999b). One problem is that the treatment of inherited methods is ambiguous. Another problem is that CBO does not satisfy property 5, disjoint module additivity, one of the coupling properties given in Section 9.1.3. If class A is coupled to classes B and C, both classes B and C have CBO = 1, assuming no other coupling connections. Now assume that classes B and C are merged creating new class D. Class D will have CBO = 1, as it is only coupled to class A. Property 5 is not satisfied because the original classes A and B were disjoint, and the property requires that the coupling of the merged classes be the sum of their coupling. Property 5 would be satisfied if the measure counted the number of connections rather than the number of classes that the class of interest is coupled to.

EXAMPLE 9.30

Chidamber and Kemerer defined another coupling measure, *response for class (RFC)*, which is metric 5 in their suite of metrics (Chidamber and Kemerer 1994). RFC captures the size of the *response* set of a class. The response set of a class consists of all the methods called by local methods. RFC is the number of local methods plus the number of external methods called by local methods. Consider again the merging of classes B and C that use methods in class A, and assume that B and C are disjoint—they do not call each other's methods. The RFC of the merged class D will not be the sum of the RFC of B and C if classes B and C used one or more of the same methods in class A. If RFC was measured by counting the number of uses of external methods rather than counting the number of methods invoked, property 5 would be satisfied.

EXAMPLE 9.31

Message passing coupling (MPC) was introduced by Li and Henry and formalized by Briand, Daly, and Wüst (Li and Henry 1993, Briand et al. 1999b). The MPC value of a class is a count of the number of static invocations (call statements) of methods that are external to the class. MPC satisfies all of the properties in Section 9.1.3 including property 5.

EXAMPLE 9.32

Robert C. Martin defines two package-level coupling measures (Martin 2003). These indicate the coupling of a package to classes in other packages:

1. Afferent coupling (C_a): "The number of classes from other packages that depends on the classes within the subject package." Only "class relationships" are counted, "such as inheritance and association." C_a is really the *fan-out* (see Section 9.3.8) of a package.
2. Efferent coupling (C_e): "The number of classes in other packages that the classes in the subject package depend on" via class relationships. C_e is a form of the *fan-in* of a package.

These measures satisfy all of the properties in Section 9.1.3, including property 5 as long as a class can only belong to a single package.

Martin suggests that high efferent coupling makes a package *unstable* as it depends on too many imported classes. He defines the following *instability (I) metric:*

$$I = \frac{C_e}{C_a + C_e}$$

Thus, a package becomes unstable as its efferent (fan-in) increases. This is because the package depends on a relatively greater number of imported classes, which makes the package more prone to problems due to changes, faults, etc. in imported classes. Martin does not provide an empirical relation system or set of properties that we can use to evaluate the intuition behind the instability metric.

9.4.2 Measuring Cohesion in Object-Oriented Systems

As described in Section 9.1.4, the cohesion of a module depends on the extent that related entities are contained within a module. Depending on the level of abstraction of interest, a module may be a method, class, or package. *Method cohesion* is conceptually the same as the cohesion of an individual function or procedure. Thus, you can measure method cohesion using the techniques and measures described in Section 9.3.7.

Class cohesion is an intraclass attribute. It reflects the degree to which the parts of a class—methods, method calls, fields, and attributes belong together. A class with high cohesion has parts that belong together because they contribute to a unified purpose. Most of the proposed cohesion metrics are class-level metrics.

Similar to their framework for measuring coupling, Briand et al. (1998) also developed a framework for measuring and analyzing cohesion in object-oriented systems. The framework identifies options for structuring classes and their effects on cohesion metrics. For example, methods, attributes, and types (or Java interfaces) can be declared locally, imported, or inherited; the declarations can be inherited and implemented locally. Connections between entities can be through invocations, and references to common attributes. The connections can be direct or indirect. Indirect method connections can be through a series of method calls, or through inherited entities. A measure can be analyzed in terms of (1) its definition (is the definition complete, operational, and objective?), (2) its scale type (can it be used during early phases such as analysis and design?), (3) its dependence on a particular language, and (4) its validation, which can be analytical or empirical. The analytical evaluation focuses on a measure's consistency with the cohesion properties in Section 9.1.4. Only three of the 10 cohesion measures analyzed by Briand, Daly, and Wüst satisfy all of the cohesion properties.

EXAMPLE 9.33

Lack of cohesion metric (LCOM) is metric 6 in the Chidamber and Kemerer suite of metrics (Chidamber and Kemerer 1994). Here, the cohesion of a class is characterized by how closely the local methods are related to the local instance variables in the class. LCOM is defined as the number of disjoint (i.e., nonintersecting) sets of local methods. Two methods in a class intersect if they reference or modify common local instance variables. LCOM is an inverse cohesion measure; higher values imply lower cohesion. Briand, Daly, and Wüst found that LCOM violates property 1 of Section 9.1.4, as it is not normalized (Briand et al. 1998). Since LCOM indicates inverse cohesion, properties 2 through 4 are also not satisfied.

EXAMPLE 9.34

Tight class cohesion (TCC) and *loose class cohesion (LCC)* are based on connections between methods through instance variables (Bieman and Kang 1995). Two or more methods have a *direct connection* if they read or write to the same instance variable. Methods may also have an *indirect connection* if one method uses one or more instance variables directly and the other uses the instance variable indirectly by calling another method that uses the same instance variable. *TCC* is based on the relative number of direct connections:

$$TCC(C) = NDC(C)/NP(C)$$

where $NDC(C)$ is the number of direct connections in class C and $NP(C)$ is the maximum number of possible connections. LCC is based on the relative number of direct and indirect connections:

$$LCC(C) = (NDC(C) + NIC(C))/NP(C)$$

where $NIC(C)$ is the number of indirect connections. The measures do not include constructor and destructor methods in the computation, since they tend to initialize and free all instance variables and will thus artificially increase the measured cohesion. Both TCC and LCC satisfy all four cohesion properties in Section 9.1.4.

EXAMPLE 9.35

Ratio of cohesive interactions (*RCI*) is defined in terms of *cohesive interactions* (*CIs*), which include interactions between public data declarations and interactions between method parameters and return types in public method interfaces (Briand et al. 1998). *RCI* is the relative number of *CIs*:

$$RCI(C) = NCI(C)/NPCI(C)$$

where $NCI(C)$ is the number of actual *CIs* in class C and $NPCI(C)$ is the maximum possible number of *CIs*. *RCI* satisfies all four cohesion properties in Section 9.1.4.

Package cohesion concerns the degree to which the elements of a package—classes and interfaces—belong together.

EXAMPLE 9.36

Robert C. Martin defines the cohesion of a package in a manner similar to that of class cohesion (Martin 2003):

Package relational cohesion $RC(P) = (R(P) + 1)/N(P)$

where $R(P)$ is the number of relations between classes and interfaces in a package and $N(P)$ is the number of classes and interfaces in the package. A one is added to the numerator so that a package with one class or interface will have $H = 1$. H does not satisfy property 1 of the cohesion properties in Section 9.1.4 because H is not normalized between zero and one. Property 1 can easily be satisfied by changing the denominator to the number of possible relations, which is $N(P) \times (N(P) - 1)$, assuming a maximum of one relation between two classes or interfaces. *H* also does not satisfy property 2 of the cohesion properties since *H* cannot be zero. However, it can approach zero as more unrelated classes are added. Thus, we revise *package relational cohesion* as follows:

$$RC'(P) = (R(P) + 1)/NP(P)$$

where *NP(P)* is the number of possible relations between classes and interfaces in the package

Martin's package relational cohesion measure captures the notion of cohesion in a manner very similar to that of class cohesion. However, this model may not reflect the desired structuring for a package. Interdependency may not be the primary reason for placing modules into the same package. We may want modules that will often be reused by the same clients to reside in the same package. Thus, we may include an analysis of the context in which a package is used to evaluate package coupling (Ponisio and Nierstrasz 2006, Zhou et al. 2008).

In structuring a package, we may really seek *logical cohesion* as defined by Yourdon and Constantine's as an ordinal scale measure and is described as a weak form of cohesion (Yourdon and Constantine 1979). Classes and interfaces are placed in the same package because they perform similar functions, but with different properties and features. Another concern for evaluating package cohesion is determining the package boundaries. For example, in Java, packages are arranged in a hierarchy that reflects the directory structure of the program files—classes and interfaces. We could evaluate package cohesion in terms of a local directory, or include containing directories. The choice would depend on the goal for measurement.

9.4.3 Object-Oriented Length Measures

Generally, length measures indicate the distance from one element to another. In object-oriented systems, distances depend on the perspective and the model representing an appropriate view of the system. One common application of distance measures involves inheritance trees as depicted in UML class diagrams.

EXAMPLE 9.36

The *depth of inheritance tree (DIT)* is metric 3 in the Chidamber and Kemerer suite of object-oriented metrics (Chidamber and Kemerer 1994). Inheritance in a class diagram is represented as a hierarchy or tree of classes. The nodes in the tree represent classes, and for each such class, the DIT metric is the length of the maximum path from the node to the root of the tree. Chidamber and Kemerer claim that DIT is a measure of how many ancestor classes can potentially affect this class. This claim could only pertain to effects due to inheritance relations, and would not be accurate due to multiple inheritance in C++ or the use of hierarchies of interfaces in Java (which were unknown in 1994).

The DIT metric tells us how deep a class is in an inheritance hierarchy. We also learn something about inheritance by determining the distribution of DIT values for all classes in a system, subsystem, and package.

EXAMPLE 9.37

Bieman and Zhao studied inheritance in 19 C++ systems containing 2744 classes (Bieman and Zhao 1995). The systems included language tools, GUI tools, thread software, and other systems. They found that the median DIT of the classes in 11 of the systems was either 0 or 1. Classes in the studied GUI tools used inheritance the most with a mean class DIH of 3.46 and median class DIH values ranging from 0 to 7. The maximum measured DIH for all GUI tool classes was 10.

We can also measure the distance between connected objects following other connections in class diagrams. Distances between classes through associations can tell us about the potential configurations of objects that may be built at run time. Distances in sequence diagrams indicate the number of object connections required to complete a task requiring collaboration between objects.

9.4.4 Object-Oriented Reuse Measurement

One of the key benefits of object-oriented development is its support for reuse through data abstraction, inheritance, encapsulation, etc. This support helps developers to reuse existing software components in several ways. Depending on the development language, they can reuse existing packages and classes in a *verbatim* fashion without changes. Developers can also reuse existing packages and classes as well as interfaces, types, generics, and templates in a *leveraged* fashion by overriding and overloading inherited methods, by implementing interfaces, instantiating generic classes or templates. Measurement of reuse involves an analysis of the structures and models used to design and implement an object-oriented system. In addition, you can measure reuse from one or both of two perspectives: (1) *client perspective:* the perspective of a new system or system component that can potentially reuse existing components, and (2) *server perspective:* the perspective of the existing components that may potentially be reused, for example, a component library or package (Bieman and Karunanithi 1995).

From the client perspective, the potential reuse measures include the number of direct and indirect server classes and interfaces reused. An indirect server class would include the classes that direct servers use either directly or indirectly—this is essentially the number of ancestor classes in an inheritance hierarchy. Another aspect of reuse involves the structure of the connections between clients and servers. To quantify this structure, we can measure the number and length of paths through a UML diagram that connects a client to indirect servers.

From the server perspective, we are concerned with the way a particular entity is being reused by clients.

EXAMPLE 9.38

The *number of children (NOC)* is metric 3 in the Chidamber and Kemerer suite of object-oriented metrics (Chidamber and Kemerer 1994). This metric relates to a node (class or interface) in an inheritance tree or UML class diagram. NOC is computed by counting the number of immediate successors (subclasses or subinterfaces) of a class or interface. NOC is a direct server reuse measure.

The NOC measure defined in Example 9.38 only gives us partial information about the reuse of a class or interface. We really learn how useful a server class is by taking the relevant indirect measure, which is the number of descendants via inheritance in the inheritance tree or UML class diagram.

9.4.5 Design Pattern Use

Object-oriented design patterns describe commonly used solutions to recurring software design problems. The book by Gamma, Helm,

Johnson, and Vlissides describes 23 creational, structural, and behavioral patterns (Gamma et al. 1994). This book was followed by numerous other design pattern books, papers, websites, etc. When used, design patterns impact the structure of a system at various levels of abstraction. An architectural pattern such as model–view–controller imparts an overall structure to a system where the functional (model) portion of the system is separated from the user interface (views and controllers) portion of the system. Other patterns, such as an adaptor, create a *microarchitecture* in a system. It affects how a cluster of classes and associated objects are arranged.

The use of design patterns provides a way to understand the structure of a system and the evolution of system structure.

EXAMPLE 9.39

Izurieta and Bieman studied the evolution and potential decay of three open-sources systems (Izurieta and Bieman 2013). They identified design pattern realizations in early versions and determined the effects of changes on the functionality, adaptability, and testability of the design pattern code. The changes did not break the functional or structural integrity of the patterns. However, the changes did introduce *grime*, which is nonpattern-related code that can obscure the pattern. The results show that the introduced grime increased the dependencies between pattern components causing an increase in both coupling and the number of test requirements.

The commonly used design structures may actually be detrimental. They may make it more difficult to test and adapt a system.

EXAMPLE 9.40

Brown, Malveau, McCormick, and Mowbray describe numerous object-oriented *antipatterns*, which are "a commonly occurring solution to a problem that generates decidedly negative consequences" (Brown et al. 1998). The *Swiss Army Knife* (also known as the *kitchen sink*) is one antipattern that is identified by an overly complicated class interface with many method signatures that support as many options as possible. These antipatterns "are prevalent in commercial software interfaces" and can be difficult to understand, maintain, and test.

EXAMPLE 9.41

Several antipatterns reduce the testability of a system. For example, the *concurrent use* antipattern occurs when transitive dependencies occur due to an inheritance hierarchy creating several paths to a class (Baudry et al. 2002). Another testing antipattern is called the *self-use relationship,* which occurs when classes reference themselves through transitive use dependencies (Baudry and Sunye 2004).

Izurieta and Bieman found that instances of the Swiss Army Knife, concurrent use, and self-use relationship design antipatterns appeared in software implementations as they age (Izurieta and Bieman 2008, Izurieta and Bieman 2013).

To measure properties in software relevant to design patterns, you must be able to identify them in a design or code. Izurieta and Bieman used a very lightweight method; they searched code and documentation for known pattern names and then confirmed that the patterns were genuine through inspection. More sophisticated pattern identification methods may employ a variety of approaches, including explanation constraint programming (Gueheneuc and Antoniol 2008) and feature-based searches (Rasool and Mader 2011).

9.5 NO SINGLE OVERALL "SOFTWARE COMPLEXITY" MEASURE

We have discussed the need for viewing overall "software complexity" as the combination of several attributes, and we have shown the importance of examining each attribute separately, so that we can understand exactly what it is that is responsible for the overall "complexity." Nevertheless, practitioners and researchers alike find great appeal in generating a single, comprehensive measure to express overall "software complexity." The single measure is expected to have powerful properties, being an indicator of such diverse notions as comprehensibility, correctness, maintainability, reliability, testability, and ease of implementation. Thus, a high value for "complexity" should indicate low comprehensibility, low reliability, etc. Sometimes, this measure is called a "quality" measure, as it purportedly relates to external product attributes such as reliability and maintainability. Here, a high "quality" measure suggests a low-quality product.

The danger in attempting to find measures to characterize a large collection of different attributes is that often the measures address conflicting goals, counter to the representational theory of measurement.

EXAMPLE 9.42

Suppose we define a measure, M, that characterizes the quality of people. If M existed, it would have to satisfy at least the following conditions:

$M(A) > M(B)$ whenever A is stronger than B and

$M(A) > M(B)$ whenever A is more intelligent than B

The fact that some highly intelligent people are very weak physically ensures that no M can satisfy both these properties.

Example 9.42 illustrates clearly why single-valued measures of "software complexity" are doomed to failure. Consider, for example, the list of properties in Table 9.2. Proposed by Weyuker, they are described as properties that should be satisfied by any good "software complexity" metric (Weyuker 1988). In Table 9.2, P, Q, and R denote any program blocks.

Zuse has used the representational theory of measurement to prove that some of these properties are contradictory; they can never all be satisfied by a single-valued measure (Zuse 1992).

TABLE 9.2 Weyuker's Properties for Any Software Complexity Metric M

Property 1: There are programs P and Q for which $M(P) \neq M(Q)$.

Property 2: If c is a nonnegative number, then there are only finitely many programs P for which $M(P) = c$.

Property 3: There are distinct programs P and Q for which $M(P) = M(Q)$.

Property 4: There are functionally equivalent programs P and Q for which $M(P) \neq M(Q)$.

Property 5: For any program bodies P and Q, we have $M(P) \leq M(P;Q)$ and $M(Q) \leq M(P;Q)$.

Property 6: There exist program bodies P, Q, and R such that $M(P) = M(Q)$ and $M(P;R) \neq M(Q;R)$.

Property 7: There are program bodies P and Q such that Q is formed by permuting the order of the statements of P and $M(P) \neq M(Q)$.

Property 8: If P is a renaming of Q, then $M(P) = M(Q)$.

Property 9: There exist program bodies P and Q such that $M(P) + M(Q) < M(P;Q)$.

EXAMPLE 9.43

We can see intuitively why properties 5 and 6 in Table 9.2 are mutually incompatible. Property 5 asserts that adding code to a program cannot decrease its complexity. This property reflects the view that program *size* is a key factor in its complexity. We can also conclude from property 5 that low comprehensibility is *not* a key factor in complexity. This statement is made because it is widely believed that, in certain cases, we can understand a program *more* easily as we see more of it. Thus, while a complexity measure M based primarily on size should satisfy property 5, a complexity measure M based primarily on comprehensibility cannot satisfy property 5.

On the other hand, property 6 asserts that we can find two program bodies of equal complexity, which, when separately concatenated to a same third program, yield programs of different complexity. Clearly, this property has much to do with comprehensibility and little to do with size.

Thus, properties 5 and 6 are relevant for very different, and incompatible, views of complexity. They cannot both be satisfied by a single measure that captures notions of size *and* low comprehensibility. The above argument is not formal. However, Zuse reinforces the incompatibility; he proves formally that while property 5 explicitly requires the ratio scale for M, property 6 explicitly excludes the ratio scale.

Cherniavsky and Smith also offer a critique of Weyuker's properties (Cherniavsky and Smith 1991). They define a code-based "metric" that satisfies all of Weyuker's properties but, as they rightly claim, is not a sensible measure of complexity. They conclude that axiomatic approaches may not work.

Cherniavsky and Smith have correctly found a problem with the Weyuker axioms. But they present no justification for their conclusion about axiomatic approaches in general. They readily accept that Weyuker's properties are not claimed to be complete. But what they fail to observe is that Weyuker did not propose that the axioms were *sufficient*; she only proposed that they were necessary. The Cherniavsky and Smith "metric" is not a real measure in the sense of measurement theory, since it does not capture a specific attribute. Therefore, showing that the "metric" satisfies a set of axioms necessary for any measure proves nothing about real measures.

These problems could have been avoided by heeding a basic principle of measurement theory: defining a numerical mapping does not in itself constitute measurement. Software engineers often use the word "metric" for any number extracted from a software entity. But while every measure is a metric, the converse is certainly not true. The confusion in analyses such as Cherniavsky and Smith's or Weyuker's arises from wrongly equating these two concepts, and from not viewing the problem from a measurement theory perspective.

The approach in this chapter is to identify necessary, but not sufficient properties for notions of complexity, coupling, cohesion, and length in Section 9.1, as proposed by Briand et al. (1996). We do not claim that measures of any of these attributes capture the overall "software complexity" of a system. No single measure can, with a single number, indicate how difficult it will be to understand, maintain, and test a software system. However, narrowly defined measures can quantify specific attributes.

9.6 SUMMARY

It is widely believed that a well-designed software product is characterized largely by its internal structure. Indeed, the rationale behind most software engineering methods is to ensure that software products are built with certain desirable structural attributes. Thus, it is important to know how to recognize and measure these attributes, since they may provide important indicators of key external attributes, such as maintainability, testability, reusability, and even reliability.

We have described how to perform measurements of what are generally believed to be key internal structural attributes, including structural complexity, coupling, cohesion, length, modularity, tree impurity, reuse, and information flow. These attributes are relevant for design documents and their models as well as code. Indeed, knowing these attributes, we can identify components that are likely to be difficult to implement, test, and maintain.

We looked in detail at control flow attributes of function, procedure, and method bodies. We showed how a program unit is built up in a unique way from the so-called prime structures, which are the building blocks of function, procedure, and method bodies. A program unit body can therefore be characterized objectively in terms of its prime decomposition, which may be automatically computed. Many measures of internal attributes of programs (including test coverage measures) can be computed easily once we know a program unit's prime decomposition. These measures are called hierarchical. The prime decomposition is a definitive representation of the control structure of a flowgraph; it can also be used as the basis for optimal restructuring of code, and hence as a *reverse engineering* tool.

Object-orientation affects the structure of most elements of a software implementation, design, and requirements. In particular, we examined how coupling and cohesion can be measured in object-oriented systems at various levels of abstraction. We applied specific coupling and cohesion measurement properties, and examined the measurement of inheritance, object-oriented reuse, and measurement involving design pattern use and misuse.

It is unrealistic to expect that general complexity measures (of either the code or system design) will be good predictors of many different attributes. Complexity is an intuitive attribute that includes the specific internal attributes discussed in this chapter. Rather than seek a single measure, we should identify specific attributes of interest and obtain accurate measures of them; in combination, we can then paint an overall picture of complexity.

EXERCISES

1. Show that there are valid flowgraphs in which not every node lies on a *simple* path from start to stop. (*Hint*: Consider one of the flowgraphs in Figure 9.3.)

2. The following sets of edges form flowgraphs on five nodes, labeled from 1 to 5. Draw the flowgraphs using the conventions of Figure 9.2. Which of the flowgraphs are primes? For those that are not, represent the flowgraphs by expressions using sequencing and nesting applied to the flowgraphs of Figure 9.2.

   a. Edges $(1, 2)$, $(2, 3)$, $(3, 4)$, $(4, 3)$, $(3, 5)$

   b. Edges $(1, 2)$, $(2, 3)$, $(1, 4)$, $(3, 4)$, $(4, 3)$, $(3, 5)$

   c. Edges $(1, 2)$, $(2, 3)$, $(3, 2)$, $(3, 4)$, $(4, 3)$, $(2, 5)$

   d. Edges $(1, 2)$, $(2, 3)$, $(2, 5)$, $(3, 4)$, $(4, 1)$

3. Draw the decomposition trees for each flowgraph in Exercise 2.

4. Draw three flowgraphs that have equal cyclomatic number but which seem intuitively to rank differently in terms of structural complexity.

What actual structural attributes are contributing to "complexity" in your examples? Find hierarchical measures that capture these attributes. (*Hint*: Example 9.7 and the largest prime in Figure 9.11 are likely candidates.)

5. For each hierarchical measure in Figure 9.11, calculate the value of the measure for the flowgraph in Figure 9.15.

6. Look at the program in Figure 9.14. How many feasible paths are there for this program? Define a set of test cases that give you 100% coverage of all the feasible paths.

7. Use the tables in Appendix A.2. to compute the branch coverage measure for the flowgraph in Figure 9.15.

8. Show that the measure m defined in Example 9.25 satisfies the four properties of tree impurity. Compute the measure m for the graphs G_1, G_5, and G_6 in Figure 9.21.

9. A measure of tree impurity proposed by Ince and Hekmatpour is given by Ince and Hekmatpour (1988):

$$m(G) = \sum_{n \in G} (id(n))^2$$

where n represents a node of G, and $id(n)$ is the in-degree of node n. Show that this measure fails to satisfy properties 1 and 3.

10. Yin and Winchester define a family of system design measures, $C_i = e_i - n_i + 1$, where e_i is the number of arcs up to level i, and n_i is the number of nodes up to level i. Taking i to be the last level, they define tree impurity m_2 of the whole design G by $m_2(G) = e - n + 1$, where e is the total number of edges and n is the total number of nodes of G. Show that properties 1 and 2 are satisfied by m_2 but not property 3 (Yin and Winchester 1978).

11. Explain why the converse of property 3 in Section 9.3.4 is reasonable for private reuse. Show that the Yin and Winchester measure r satisfies the four proposed properties.

12. Compute the Yin and Winchester measure for the graph in Figure 9.20.

13. What is the rationale behind squaring the product of fan-in and fan-out in the Henry–Kafura (Henry and Kafura 1981) measure? (*Hint:* Examine the effect of squaring in the example in Figure 9.23.)

14. Consider the following proposed measure of object-oriented *design pattern realization complexity*: $DPRC(p) = NC + NI + NA$, where NC is the number of classes in pattern realization p, NI is the number of inheritance links in pattern realization p, and NA is the number of associations in pattern realization p. Using simple UML class diagrams representing design pattern realizations, demonstrate why this proposal may be unreasonable.

15. Consider a new complexity measure, *Cnew*. The *Cnew* complexity of a module m is defined as follows: $Cnew = length_m \times (fan\text{-}in_m - 0.5 \times fan\text{-}out_m)$. The rationale is that high fan-out aids understanding of a module because fan-out is associated with delegation of responsibility. However, fan-out does not lower the complexity as much as length and fan-in add to complexity. Therefore, fan-out is multiplied by the weight 0.5. The *Cnew* complexity of a system composed of modules 1, 2,…, n is defined as follows:

$$CnewSyst = \sum_{i=1}^{n} length_i \times \left(\sum_{i=1}^{n} fan\text{-}in_i - 0.5 \times \sum_{i=1}^{n} fan\text{-}out_i \right)$$

Show which of the Briand et al. properties for complexity measures (described in Section 9.1.1) are satisfied or dissatisfied by the *Cnew* and *CnewSyst* measures and why.

16. Below is a list of measures that can be applied to program bodies. In each case, determine which of the Weyuker properties hold:

- *LOC*
- Cyclomatic number
- Is-*D*-structured
- Henry–Kafura measure
- Function points

17. "A good design should exhibit high module cohesion and low module coupling." Briefly describe what you understand this assertion to mean.

18. McCabe's cyclomatic number is a classic example of a software metric. Which software entity and attribute do you believe it really measures?

19. Explain briefly the principles behind the statement coverage testing criterion.

20. Explain why exhaustive path testing is generally infeasible by giving an example using a flowgraph.

21. What is a prime flowgraph? Give examples of two prime flowgraphs that are not building blocks of the traditional *D*-structured programming, but that are (in a sense you should explain briefly) natural control constructs.

22. By considering the formal definition of *D*-structured programs in terms of prime flowgraphs, deduce that the following procedure is *D*-structured:

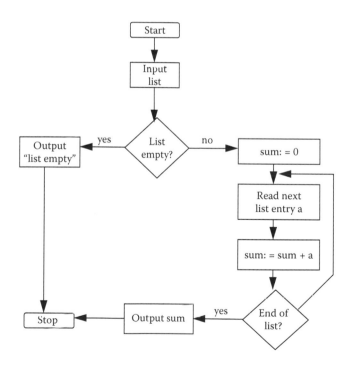

23. The following algorithm describes a simple control system. What can you say about the structuredness of this algorithm?

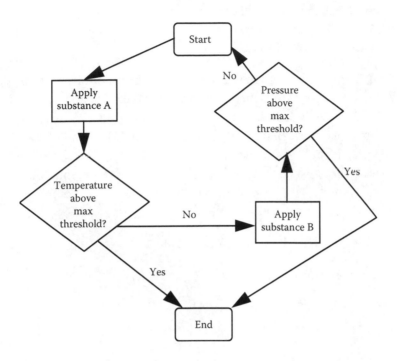

APPENDICES TO CHAPTER 9

A.1 McCabe's Testing Strategy

A.1.1 Background

In a *strongly connected graph G* for any nodes *x,y* there is a path from *x* to *y* and vice versa. For example, in the graph G of Figure 9.24 each path can be represented as a 6-tuple (vector with six components):

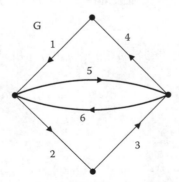

FIGURE 9.A.24   A strongly connected graph.

2.5in <1,2,3,4> = (1 1 1 1 0 0)
<1,5,6,2,3,4,1 > = (2 1 1 1 1 1)
<1,5,4> = (1 0 0 1 1 0), etc.

That is, the ith position in vector is the number of occurrences of edge i.

A *circuit* is a path which begins and ends at the same node, for example, <1,2,3,6,5,4>. A *cycle* is a circuit with no node (other than the starting node) included more than once, for example, <1,2,3,4> and <5,6>.

A path p is said to be a *linear combination* of paths $p_1, \ldots, p_n$ if there are integers $a_1, \ldots, a_n$ such that $p = \sum a_i p_i$ in the vector representation.

For example, path <1,2,3,4,5,6,2> is a linear combination of paths <1,2,3,4> and <5,6,2> since

$$(1\ 2\ 1\ 1\ 1\ 1) = (1\ 1\ 1\ 1\ 0\ 0) + (0\ 1\ 0\ 0\ 1\ 1)$$

As another example, let:

a = <1,2,3,4> = (1 1 1 1 0 0)
b = <5,6> = (0 0 0 0 1 1)
c = <1,5,4> = (1 0 0 1 1 0)
d = <2,3,6> = (0 1 1 0 0 1)

Then $a + b - c = d$ (*)

A set of paths is *linearly independent* if no path in the set is a linear combination of any other paths in the set. Thus {a,b,c} is linearly independent, but {a,b,c,d} is not by virtue of (*).

A *basis* set of cycles is a maximal linearly independent set of cycles. In a graph of e edges and n nodes the basis has $e - n + 1$ cycles. Although the size of the basis is invariant, its content is not. For example, for the graph G above:

$$\{a,b,c\},\{a,b,d\},\{b,c,d\},\{a,c,d\}$$

are different basis sets of cycles. *Every path is a linear combination of basis cycles.*

A.1.2 The Strategy

Any flowgraph can be transformed into a strongly connected graph by adding an edge from stop node to start node. Figure 9.25 shows how we transform a flowgraph to obtain the same graph as that in Figure 9.24.

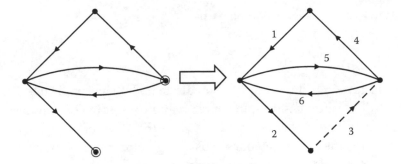

FIGURE 9.25   Transforming a flowgraph into a strongly connected graph.

McCabe's test strategy is based on choosing a basis set of cycles in the resulting graph G.

As we have seen above, the number of these, $v(G)$ – McCabe's "cyclomatic complexity" satisfies

$$v(G) = e - n + 1.$$

Since the original flowgraph has had one edge added, the formula for computing v (G) for an arbitrary flowgraph G with e edges and n nodes is

$$v(G) = e - n + 2.$$

The idea is that these cycles are "representative" of all the paths since every path is a linear combination of basis cycles. In the example, one possible basis set of cycles is

$$\{<4,1,2,3>, <6,5>, <6,2,3>\}.$$

Unfortunately these do not correspond to paths through the flowgraph (because of the artificial introduction of edge 3) which is what we need for a structural testing strategy. We thus have to derive the "smallest" associated paths. These are:

$$<4,1,2>, <6,5,6,2>, <6,2>.$$

Note that a different basis set of cycles, for example,

$$\{<6,5>, <6,2,3>, <4,1,5>\}$$

leads to a different set of testing paths:

$$< 6,5,6,2>, <6,2>, <4,1,5,6,2>.$$

In fact, we can show that if all predicate nodes have outdegree 2 then $v(G) = d + 1$ where d is the number of predicate nodes in G. For if there are p procedure nodes, then $n = p + d + 1$ since the nodes of G are the procedure nodes, the predicate nodes, plus a single stop node. Now, $e = p + 2d$ since each procedure node contributes 1 to the total number of edges and each predicate node contributes 2. Thus, $e - n + 2 = (p + 2d) - (p + d + 1) + 2 = d + 1$.

A.2 Computing Test Coverage Measures

	Measurement Values for Primes							
Test Strategy	P_1	D_0	D_1	C_n	D_2	D_3	D_4	L_2
All-path coverage	1	2	2	n	—	—	—	—
Branch coverage	1	2	2	n	1	1	1	2
Statement coverage	1	1	2	n	1	1	1	1

	Sequencing Function
Test Strategy	$F_1; \ldots F_n$
All-path coverage	$\prod_{i=1}^{n} \mu(F_i)$
Branch coverage	$max(\mu(F_1), \ldots, \mu(F_n))$
Statement coverage	$max(\mu(F_1), \ldots, \mu(F_n))$

	Nesting Function			
Test Strategy	$D_1(F_1,F_2)$	$C_n(F_1, \ldots, F_n)$	$D_0(F)$	$D_2(F)$
All – path coverage	$\mu(F_1) + \mu(F_2)$	$\sum_{i=1}^{n} \mu(F_i)$	$\mu(F) + 1$	—
Branch coverage	$\mu(F_1) + \mu(F_2)$	$\sum_{i=1}^{n} \mu(F_i)$	$\mu(F) + 1$	1
Statement coverage	$\mu(F_1) + \mu(F_2)$	$\sum_{i=1}^{n} \mu(F_i)$	$\mu(F)$	1

Test Strategy	$D_3(F)$	$D_4(F_1,F_2)$	$L_2(F_1,F_2)$
All-path coverage	—	—	—
Branch coverage	1	1	2
Statement coverage	1	1	1

FURTHER READING

Many of the measures introduced in this chapter rely on an understanding of graph theory. Wilson's book provides a standard and accessible reference to graph theory concepts and techniques.

Wilson R.I., *Introduction to Graph Theory*, 5th Edition, Prentice-Hall, Harlow, New York, 2010.

More detailed accounts of our particular approach to prime decomposition may be found in these references:

Fenton N.E. and Kaposi A.A., Metrics and software structure, *Journal of Information and Software Technology*, 29, 301–320, July 1987.
Fenton N.E. and Whitty R.W., Axiomatic approach to software metrication through program decomposition, *Computer Journal*, 29(4), 329–339, 1986.

A comprehensive treatment of the generalized theory of structuredness can be found in these publications:

Fenton N.E. and Hill G., *Systems Construction and Analysis: A Mathematical and Logical Approach*, McGraw-Hill, New York, 1992.
Fenton N.E., Whitty R.W., and Kaposi A.A., A generalized mathematical theory of structured programming, *Theoretical Computer Science*, 36, 145–171, 1985.

Van den Broek and van den Berg generalize the prime decomposition of flow-graphs by allowing arbitrary decomposition operations. They also describe a flowgraph-type model for functional programs and apply the theory of decomposition as described in this chapter.

Berg van den K.G. and Broek van den P.M., Static analysis of functional programs, *Information and Software Technology*, 37(4), 213–224, 1995.
Broek van den P.M. and Berg van den K.G., Generalised approach to software structure metrics, *Software Engineering Journal*, 10(2), 61–68, 1995.

The following papers discuss axiomatic-type properties that should be satisfied by so-called software complexity metrics. Melton and his colleagues restrict their discussion to control flow structure and adopt a measurement theory-type approach by defining a partial order on flowgraphs. The partial order preserves an intuitive notion of the relation "more complex than." Lakshmanan and colleagues also restrict their properties to control flow measures.

Lackshmanan K.B., Jayaprakesh S., and Sinha P.K., Properties of control-flow complexity measures, *IEEE Transactions on Software Engineering*, 17(12), 1289–1295, 1991.
Melton A.C., Bieman J.M., Baker A., and Gustafson D.A., Mathematical perspective of software measures research, *Software Engineering Journal*, 5(5), 246–254, 1990.

Weyuker E.J., Evaluating software complexity measures, *IEEE Transactions on Software Engineering, SE*, 14(9), 1357–1365, 1988.

Briand, Morasca, and Basili propose properties that any measure of well-defined notions of size, complexity, coupling, cohesion, and length should satisfy.

Briand L.C., Morasca S., and Basili V.R., Property-based software engineering measurement. *IEEE Transactions on Software Engineering*, 22(1), 68–86, January 1996.

Bieman and Ott define a range of intramodular measures of functional cohesion based on the notion of data slices. This paper is also especially interesting from our perspective, because the authors discuss the scale properties of their measures using measurement theory. Bieman and Kang derive a measure of cohesion at the design level focusing on a module's interface.

Bieman J.M. and Kang B.-K., Measuring design level cohesion, *IEEE Transactions on Software Engineering*, 24(2), 111–124, February 1998.
Bieman J.M. and Ott L.M., Measuring functional cohesion, *IEEE Transactions on Software Engineering*, 20(8), 644–657, 1994.

Briand, Daly, and Wüst develop comprehensive frameworks for measuring coupling and cohesion in object-oriented systems. Bieman and Kang define and apply an object-oriented cohesion measure that satisfies the requirements of the cohesion framework.

Bieman J.M. and Kang B.-K., Cohesion and reuse in an object-oriented system, *Proceedings of the ACM Symposium Software Reusability (SSR'95)*, Seattle, Washington, 259–262, 1995.
Briand L.C., Daly J.W., and Wüst J.K., Unified framework for coupling measurement in object-oriented systems, *IEEE Transactions on Software Engineering*, 25(1), 91–121, Jan/Feb 1999.
Briand L.C., Daly J.W., and Wüst J.K., Unified framework for cohesion measurement in object-oriented systems, *Empirical Software Engineering*, 3, 65–117, 1998.

Zuse provides a comprehensive review of almost every control structure measure appearing in the literature from the viewpoint of measurement theory.

Zuse H., *Software Complexity: Measures and Methods*, De Gruyter, Berlin, 1991.

Bertolino and Marre have written an excellent paper that explains the practical problems with the branch coverage metric and proposes an alternative metric for branch coverage. The theoretical minimum number of paths for this strategy is more closely aligned to minimum number of practically feasible paths. The paper also shows how the metric can be formulated using the prime decomposition approach.

Bertolino A. and Marre M., How many paths are needed for branch testing? *Journal of Systems and Software*, 35(2), 95–106, 1995.

Ammann and Offutt's book on testing provides a thorough account of many model-based testing strategies. This book gives a comprehensive account of both control-flow and data-flow-oriented test coverage criteria along with details concerning the specification and identification of test requirements and test paths. This book introduced the notion of prime path coverage discussed in this chapter.

Ammann P. and Offutt J., *Introduction to Software Testing*, Cambridge University Press, 2008.

Rapps and Weyuker provide an extensive account of data-flow testing strategies. An excellent account of these strategies may also be found in Bieman and Schultz.

Bieman J.M. and Schultz J.L., An empirical evaluation (and specification) of the all-du-paths testing criterion, *Software Engineering Journal*, 7(1), 43–51, 1992.
Rapps S. and Weyuker E.J., Selecting software test data using data flow information, *IEEE Transactions on Software Engineering*, 11(4), 367–375, 1985.

Measuring External Product Attributes

A PRINCIPAL OBJECTIVE OF SOFTWARE engineering is to improve the quality of software products. But quality, like beauty, is very much in the eyes of the beholder. In the philosophical debate about the meaning of software quality, proposed definitions include:

- Fitness for purpose

- Conformance to specification

- Degree of excellence

- Timeliness

However, from a measurement perspective, we must be able to define quality in terms of specific software product attributes of interest to the user. That is, we want to know how to measure the extent to which these attributes are present in our software products. This knowledge will enable us to specify (and set targets for) quality attributes in measurable form.

In Chapter 3, we defined external product attributes as those that can be measured only with respect to how the product relates to its environment. For example, if the product is software code, then its reliability (defined in terms of the probability of failure-free operation) is an external attribute; it is dependent on both the machine environment and the user. Whenever we think of software code as our product and we investigate an external attribute that is dependent on the user, we inevitably are dealing with an attribute synonymous with a particular view of quality (i.e., a *quality attribute*). Thus, it is no coincidence that the attributes considered in this chapter relate to some popular views of software quality.

In Chapter 9, we considered a range of internal attributes believed to affect quality in some way. Many practitioners and researchers measure and analyze internal attributes because they may be predictors of external attributes. There are two major advantages to doing so. First, the internal attributes are often available for measurement early in the life cycle, whereas external attributes are measurable only when the product is complete (or nearly so). Second, internal attributes are often easier to measure than external ones.

The objective of this chapter is to focus on a small number of especially important external attributes and consider how they may be measured. Where relevant, we indicate the relationships between external and internal attributes. We begin by considering several general software quality models, each of which proposes a specific set of quality attributes (and internal attributes) and their interrelationships. We use the models to identify key external attributes of interest, including *reliability, maintainability, usability,* and *security*. Given the increasing use of software in systems that are crucial to our life and health, software reliability is particularly important. Because reliability has received a great deal of scrutiny from practitioners and researchers, spawning a rich literature about its measurement and behavior, we postpone our discussion of it until Chapter 11, where we present a detailed account. In this chapter, we focus primarily on how usability, maintainability, and security may be measured.

10.1 MODELING SOFTWARE QUALITY

Because quality is really a composite of many characteristics, the notion of quality is usually captured in a model that depicts the composite characteristics and their relationships. Many of the models blur the distinction between internal and external attributes, making it difficult for us to understand exactly what quality is. Still, the models are useful in articulating what people think is important, and in identifying the commonalities of view. In this section, we look at some very general models of software quality that have gained acceptance within the software engineering community. By extracting from them several common external attributes of general interest, we show how the models and their derivatives may then be tailored for individual purpose.

In Chapter 1, we introduced the notion of describing quality by enumerating its component characteristics and their interrelationships. Figure 1.2 presented an example of such a quality model. Let us now take a closer look at this type of model to see how it has been used by industry and what we can learn from the results.

10.1.1 Early Models

Two early models described quality using a decomposition approach (McCall et al. 1977; Boehm et al. 1978). Figure 10.1 presents the Boehm et al. view of quality's components, while Figure 10.2 illustrates the McCall et al. view.

In models such as these, the model-builders focus on the final product (usually the executable code), and identify key attributes of quality from the user's perspective. These key attributes, called *quality factors*, are normally high-level external attributes like *reliability, usability,* and *maintainability.* But they may also include several attributes that arguably are internal, such as *testability* and *efficiency.* Each of the models assumes that the quality factors are still at too high a level to be meaningful or to be measurable directly. Hence, they are further decomposed into lower-level attributes called *quality criteria* or *quality subfactors.*

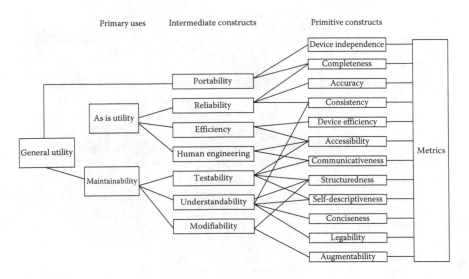

FIGURE 10.1 Boehm software quality model.

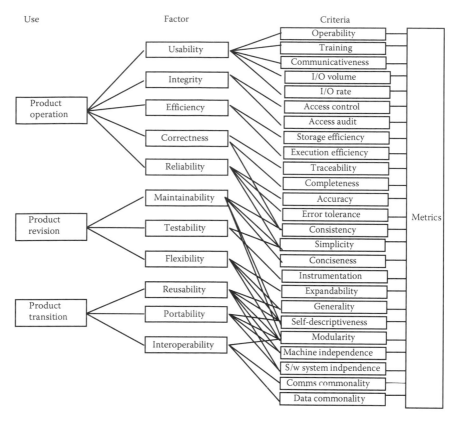

FIGURE 10.2   McCall software quality model.

EXAMPLE 10.1

In McCall's model, the factor *reliability* is composed of the criteria (or subfactors) *consistency, accuracy, error-tolerance,* and *simplicity.*

Sometimes the quality criteria are internal attributes, such as *structuredness* and *modularity,* reflecting the developers' belief that the internal attributes have an affect on the external quality attributes. A further level of decomposition is required, in which the quality criteria are associated with a set of low-level, directly measurable attributes (both product and process) called *quality metrics.* For instance, Figure 10.3 shows how maintainability can be described by three subfactors and four metrics, forming a complete decomposition. (This structure has been adapted from an IEEE standard for software quality metrics methodology, which uses the term *subfactor* rather than *criteria* (IEEE Standard 1061 2009).)

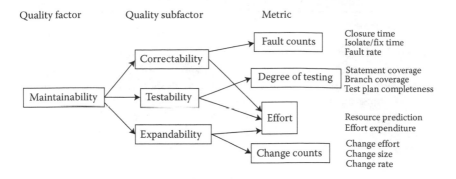

Quality factor   Quality subfactor   Metric

FIGURE 10.3   A decomposition of maintainability.

This presentation is helpful, as we may use it to monitor software quality in two different ways:

1. *The fixed model approach*: We assume that all important quality factors needed to monitor a project are a subset of those in a published model. To control and measure each attribute, we accept the model's associated criteria and metrics and, most importantly, the proposed relationships among factors, criteria, and metrics. Then, we use the data collected to determine the quality of the product.

2. *The "define your own quality model" approach*: We accept the general philosophy that quality is composed of many attributes, but we do not adopt a given model's characterization of quality. Instead, we meet with prospective users to reach a consensus on which quality attributes are important for a given product. Together, we decide on a decomposition (possibly guided by an existing model) in which we agree on specific measures for the lowest-level attributes (criteria) and specific relationships between them. Then, we measure the quality attributes objectively to see if they meet specified, quantified targets.

The Boehm and McCall models are typical of fixed quality models. Although it is beyond the scope of this book to provide a detailed and exhaustive description of fixed model approaches, we present a small picture of how such a model can be used.

EXAMPLE 10.2

The McCall model, depicted in Figure 10.2, includes 41 metrics to measure the 23 quality criteria generated from the quality factors. Measuring any factor requires us first to consider a checklist of conditions that may apply to the requirements (R), the design (D), and the implementation (I). The condition is designated "yes" or "no," depending on whether or not it is met. To see how the metrics and checklists are used, consider measuring the criterion *completeness* for the factor *correctness*. The checklist for *completeness* is:

1. Unambiguous references (input, function, output) [R,D,I].
2. All data references defined, computed, or obtained from external source [R,D,I].
3. All defined functions used [R,D,I].
4. All referenced functions defined [R,D,I].
5. All conditions and processing defined for each decision point [R,D,I].
6. All defined and referenced calling sequence parameters agree [D,I].
7. All problem reports resolved [R,D,I].
8. Design agrees with requirements [D].
9. Code agrees with design [I].

Notice that there are six conditions that apply to requirements, eight to design and eight to implementation. We can assign a 1 to a yes answer and 0 to a no, and we can compute the completeness metric in the following way to yield a measure that is a number between 0 and 1:

$$\frac{1}{3}\left(\frac{\text{Number of yes for R}}{6} + \frac{\text{Number of yes for D}}{8} + \frac{\text{Number of yes for I}}{8}\right)$$

Since the model tells us that *correctness* depends on *completeness, traceability,* and *consistency,* we can calculate analogous measures for the latter two. Then, the measure for *correctness* is the mean of their measures:

$$\text{Correctness} = \frac{x + y + z}{3}$$

That is, *x, y,* and *z* are the metrics for *completeness, traceability,* and *consistency,* respectively.*

---

* In this example, all of the factors are weighted the same. However, it is possible to use different weightings, so that the weights reflect importance, cost, or some other consideration important to the evaluator.

The McCall model was originally developed for the U.S. Air Force, and its use was promoted within the U.S. Department of Defense for evaluating software quality. But, many other standard (and competing) measures have been employed in the Department of Defense; no single set of measures has been adopted as a department-wide standard.

10.1.2 Define-Your-Own Models

Gilb, Kitchenham, and Walker pioneered the define-your-own-model approach (Gilb 1976, 1988; Kitchenham and Walker 1989). Gilb's method can be thought of as "design by measurable objectives"; it complements his philosophy of *evolutionary development*. The software engineer delivers the product incrementally to the user, based on the importance of the different kinds of functionality being provided. To assign priorities to the functions, the user identifies key software attributes in the specification. These attributes are described in measurable terms, so the user can determine whether measurable objectives (in addition to the functional objectives) have been met. Figure 10.4 illustrates an example of this approach. This simple but powerful technique can be used to good effect on projects of all sizes, and can be applied within agile processes.

10.1.3 ISO/IEC 9126-1 and ISO/IEC 25010 Standard Quality Models

For many years, the user community sought a single model for depicting and expressing quality. The advantage of a universal model is clear: it makes it easier to compare one product with another. In 1992, a derivation of the McCall model was proposed as the basis for an international standard for software

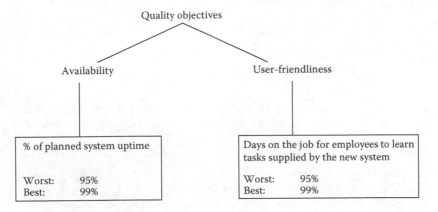

FIGURE 10.4   Gilb's attribute expansion approach.

quality measurement and adopted. It evolved into *ISO/IEC Standard 9126-1* (ISO/IEC 9126-1 2003). In 2011, ISO/IEC 25010 "Systems and Software Engineering—Systems and Software Quality Requirements and Evaluation (SQuaRE)" replaced ISO/IEC 9126-1 (ISO/IEC 25010 2011). In the ISO/IEC 25010 standard, software quality is defined to be the following: "The degree to which a software product satisfies stated and implied needs when used under specified conditions."

Then quality is decomposed into eight characteristics:

1. Functional suitability

2. Performance efficiency

3. Compatibility

4. Usability

5. Reliability

6. Security

7. Maintainability

8. Portability

The standard claims that these eight are comprehensive; that is, any component of software quality can be described in terms of some aspect of one or more of the eight characteristics. In turn, each of the eight is defined in terms of other attributes on a relevant aspect of software, and each can be refined through multiple levels of subcharacteristics.

EXAMPLE 10.3

In ISO/IEC 25010:2011, *reliability* is defined as the

> degree to which a system, product or component performs specified functions under specified conditions for a specified period of time...

while *portability* is defined as the

> degree of effectiveness and efficiency with which a system, product or component can be transferred from one hardware, software or other operational or usage environment to another.

ISO/IEC 25010:2011 contains definitions of numerous subcharacteristics and metrics; many of these were derived from those in three technical reports issued as part of ISO/IEC 9126 (ISO/IEC 9126-2 2003, ISO/IEC 9126-3 2003, ISO/IEC 9126-4 2004).

Another standards document describes an evaluation process for evaluating software product quality (ISO/IEC 25040:2011 2011). The process is described in terms of five major stages:

1. Establish evaluation requirements by determining the evaluation objectives, quality requirements, and the extent of the evaluation.

2. Specify the evaluation by selecting measures, criteria for measurement, and evaluation.

3. Design the evaluation activities.

4. Conduct the evaluation.

5. Conclude the evaluation by analyzing results, preparing reports, providing feedback, and storing results appropriately.

The standard also describes the roles of various stakeholders involved in software quality evaluations.

10.2 MEASURING ASPECTS OF QUALITY

Many software engineers base their quality assessments on measures defined for a specific purpose, separate from any formal quality model. These definitions often reflect the use to which the software will be put, or the realities of testing a system. For example, certain systems have stringent requirements for *portability* (the ability to move an application from one host environment to another) and *integrity* (the assurance that modifications can be made only by authorized users). The user or practitioner may offer simple definitions of these terms, such as

$$\text{Portability} = 1 - \frac{ET}{ER}$$

where ET is a measure of the resources needed to move the system to the target environment, and ER is a measure of the resources needed to create the system for the resident environment. Gilb recommends that the onus for setting measurable targets for these attributes should lie with the user.

Measuring many of the quality factors described in formal models, including McCall's, Boehm's, and those given in various standards documents are dependent on subjective ratings. Although objective measures are preferable, subjectivity is better than no measurement at all. However, those performing the rating should be made aware of the need for consistency, so that variability is limited wherever possible.

10.2.1 Defects-Based Quality Measures

Software quality measurement using decomposition approaches clearly requires careful planning and data collection. Proper implementation even for a small number of quality attributes uses extra resources that managers are often reluctant to commit. In many situations, we need only a rough measure of overall software quality based on existing data and requiring few resources. For this reason, many software engineers think of software quality in a much narrower sense, where quality is considered only to be a lack of defects. Here, "defect" is interpreted to mean a known error, fault, or failure, as discussed in Chapter 5.

10.2.1.1 Defect Density Measures

A de facto standard measure of software quality is *defect density*. For a given product (i.e., anything from a small program function to a complete system), we can consider the defects to be of two types: the *known defects* that have been discovered through testing, inspection, and other techniques, and the *latent defects* that may be present in the system but of which we are as yet unaware. Then, we can define the *defect density* as the following:

$$\text{Defect density} = \frac{\text{Number of known defects}}{\text{Product size}}$$

Product size is usually measured in terms of lines of code or one of the other length measures described in Chapter 8; some organizations (notably in the financial community) use function points as their size measure. The defect density measure is sometimes incorrectly called a defect *rate*.

EXAMPLE 10.4

Coverity reports the number of defects per thousand source statements (KNCSS) in three well-known open source systems: Linux 2.6 (6849 KLOC),

PHP 5.3 (538 KLOC), and PostreSQL 9.1 (1106 KLOC). Figure 10.5 shows how the defect density can vary (Coverity 2011). Note that the defect densities reported by Coverity in these three open source systems are approximately 100 times lower than defect densities reported in commercial systems 25 years ago (Grady and Caswell 1987).

Defect density is certainly an acceptable measure to apply to your projects, and it provides useful information. However, the limitations of this metric were made very clear in prior chapters. Before using it, either for your own internal quality assurance purposes or to compare your performance with others, you must remember the following:

1. As discussed in Chapter 5, there is no general consensus on what constitutes a defect. A defect can be either a *fault* discovered during review and testing (which may *potentially* lead to an operational failure), or a *failure* that has been observed during software operation. In published studies, defect counts have included

    a. Post-release failures

    b. Residual faults (i.e., all faults discovered after release)

    c. All known faults

    d. The set of faults discovered after some arbitrary fixed point in the software life cycle (e.g., after unit testing)

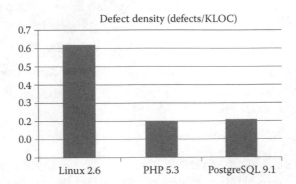

FIGURE 10.5 Reported defect densities (Defects/KLOC) in three open source systems (Coverity 2011).

The terminology differs widely among studies; fault rate, fault density, and failure rate are used almost interchangeably. Thus, to use defect density as a comparative measure, you must be sure that all parties are counting the same things in the same ways.

2. The implication of the phrase "defect rate" is that the number of defects is being recorded with respect to a measure of time (such as operational time or clock time). This measure can be very important. For example, when recording information about operational failures, a defect rate can be calculated based on interfailure times. In this case, the defect rate, defined with respect to time, is an accurate measure of *reliability*. However, many studies capture size information but present it as part of a defect rate. Here, *size* is being used as a surrogate measure of *time* (usually when time is considered to be too difficult to record). Be sure that when you evaluate a study's results, you separate the notion of defect rate from defect density.

3. As discussed in Chapter 8, there is no consensus about how to measure software size in a consistent and comparable way. Unless defect densities are consistently calculated using the same definition of size, the results across projects or studies are incomparable.

4. Although defect density is a product measure in our sense, it is derived from the process of finding defects. Thus, defect density may tell us more about the quality of the defect-finding and defect-reporting process than about the quality of the product itself. Chapter 7 explained how causal Bayesian network models enable us to properly incorporate the impact such process factors.

5. Even if we were able to know exactly the number of residual faults in our system, we would have to be extremely careful about making definitive statements about how the system will operate in practice. Our caution is based on two key findings:

   a. It is difficult to determine in advance the seriousness of a fault.

   b. There is great variability in the way systems are used by different users, and users do not always use the system in the ways expected or intended. Thus, it is difficult to predict which faults are likely to lead to failures, or to predict which failures will occur often.

Adams, who examined IBM operating system data, has highlighted the dramatic difference in rate of failure occurrence.

EXAMPLE 10.5

Ed Adams at IBM examined data on nine software products, each with many thousands of years of logged use worldwide (Adams 1984). He recorded the information in Table 10.1, relating detected faults to their manifestation as observed failures. For example, Table 10.1 shows that for product 4, 11.9% of all known defects led to failures that occur on average every 160–499 years of use.

Adams discovered that about a third of all detected faults lead to the "smallest" types of failures, namely, those that occur on average every 5000 years (or more) of run-time. Conversely, a small number of faults (<2%) cause the most common failures, namely those occurring at least once every 5 years of use. In other words, a very small proportion of the faults in a system can lead to most of the observed failures in a given period of time; conversely, most faults in a system are benign, in the sense that in the same given period of time they will not lead to failures. In addition, less than 2% of the failures were classified as "important failures." Figure 10.6 summarizes the relationship between faults, failures, and the distribution of the severity of the failures.

It is quite possible to have products with a very large number of faults failing very rarely, if at all. Such products are certainly high quality, but their quality is not reflected in a measure based on fault counts. It follows

TABLE 10.1 Adams Data: Fitted Percentage Defects—Mean Time to Problem Occurrence in Years

Product	1.6 Years	5 Years	16 Years	50 Years	160 Years	500 Years	1600 Years	5000 Years
1	0.7	1.2	2.1	5.0	10.3	17.8	28.8	34.2
2	0.7	1.5	3.2	4.5	9.7	18.2	28.0	34.3
3	0.4	1.4	2.8	6.5	8.7	18.0	28.5	33.7
4	0.1	0.3	2.0	4.4	11.9	18.7	28.5	34.2
5	0.7	1.4	2.9	4.4	9.4	18.4	28.5	34.2
6	0.3	0.8	2.1	5.0	11.5	20.1	28.2	32.0
7	0.6	1.4	2.7	4.5	9.9	18.5	28.5	34.0
8	1.1	1.4	2.7	6.5	11.1	18.4	27.1	31.9
9	0.0	0.5	1.9	5.6	12.8	20.4	27.6	31.2

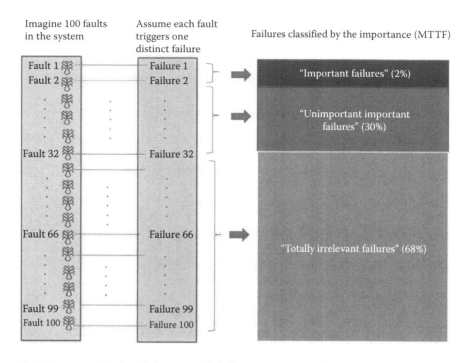

Imagine 100 faults in the system | Assume each fault triggers one distinct failure | Failures classified by the importance (MTTF)

FIGURE 10.6 Faults, failures, and failure severity. (Adapted from Adams E., *IBM Journal of Research and Development*, 28(1), 2–14, 1984.)

that finding (and removing) large numbers of faults may not necessarily lead to improved *reliability*, as reliability measures are based on failure data, not fault data. It also follows that a very accurate residual fault density (or even rate) prediction may be a very poor predictor of operational reliability.

EXAMPLE 10.6

Mockus and Weiss investigated the external attribute of customer *perceived quality*, which is measured using *interval quality*—"the probability that a customer will observe a failure within a certain interval after software release." In a 4-year study of telecommunications software developed at Avaya, they found that defect density was inversely related to perceived quality (Mockus and Weiss 2008).

Despite these and other serious problems with using defect density, we understand the need for such a measure and the reason it has become a de facto standard in industry. Commercial organizations argue that they avoid many of the problems with the measure by having formal definitions that are understood and applied consistently in their own environment. Thus, what works for a particular organization may not transfer to other organizations, so cross-organizational comparisons are dangerous. Nevertheless, organizations are hungry both for benchmarking data and for predictive models of defect density. To meet these goals, we must make cross-project comparisons and inferences. We have seen in Chapter 4 that good experimental design can help us to understand when such comparison makes sense.

Benchmarking is also performed using function points as a size measure, rather than lines of code. Jones reports an average of 5.87 defect potentials per function point, where *defect potentials* are the sum of all defects discovered during development and by users after delivery. This result is based on data from a variety of commercial sources from many application domains (Jones 2008).

10.2.1.2 Other Quality Measures Based on Defect Counts
Defect density is not the only useful defect-based quality measure. For many years, Japanese companies have defined quality in terms of *spoilage*. Specifically, they compute it as

$$\text{System spoilage} = \frac{\text{Time to fix post} - \text{Release defects}}{\text{Total system development time}}$$

For example, Tajima and Matsubara described quality improvements in the 1970s at Hitachi in terms of this measure. The results are displayed in Figure 10.7 (Tajima and Matsubara 1981).

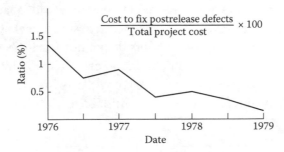

FIGURE 10.7   Quality improvements at Hitachi in the 1970s.

Inglis describes a set of "standard software quality measures used at AT&T Bell Laboratories," all of which are derived from defects data collection (Inglis 1985):

- Cumulative fault density—faults found internally

- Cumulative fault density—faults found by customers

- Total serious faults found

- Mean time to close serious faults

- Total field fixes

- High-level design review errors per thousand NCLOC

- Low-level design errors per thousand NCLOC

- Code inspection errors per inspected thousand NCLOC

- Development test and integration errors found per thousand NCLOC

- System test problems found per developed thousand NCLOC

- First application test site errors found per developed thousand NCLOC

- Customer found problems per developed thousand NCLOC

Clearly, some of these measures require careful data collection.

Assuming that appropriate data-collection procedures are in place, all of the defect-based measures we have discussed may be used for general monitoring purposes and for establishing baselines. In the absence of measures of specific quality attributes, we strongly recommend that these kinds of measures be used. However, it is also important to be aware of the dangers of misinterpreting and misusing these measures. The drawback with using low "defect" rates as if they were synonymous with quality is that, in general, the presence of defects may not lead to subsequent system quality problems. In particular, software faults do not necessarily lead to software failures. Only failures are seen by users, so perceived quality reflects only failure information. As we saw with startling clarity from the Adams study, a preoccupation with faults can paint a misleading picture of quality.

10.3 USABILITY MEASURES

As developers, we sometimes focus on implementing functionality, without much regard for how the user will actually be interacting with the system. The *usability* of the system plays a big role, not only in customer satisfaction but also in terms of additional functionality and life-cycle costs. The ISO/IEC 25010 standard defines usability as follows:

> *Usability* is the degree to which a product or system can be used by specified users to achieve specified goals with effectiveness, efficiency and satisfaction in a specified context of use.

ISO/IEC 25010 2011

This standard definition clearly specifies that *usability* is an external attribute. The usability of a system is a function of the system, users, and the context of use. The common intuitive notion of usability is often called *user-friendliness,* which includes many characteristics or subattributes. We need to know how easy it is to learn to use a system, how efficient we can be when using it, how well we remember how to use it, and how frequently we make errors (Holzinger 2005). Usability also includes the subjective characteristic of user satisfaction, which clearly depends on the skills, knowledge, and personal preferences of users. We can say that usability is an indirect measure, because it cannot be directly measured in terms of only one attribute (Hornbaek 2006).

One can develop models to predict usability based on measures of internal attributes of the structure of the interface (e.g., the number of button clicks required to complete a use case). However, these models can be used only to predict usability, not to measure it. To measure usability, we need to include the user and the context of use in the measurement system.

10.3.1 External View of Usability

Hornbaek reviewed 180 published studies that involved various measures of usability (Hornbaek 2006). He classified the measures used in the studies into the three areas specified in ISO/IEC 25010 definition of usability: effectiveness, efficiency, and satisfaction.

1. *Effectiveness* measures indicate the degree to which users can correctly complete tasks. Thus, counts or percentages of completed tasks, as well as errors made can measure effectiveness. These errors can be mental errors—misunderstanding a system option—or physical errors—problems in accurately pointing and selecting an option. We are also concerned with the completeness of the solutions generated by the users' interactions. User interface effectiveness also includes user recall, which indicates whether a user can remember

information provided by an interface. Researchers have measured users' understanding of a system through testing methods, and experts have rated users' interactions in terms of effectiveness.

2. *Efficiency* measures generally involve the time required to complete tasks. They may be concerned with the rate at which a user can input data via a keyboard, mouse, or other means. Studies have employed questionnaires, expert ratings, and users' ratings to measure the mental effort required for users to complete a task.

3. *Satisfaction* measures indicate subjective notions of the quality of interactions with a software system. It can be measured using standard questionnaires about users' experiences. Preferences between alternatives can be determined by users' rankings or by observations of user behavior. Satisfaction can be evaluated during use using biological measurements such as pulse rates and facial expressions.

The MUSiC project was specifically concerned with defining measures of software usability that include indicators of effectiveness, efficiency, and satisfaction (Bevan 1995). For instance, *task effectiveness* is defined as

$$\text{Tast effectiveness} = \frac{\text{Quantity} \times \text{Quality}}{100}\%$$

The measure tries to capture the notion that the effectiveness with which a user carries out a task has two components: *quantity* of task completed, and *quality* of the goals achieved. Both quantity and quality are measured as percentages. For example, suppose the desired goal is to transcribe a two-page document into a specified format. We could measure quantity as the proportion of transcribed words to original words, and quality as the proportion of nondeviations from the specified format. If we then manage to transcribe 90% of the document, with 3 out of 10 deviations, the task effectiveness is 63%.

MUSiC efficiency measures are

$$\text{Temporal efficiency} = \frac{\text{Effectivness}}{\text{Tast time}}$$

$$\text{Productive period} = \frac{\text{Tast time} - \text{Unproductive time}}{\text{Task time}} \times 100\%$$

$$\text{Relative user efficiency} = \frac{\text{User efficiency}}{\text{Expert efficiency}} \times 100\%$$

The MUSiC project captures user satisfaction, convenience, or ease of use through surveys of actual users that indicate the proportion who "like to work with the product." The MUSiC project developed an internationally standardized 50-item questionnaire (taking 10 min to complete) to assess user satisfaction. This *software usability measurement inventory* provides an overall assessment and usability profile.

It is possible to more directly measure attributes relevant to subjective attributes of user satisfaction, at least in a laboratory setting.

EXAMPLE 10.7

Factors that affect user satisfaction include the aesthetics of users experiences as well as users emotional responses. Through a series of experiments involving human subjects, Thuring and Mahlke (2007) found that well-designed user interfaces led to greater user satisfaction. The study used the same system with combinations of well-designed and poorly designed user interfaces, visually attractive and visually unattractive interfaces. In addition to the subjects' quality perceptions, the study measured users emotional response in terms of changes in heart rates and facial expressions. Both well-designed user interfaces and good visual aesthetics led to higher levels of user satisfaction. However, well-designed interfaces were more important than visual aesthetics.

Three important usability attributes that are not included in the ISO/IEC 25010 definition are *accessibility*, *universality*, and *trustfulness* (Seffah et al. 2006). People with disabilities can use a system with accessibility. Disabilities include visual impairments, hearing loss, or physical impairments. *Universality* is related to the affects of varying cultural norms on the use of a system—certain layouts and naming conventions may have negative (or positive) connotations in some cultures. *Trustfulness* indicates the relative level of trust that users should have in the system. We will deal further with the notion of trust in the discussion of security measurement in Section 10.5.

10.3.2 Internal Attributes Affecting Usability

Collecting early data on usability is often very difficult, especially before a system is released. One popular technique is to look for internal characteristics that we think lead to good usability. For example, we often assume that evidence of good usability includes

- Good use of menus and graphics
- Informative error messages
- Help functions
- Consistent interfaces
- Well-structured manuals

Unfortunately, we know of no explicit relationships between these internal attributes and the external notion of usability, so the artifacts of good usability do not help us in defining it.

Thus, although we can measure the internal attributes (e.g., by counting the number of help screens and error messages, or measuring the structuredness of user manuals), we cannot define them to be measures of usability, just as we could not define structural complexity measures to be measures of quality.

With the exception of the counting measures, such as number of help screens and menu options, most proposed measures of usability based on internal attributes are actually measures of text structure. These in turn are claimed to be measures of text readability or comprehensibility. Again we have an instance of measures of internal attributes being claimed to be measures of an external attribute; it is claimed that when the measures are applied to user manuals and error messages, they measure usability. However, their use is suspect, and we do not recommend them.

10.4 MAINTAINABILITY MEASURES

Most software is exercised repeatedly, and some of us are responsible for the upkeep of delivered software. Both before and during this maintenance period, software measurement can be extremely valuable. We want our software to be easy to understand, enhance, or correct; if it is, we say that our software is maintainable, and measurements during development can tell us the likelihood that we will meet this goal. Once the software is delivered, measurements can guide us during the maintenance process, so that we can evaluate the impact of a change, or assess the relative merits of several proposed changes or approaches.

The ISO/IEC 25010 standard defines maintainability as follows:

Maintainability is the degree of effectiveness and efficiency with which a product or system can be modified by the intended maintainers.

ISO/IEC 25010 2011

We noted in Chapter 5 that maintenance involves several types of changes. The change may be *corrective*, in that it is correcting a fault that has been discovered in one of the software products. Thus, *corrective maintenance* involves finding and fixing faults. Or the change may be *adaptive*: the system changes in some way (the hardware is changed, or some part of the software is upgraded), and a given product must be adapted to preserve functionality and performance. Implementing these changes is called *adaptive maintenance*. Changes also occur for *preventive* reasons; developers discover faults by combing the code to find faults before they become failures. Thus, maintainers are also involved in *preventive maintenance*, where they fix problems before the user sees them. And finally, developers sometimes make *perfective* changes, rewriting documentation or comments, or renaming a variable or routine in the hope of clarifying the system structure so that new faults are not likely to be introduced as part of other maintenance activities. This *perfective maintenance* differs somewhat from preventive, in that the maintainers are not looking for faults; they are looking for situations that may lead to misinterpretation or misuse. Perfective maintenance may also involve the addition of new functionality to a working, successful system (Pfleeger and Atlee 2006).

Maintainability is not restricted to code; it is an attribute of a number of different software products, including specification and design documents, and even test plan documents. Thus, we need maintainability measures for all of the products that we hope to maintain. As with usability, there are two broad approaches to measuring maintainability, reflecting external and internal views of the attribute. Maintainability is an external product attribute, because it is clearly dependent not only on the product, but also on the person performing the maintenance, the supporting documentation and tools, and the proposed usage of the software. The external and more direct approach to measuring maintainability is to measure the maintenance process; if the process is effective, then we assume that the product is maintainable. The alternative, internal approach is to identify internal product attributes (e.g., those relating to the structure of the product) and establish that they are predictive of the process measures. We stress once again that, although the internal approach is more practical since the measures can be gathered earlier and more easily, we can never define maintainability *solely* in terms of such measures.

10.4.1 External View of Maintainability

Suppose we seek measures to characterize the ease of applying the maintenance process to a specific product. All four types of maintenance activity (corrective, adaptive, preventive, and perfective) are concerned with making specific changes to a product; for simplicity, we refer simply to making changes, regardless of the intent of the change. Once the need for a change is identified, the speed of implementing that change is a key characteristic of maintainability. We can define a measure called *mean time to repair* (MTTR), sometimes measured alternatively as the *median* time to repair; it is the average time it takes the maintenance team to implement a change and restore the system to working order. Many measures of maintainability are expressed in terms of MTTR.

To calculate this measure, we need careful records of the following information:

- Problem recognition time

- Administrative delay time

- Maintenance tools collection time

- Problem analysis time

- Change specification time

- Change time (including testing and review)

EXAMPLE 10.8

Brewer reports that reducing MTTR is more effective in improving the availability of evolving systems providing giant-scale web services than increasing the mean-time-between-failures. He found that it is easier to reduce repair time than the failure rate in these systems (Brewer 2001).

Maintainability is also related to the number of required changes, which in turn is dependent on the number of faults or failures as described in Section 10.2.1. Other (environment-dependent) measures may be useful if the relevant data are collected and available:

- Ratio of total change implementation time to total number of changes implemented

- Number of unresolved problems

- Time spent on unresolved problems

- Percentage of changes that introduce new faults

- Number of modules modified to implement a change

All of these measures in concert paint a picture of the degree of maintenance activity and the effectiveness of the maintenance process. An actual measure of *maintainability* can be derived from these component measures, tailored to the goals and needs of the organization.

EXAMPLE 10.9

Gilb's approach to measuring quality attributes, introduced in Figure 10.4, is particularly useful for maintainability. Table 10.2 illustrates his suggestions for decomposing maintainability into seven aspects, each of which reveals useful information about overall maintainability (Gilb 1988).

10.4.2 Internal Attributes Affecting Maintainability

Numerous measures of internal attributes have been proposed as indicators of maintainability. In particular, a number of the "complexity" measures described in Chapter 9 have been correlated significantly with levels

TABLE 10.2   Gilb's Approach to Measuring Maintenance

Maintainability
$SCALE$ = minutes/NCLOC maintained/year
$TEST$ = logged maintenance minutes for system/estimated NCLOC
$WORST$ (new code, created by new process) = 0.2 min/NCLOC/year
$WORST$ (old code, existing now) = 0.5 min/NCLOC/year
$PLAN$ (new code) = 0.1 min/NCLOC/year (based on Fagan's inspection experience)
$PLAN$ (old code) = 0.3 min/NCLOC/year
$REFERENCE$ = estimate based on 5000 programs, average 5000 NCLOC/program, 70% of 250 programmers in maintenance

of maintenance effort. As we have noted previously, correlation with a characteristic does not make something a *measure* of that characteristic, so we continue to separate the structural measures from maintainability measures. Nevertheless, some of the structural measures can be used for "risk avoidance" with respect to maintainability. There is a clear *intuitive* connection between poorly structured and poorly documented products and the maintainability of the implemented products that result from them.

To determine which measures (relating to specific internal attributes) most affect maintainability, we must gather them in combination with external maintainability measures. On the basis of accumulated evidence, we may, for example, identify a particular module having measurably poor structure. We cannot say that such a module will inevitably be difficult to maintain. Rather, past experience tells us that modules with the identified profile have had poor maintainability, so we should investigate the reasons for the given module's poor structure and perhaps restructure it. (There may be other options, such as changing the documentation or enhancing the error-handling capabilities.)

EXAMPLE 10.10

Many organizations investigate the relationship between cyclomatic numbers and other structural attributes and maintenance effort. Based on past history, they develop structural guidelines for code development. The most well known of such guidelines is McCabe's; he suggests that no module be allowed a cyclomatic number above 10 (McCabe 1976). Many studies suggest that cyclomatic number correlates most closely with module size. Thus, module size can predict maintenance effort about as well as the cyclomatic number.

EXAMPLE 10.11

Menzies, Greenwals, and Frank use data mining techniques to explore the relationships between static code attributes and faults (Menzies et al. 2007). This approach includes module length, cyclomatic number, Halstead measures, and various other primitive metrics as predictors. Results show that a naive Bayes data miner was effective as a defect predictor, and that the choice of a good predictor is more important than the set of attributes.

As described in Chapter 9, design constructs that concern the interconnections between modules can potentially affect maintainability.

Object-oriented design patterns are one structuring method that is often touted as supporting adaptability (Gamma et al. 1995). However, there is not clear evidence of their beneficial effects on adaptability.

EXAMPLE 10.12

Bieman et al. studied the relationship between object-oriented design pattern use and change proneness in five systems—three commercial Java systems and two open source systems (Netbeans and JRefactory). They examined changes over multiple releases and found that pattern classes tend to be more change prone than nonpattern classes in four of the five systems (Bieman et al. 2001, 2003b).

For textual products, *readability* is believed to be a key aspect of maintainability. In turn, the internal attributes determining the structure of documents are considered to be important indicators of readability. The most well-known readability measure is Gunning's *fog index F*, defined by

$$F = 0.4 \times \frac{\text{Number of words}}{\text{Number of sentences}} + \text{Percentage of words of three or more syllables}$$

The measure is supposed to correspond roughly with the number of years of schooling a person would need in order to be able to read a passage with ease and understanding. For large documents, the measure is normally calculated on the basis of an appropriate sample of the text. The fog index is sometimes specified in contracts; documentation must be written so that it does not exceed a certain fog index level, making it relatively easy for the average user to understand (Gunning 1968).

There are other readability measures that are specific to software products, such as source code.

EXAMPLE 10.13

De Young and Kampen defined the readability R of programs as

$$R = 0.295a - 0.499b + 0.13c$$

where *a* is the average normalized length of variables (where "length of variable" is the number of characters in a variable), *b* is the number of lines containing statements, and *c* is McCabe's cyclomatic number. The formula was derived using regression analysis of data about subjective evaluation of readability (De Young and Kampen 1979).

10.5 SECURITY MEASURES

We want to be able to use software systems without fear that external attackers will hijack our computers, data files, passwords, and/or accounts. Software security measures can potentially indicate the relative level of security of computer systems from various perspectives. The ISO/IEC 25010 standard defines security as follows:

> *Security* is the degree to which a product or system protects information and data so that persons or other products or systems have the degree of data access appropriate to their types and levels of authorization.

> ISO/IEC 25010:2011

The standard definition focuses on the internal information that needs to be protected. However, in general, we need to be protected from external threats. When we measure software reliability or fault proneness, we generally accept the *competent programmer* hypothesis (Demillo et al. 1978). That is, the developers were diligent and were trying to build correct systems, programs will be nearly correct, and faults will tend to be simple in nature. However, we make no such assumption for security. We assume that there are external entities that are diligently working to overcome any security protections in place and attackers will aim to hide their activities.

Security is an external attribute primarily because of the interactions between external attackers and the systems. Attackers gain entry into a system through some security vulnerability. Although vulnerability is an internal characteristic, it is only a vulnerability with respect to particular attacks. You can remove identified vulnerabilities and/or put defense mechanisms in place once the nature of an attack is identified. A highly secure system becomes insecure when attackers develop new strategies to overcome protection mechanisms.

10.5.1 External View of Security

Berger suggests conducting security risk assessments both qualitatively and quantitatively (Berger 2003). Using either method, the security risk to an organization or individual is the product of the following factors:

- Impact

- Likelihood

- Threat

- Vulnerability

We would like to be able to assess each of the factors independently. However, the specific threat partially determines the other three factors. The impact of a successful attack depends on the nature of the threat. The likelihood of an attack depends on the prevalence and source of the threat. The vulnerability of a system depends on the existence and quality of protections against specific threats—a system may be completely protected against one type of attack, but vulnerable to another threat.

To assess security risks, we need ways to balance several interdependent factors. A widely used method, developed by the Forum of Incident Response and Security Teams (FIRST), is the Common Vulnerability Scoring System (CVSS) (Mell et al. 2007). The CVSS base metric group can have a score between zero and one, with zero representing no vulnerability and one representing the maximum vulnerability. The base metric group includes six individual measures related to the required access to exploit vulnerability and the impact of the vulnerability:

1. Access vector (AV) indicates how remote an attacker can be to exploit the vulnerability. The AV can be local (AV = 0.395), adjacent network accessible (AV = 0.646), or network accessible (AV = 1.0).

2. Access complexity (AC) indicates how complex the attack method needs to be in order to mount a successful attack. Attack complexity is rated as high (AC = 0.35), medium (AC = 0.61), or low (AC = 0.71).

3. Authentication (Au) indicates whether an attacker needs to authenticate two or more times (Au = 0.45), once (Au = 0.56), or no authentication (0.704) to exploit the vulnerability after gaining access to the system.

4. Confidentiality impact (C) indicates whether there is no impact to system confidentiality (C = 0), partial impact to confidentiality (C = 0.275), or complete impact—all files are revealed (C = 0.660).

5. Integrity impact (I) indicates whether there is no impact to system integrity (I = 0), partial impact to integrity—some system files may be modified (I = 0.275), or complete impact—all files may be corrupted (I = 0.660).

6. Availability impact (A) indicates whether there is no impact to system availability (A = 0), reduced performance (A = 0.275), or no availability—a total system shutdown (A = 0.660).

An overall base score is computed through a fairly complicated combination of the six individual measures to produce a single overall metric. In addition, the CVSS includes temporal metrics that indicate evolving attributes of a vulnerability, and environmental metrics that indicate attributes of a vulnerability that depend on characteristics of individual applications that are targets of the vulnerability. Generally, CVSS base and temporal metrics are computed by security experts and posted at sites such as the one managed by NIST.* The developers and users who are most familiar with the potential impact on their own systems determine environmental metrics.

Another way to make sense of the many factors that contribute to security or lack of security is to use multiattribute analyses.

EXAMPLE 10.14

The security attribute analysis method (SAEM) applies multiattribute analysis to rank a set of threat types in terms of their frequencies, and outcomes in terms of lost revenue, reputation, and productivity. The method supports a mixture of units—lost revenue is reported as dollars, while lost reputation uses an ordinal scale, and lost productivity uses hours. Security managers rank and assign a weight to each attribute. An overall threat index is the sum of weighted and normalized individual attribute values.

---

* http://nvd.nist.gov/nvd.cfm.

Although the procedure includes transformations that are not "meaningful" as defined in Chapter 2, the SAEM does encourage security managers to provide explicit assumptions and then allows them to view the results. It gives them a basis for selecting between security solutions (Butler 2002; Butler and Fischbeck 2002).

If you can convert the likelihood of an attack as a probability and its cost in monetary units, you can apply Bayesian analyses to evaluate security risks.

EXAMPLE 10.15

Poolsappasit et al. show that you can use the CVS metrics along with a Bayesian analysis of network states to quantify security risks. Their method is based on the notion of a Bayesian attack graph, which represents causal dependencies between network states along with probabilities of the system being in a particular state. Transitions in the graph represent the likelihood that an attacker can successfully perform an exploit. This method depends on an initial risk assessment based on analyses of subjective belief that a known threat turns into an actual attack. Risk assessments are adjusted following attack incidents using Bayesian propagation methods. This approach allows one to adjust a security plan to more efficiently reduce risks. The CVSS metrics are used to determine the probabilities of attack occurrences (Poolsappasit et al. 2012).

We might expect that security vulnerabilities are more likely if there are other problems in a system.

EXAMPLE 10.16

Alhazmi et al. developed models for predicting the number of security vulnerabilities based on the number of reported faults in multiple versions of the Windows and Linux operating systems. The models normalized the counts of faults and vulnerabilities using fault density (faults per KLOC) and vulnerability density (vulnerabilities per KLOC). The ratio of vulnerabilities to defects was generally between 1% and 5%. The study also analyzed the vulnerability discovery rates over time on these systems. The increase in the number of vulnerabilities for these systems followed a linear trend (Alhazmi et al. 2007).

10.5.2 Internal Attributes Affecting Security

Software security is clearly an external attribute, as security depends on the how vulnerable a system is to attacks from external sources. Many attacks exploit program constructs that are vulnerable to an attack. Thus we ask the question, can we assess the relative security of a system through an analysis of internal attributes of a system?

Since attacks generally come from external sources, we can expect "that functions near a source of input are most likely to contain a security vulnerability" that can be exploited (DaCosta et al. 2003). Thus, we can evaluate internal security attributes with respect to their relative proximity to external attack points.

EXAMPLE 10.17

Manadhata and Wing introduce an *attack surface metric*, which quantifies the security of an implementation independently from external threats (Manadhata and Wing 2011). The attack surface metric is calculated using the number and properties of entry points (i.e., input methods), exit points (i.e., output methods), channels, and untrusted persistent data items (i.e., a file accessed by both the system and a user) in a system implementation. The individual counts of methods, channels, and data items are multiplied by *damage potential–effort ratio (der)* values. The numerator of a *der* is based on the *privilege value* of the entity and the denominator is the entity's *access rights value*. The privilege and access rights values are determined by domain experts' judgments of the relative security risk of an entity's privilege and access rights.

Mandadhata and Wing demonstrate that the attack surface metric is internally valid by showing that it is consistent with an empirical relation system, and thus satisfies the representation condition of measurement. They also show that the measure can predict exploitable vulnerabilities found in Microsoft software. The attributes in the attack surface metric can predict damage potential and effort. The attack surface metric also correlates with the software security risk indicated by patches in Firefox and ProFTP open source systems. Thus, since the attack surface metric is internally valid and appears to be a component of a valid prediction system, we can consider it to be valid in the wide sense, as defined in Chapter 3.

10.6 SUMMARY

The external attributes that interest us are synonymous with aspects of software quality that, in combination, present a comprehensive picture of quality. Most practitioners focus on reliability, maintainability, usability, and security with debate continuing about what other components of quality may be important. However, users and developers can agree among themselves how to measure a particular attribute that interests them. Normally, this measurement involves decomposing an attribute into measurable components, sometimes guided by published quality models or standards. The models make specific assumptions about relationships among attributes. Although it is relatively easy to compute measures using models, the measures are heavily dependent on subjective assessments.

A different but popular view equates quality with few defects. Under certain circumstances, defect-based measures can be useful, but we cannot assume that they are always accurate indicators of quality as perceived by the user. One limitation is that software defects discovered during testing, review, or compilation may not lead to failures in operation. Thus, high defect levels do not always indicate low quality, and low defect levels may not mean high quality. Nevertheless, defect-based measures provide a powerful basis for baselining and monitoring quality changes.

Measuring maintainability involves identifying the number and types of changes to software components. It also involves monitoring the maintenance process, capturing process measures such as the time to locate and fix faults. Some internal attribute measures, notably structural measures described in Chapter 9, may be used as indicators of likely maintainability. Usability must involve assessing people who use the software, and we described several external approaches to usability measurement. There are no internal attributes that obviously predict usability.

Security depends on external attackers, and vulnerabilities are discovered by attacks, and then quantified in terms of their likelihood of success and potential damage. Security experts generally provide the metric values for vulnerabilities. There are internal attributes that can predict security.

EXERCISES

1. Software engineering practices are intended to lead to software products with certain desirable quality attributes. For products that might be used in safety-critical environments, dependability is a particularly important requirement. Briefly describe three key attributes of software dependability. List three other quality attributes that might generally be expected to result from good software engineering practice.

2. The most commonly used software quality measure in industry is the *number of faults per thousand lines of product source code*. Compare

the usefulness of this measure for developers and users. List some possible problems with this measure.

3. Suppose you have overall responsibility for a number of ongoing software projects (some of which are being beta-tested). There are wide quality variations among the projects, and you have available the following measures for each project:

   a. Mean time to failure.

   b. MTTR reported defects.

   c. Total number of user-reported failures.

   d. Total number of defects found during system testing.

   e. Total number of changes made during development.

   f. Maximum cyclomatic number.

   g. Total project overspend/underspend.

   h. Average number of function points produced per month of programmer effort.

Discuss the relative merits of these measures for purposes of comparison. Are there any other measures (that are relatively straightforward to collect) that might help you?

4. List five factors (not described in this chapter) that can potentially affect the maintainablility of software. For each one, name the entity and the relevant measurable attribute and propose a measure of the attribute.

5. Compute the fog index for

   a. This book

   b. A recent document of your own

6. Comment on how Gunning's interpretation of the fog index corresponds to your intuitive perception.

7. Explain why is it not possible (or at least very difficult) to measure software security directly and precisely.

FURTHER READING

Heston and Phifer provide a practical way to take advantage of the key concepts in a set of six different software quality models and standards.

Heston K.M. and Phifer W., The multiple quality models paradox: How much 'best practice' is just enough? *Journal of Software Maintenance and Evolution: Research and Practice*, 23, 517–531, 2011.

Haigh reports on the results of a survey of professionals with an MBA background. The subjects are classified into one of the following groups: (1) user and manager, (2) user and nonmanager, (3) developer and manager, and (4) developer and nonmanager. The subjects rated the importance of 11 nonfunctional quality attributes and 2 functional attributes (correctness and accuracy). There are no significant differences between the groups in their "conceptions of software quality." However, there are differences in the priorities of the groups. Users favored usability, managers of users favored accuracy, and developers favored testability and maintainability.

Haigh M., Software quality, non-functional software requirements and IT-business alignment, *Software Quality Journal*, 18, 361–385, 2010.

Riaz, Mendes, and Tempero published a systematic literature review that identifies 15 published research papers that report on empirical validation of methods for predicting software maintainability. They conclude that there is little evidence that the methods are effective.

Riaz M., Mendes E., and Tempero E., A systematic review of software maintainability prediction and metrics, *Proceedings of the Third International Symposium on Empirical Software Engineering and Measurement*, Lake Buena Vista, Florida, pp. 367–377, 2009.

CHAPTER 11

Software Reliability

Measurement and Prediction[*]

ALL THE SOFTWARE-QUALITY MODELS discussed in Chapter 10 identified software reliability as a key high-level attribute. In Chapter 7, although we did not refer to reliability explicitly, the main examples treated software "quality" as a surrogate for software reliability. So, it is not surprising that software reliability has been the most extensively studied of all the quality attributes. Quantitative methods for its assessment date back to the early 1970s, evolving from the theory of hardware reliability. In this chapter, we describe these methods, highlighting their limitations as well as their benefits. Building on this work, we describe an approach to software reliability assessment that can provide us with truly accurate predictive measures, providing we have been able to collect data about past failures. The approach to measuring software reliability described in this chapter can also be incorporated with the causal modeling approach described in Chapter 7 to achieve improved decision making by software quality and test managers, even when there is minimal data available.

We begin by introducing the basics of reliability theory. Then we address what has become known as the *software reliability growth problem*: estimating and predicting the reliability of a program as faults are identified and attempts are made to fix them. In the approach described here, no individual technique is singled out for unreserved recommendation from

---

[*] Including contributions from prior editions by Bev Littlewood (City University, London).

many that have been proposed over the years. We explain in some detail that we reserve judgment, because empirical observation suggests that no such technique has been able to consistently give accurate results over different data sources. Instead, we emphasize the need to examine the accuracy of the actual reliability measures, obtained from several techniques in a particular case, with a view to selecting the one (if any) that yields trustworthy results.

It is important to note that no current methods can feasibly assure software systems with ultra-high reliability requirements. However, the techniques we describe apply to the vast majority of the systems we build: those with relatively modest reliability requirements. Our suggested approach is more computationally intensive than simply adopting a single technique, and it involves some novel statistical techniques. However, we explain the new methods intuitively, and you can apply the methods in practice by using commercially available software tools. The result is reliability measures that are known to be trustworthy; so, this approach is well worthwhile.

11.1 BASICS OF RELIABILITY THEORY

The theory of software reliability has its roots in the more general theory of systems and hardware reliability; the hardware approaches are described in many textbooks (e.g., Rausand and Hoyland 2004; Modarres et al. 2010; Birolini 2007). We apply many of the basic concepts of this general theory to software reliability problems in the discussion in this chapter.

The basic problem of reliability theory is to predict when a system will eventually fail. In hardware reliability, we are normally concerned with component failures due to physical wear; they can be caused, for example, by corrosion, shock, or overheating. Such failures are probabilistic in nature; that is, we usually do not know exactly when something will fail, but we know that the product eventually will fail; so, we can assign a probability that the product will fail at a particular point in time. For example, suppose we know that a hose will eventually dry out, and that the average usage time for the hose is 3 years. In other words, we can expect that some time around the 3-year mark, the hose will begin to leak. The probability of failure of the hose does not go from 0 on day 1094 (1 day short of 3 years) to 1 on day 1095; rather, the probability may start at 0 on day 1 and then increase slowly as we approach the 3-year mark. We can graph these daily probabilities over time, and the shape of the curve depends on the characteristics of the hose that affect the failure: the materials, pressure, usage, etc. In this way, we build a model to describe the likely failure.

The same approach applies in software. We build a basic model of component reliability and create a *probability density function* (pdf) *f* of time *t* (written as $f(t)$) that describes our uncertainty about when the component will fail.

EXAMPLE 11.1

Suppose we know that a component has a maximum life span of 10 h. In other words, we know it is certain to fail within 10 h of use. Suppose also that the component is equally likely to fail during any two time periods of equal length within 10 h. Thus, for example, it is just as likely to fail in the first 2 min as in the last 2 min. Then we can illustrate this behavior with the pdf $f(t)$ shown in Figure 11.1. The function $f(t)$ is defined to be 1/10 for any *t* between 0 and 10, and 0 for any $t > 10$. We say it is *uniform* in the interval of time from $t = 0$ to $t = 10$. (Such an interval is written as [0,10].) In general, for any *x*, we can define the *uniform pdf* over the interval [0,*x*] to be 1/*x* for any *t* in the interval [0,*x*] and 0 elsewhere. Of special interest (for technical reasons) is the pdf that is uniform on the interval [0,1].

The uniform distribution in Example 11.1 has a number of limitations for reliability modeling. For example, it applies to components only where the failure time is bounded (and where the bound is known). In many situations, no such bound exists, and we need a pdf that reflects the fact that there may be an arbitrarily long time to failure.

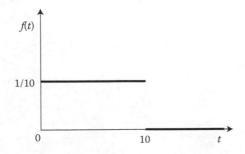

FIGURE 11.1   Uniform pdf.

EXAMPLE 11.2

Figure 11.2 illustrates an unbounded pdf that reflects the notion that the failure time occurs purely randomly (in the sense that the future is statistically independent of the past). The function is expressed as the exponential function

$$f(t) = \lambda e^{-\lambda t}$$

In fact, the exponential function follows inevitably from the randomness assumption. As you study reliability, you will see that the exponential is central to most reliability work.

Having defined a pdf $f(t)$, we can calculate the probability that the component fails in a given time interval $[t_1, t_2]$. Recall from calculus that this probability is simply the area under the curve between the endpoints of the interval. Formally, we compute the area by evaluating the integral:

Probability of failure between time t_1 and t_2 $= \int_{t_2}^{t_1} f(t)dt.$

EXAMPLE 11.3

For the pdf in Example 11.1, the probability of failure from time 0 to time 2 h is 1/5. For the pdf in Example 11.2, the probability of failure during the same time interval is

$$\int_0^2 \lambda e^{-\lambda t}\, dt = \left[-e^{-\lambda t} \right]_0^2 = 1 - e^{-2\lambda}$$

When $\lambda = 1$, this value is equal to 0.63; when $\lambda = 3$, it is equal to 0.998.

It follows from our definition that it does not make sense to consider the probability of failure at any *specific* instance of time t because this is always

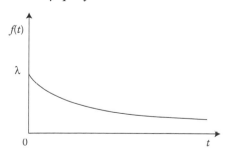

FIGURE 11.2   Pdf $f(t) = \lambda e^{\lambda t}$.

equal to 0 (because the time interval has length 0). Instead, we always consider the probability of failure during some nonzero time interval.

Usually, we want to know how long a component will behave correctly before it fails. That is, we want to know the probability of failure from time 0 (when the software first begins operation) to a given time t. The *distribution function* (also called the *cumulative density function*) $F(t)$ is the probability of failure between time 0 and t, expressed as

$$F(t) = \int_0^t f(t)dt$$

We say that a component *survives* until it fails for the first time, so that we can think of survival as the opposite concept to failure. Thus, we define the *reliability function* (also called the *survival function*) $R(t)$ as

$$R(t) = 1 - F(t)$$

This function generates the probability that the component will function properly (i.e., without failure) up to time t. If we think of t as the "mission" time of a component, then $R(t)$ is the probability that the component will survive the mission.

EXAMPLE 11.4

Consider the pdf that is uniform over the interval [0,1] (as described in Example 11.1). Then $f(t) = 1$ for each t between 0 and 1, and

$$F(t) = \int_0^t f(t)dt = \int_0^t 1dt = [t]_0^t = t$$

for each t between 0 and 1. The graphs of both $F(t)$ and $R(t)$ are shown in Figure 11.3.

EXAMPLE 11.5

The distribution function $F(t)$ for the pdf of Example 11.2 is

$$F(t) = \int_0^t \lambda e^{-\lambda t}dt = \left[-e^{-\lambda t}\right]_0^t = 1 - e^{-\lambda t}$$

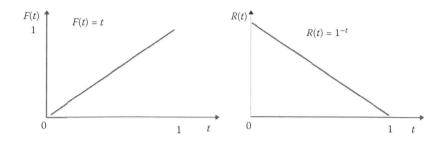

FIGURE 11.3 Distribution function and Reliability function for uniform [0,1] density function.

Thus,

$$R(t) = e^{-\lambda t}$$

which is the familiar exponential reliability function. Both $F(t)$ and $R(t)$ are shown in Figure 11.4.

Clearly, any one of the functions $f(t)$, $F(t)$, or $R(t)$ may be defined in terms of the others. If T is the random variable representing the yet-to-be-observed time to failure, then any one of these functions gives a complete description of our uncertainty about T. For example,

$$P(T > t) = R(t) = 1 - F(t)$$

where P stands for the probability function. The equation tells us that the probability that the actual time to failure will be greater than a given time t is equal to $R(t)$ or $1 - F(t)$. Thus, having any one of these functions allows us to compute a range of specific reliability measures:

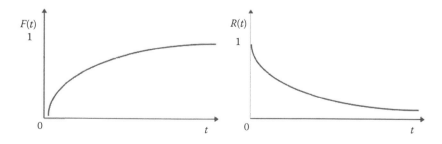

FIGURE 11.4 Distribution function and reliability function for exponential pdf.

The *mean time to failure (MTTF)* is the mean of the pdf, also called the *expected value* of T (written as $E(T)$). We can compute the mean of the pdf $f(t)$ as

$$E(t) = \int tf(t)dt$$

EXAMPLE 11.6

For the pdf in Example 11.1, the MTTF is 5 h. The MTTF for the pdf in Example 11.5 is $1/\lambda$.

The *median time to failure* is the point in time t at which the probability of failure after t is the same as the probability of failure before t. It follows that we can calculate its value by finding the m satisfying $F(m) = 1/2$.

EXAMPLE 11.7

For the pdf of Example 11.1, the median time to failure is 5 h. For the pdf of Example 11.2, we find the median time to failure by solving the following equation for m:

$$\frac{1}{2} = F(m) = \int_0^m \lambda e^{-\lambda t} \, dt = 1 - e^{-\lambda m}$$

Rearranging this equation gives us:

$$m = \frac{1}{\lambda} \log_e 2$$

So, for example, when $\lambda = 1$, we have $m = 0.69$; when $\lambda = 2$, we have $m = 0.35$.

In some sense, the median time to failure gives us a "middle value" that splits the interval of failure possibilities into two equal parts. We can also consider a given interval, and calculate the probability that a component will fail in that interval. More formally, we define the *hazard rate h(t)* as

$$h(t) = \frac{f(t)}{R(t)}$$

$h(t)\delta t$ is the probability that the component will fail during the interval $[t, t + \delta t]$, given that it had not failed before t.

EXAMPLE 11.8

The hazard rate for the important exponential pdf of Example 11.2 is λ.

So far, we have been concerned only with the uncertainty surrounding the time at which the system fails *for the first time*. But, in many cases, systems fail repeatedly (not always from the same cause), and we want to understand the behavior of all these failures collectively. Thus, suppose that a system fails at time t_1. We attempt to fix it (e.g., by replacing a particular component that has failed), and the system runs satisfactorily until it fails at time t_2. We fix this new problem, and again, the system runs until the next failure. After a series of $i - 1$ failures, we want to be able to predict the time of the ith failure. This situation is represented in Figure 11.5.

For each i, we have a new random variable t_i representing the time of the ith failure. Each t_i has its own pdf f_i (and so, of course, also its own F_i and R_i). In classical hardware reliability, where we are simply replacing failed components with identical working components, we might expect the series of pdfs to be identical. However, sometimes, we can replace each failed component with one of superior quality. For example, we may be able to make a design change to minimize the likelihood of recurrence of the fault that caused the previous one to fail. Here, we expect the pdf of t_{i+1} to be different from that of the pdf of t_i. In particular, we would expect

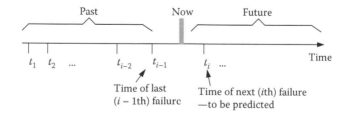

FIGURE 11.5 Reliability problem for the scenario of attempting to fix failures after each occurrence.

the mean of f_{i+1} to be *greater* than that of f_i; in other words, the new component should fail less often than the old one. In such a situation, we have *reliability growth*: successive observed failure times tend to increase. This situation is not normally considered in hardware reliability, but reliability growth is a goal of software maintenance; so, we assume it in our study of software reliability.

In this scenario, there are several other measures that may be useful to us. The hazard rate helped us to identify the likely occurrence of a first failure in an interval. We can define a similar measure to tell us the likelihood of *any* failure, whether or not it is the first one. The *rate of occurrence of failures (ROCOFs)* $\lambda(\tau)$ is defined so that $\lambda(\tau)\delta t$ is the probability of a failure, not necessarily the first, in the interval $[t, t + \delta t]$. ROCOF can be quite difficult to compute in practice. For this reason, ROCOF is often crudely (but wrongly) defined as the number of failures in a unit interval.

A system runs successfully for a time, and then it fails. The measures we have introduced so far have focused on the interruption to successful use. However, once a failure occurs, there is additional time lost as the faults causing the failure are located and repaired. Thus, it is important to know the *mean time to repair (MTTR)* for a component that has failed. Combining this time with the mean time to failure tells us how long the system is unavailable for use: the *mean time between failures (MTBF)* is simply

$$MTBF = MTTF + MTTR$$

These measures tell us about the system's availability for use at a given point in time. In particular, Pressman (2010) and others define availability as

$$\text{Availability} = \frac{MTTF}{MTTF + MTTR} \times 100\%$$

However, this formulation is not always meaningful for software.

In the discussion so far, we have assumed that time is measured as *continuous* operational time. Although we do not address it here, there is an analogous theory in which operational time is treated in discrete units that are normally regarded as *demands* on the system.[*] For example, any control device (such as a simple thermostat) is called into action only

---

[*] We can justify addressing only the continuous case; in practice, we can treat *counts* of demands as continuous variables with little error, since these interfailure counts are usually very large.

when certain environmental conditions prevail. Every time this happens, we have a demand on the device, which is expected to operate in a certain way. It will either succeed or fail. In such situations, the most important measure of reliability may be the *probability of failure on demand*. In software, the discrete scenario is highly relevant when a program is regarded as executing a sequence of inputs or transactions.

11.2 THE SOFTWARE RELIABILITY PROBLEM

There are many reasons for software to fail, but none involves wear and tear. Usually, software fails because of a design problem, that is, the introduction of a software fault into the code (as explained in Chapter 5), or when changes to the code are introduced to a working system. These faults, resulting from the existing, new or changed requirements, revised designs, or corrections of existing problems, do not always create failures immediately (if they do at all); as explained in Chapter 5, the failures are triggered only by certain states and inputs. Ideally, we want our changes to be implemented without introducing new faults, so that by fixing a known problem, we increase the overall reliability of the system. If we can fix things cleanly in this way, we have *reliability growth*.

In contrast, when hardware fails, the problem is fixed by replacing the failed component with a new or repaired one, so that the system is restored to its previous reliability. Rather than growing, the reliability is simply maintained. Thus, the key distinction between software reliability and hardware reliability is the difference between intellectual failure (usually due to design faults) and physical failure. Although hardware can also suffer from design faults, the extensive theory of hardware reliability does not deal with them. For this reason, the techniques we present represent a theory of *design reliability*, equally applicable to software and hardware. However, we discuss reliability solely in terms of software.

In the discussion that follows, we make two assumptions about the software whose reliability we wish to measure:

1. The software is operating in a real or simulated user environment.

2. When software failures occur, attempts are made to find and fix the faults that caused them.

In the long run, we expect to see the reliability improve; however, there may be short-term decreases caused by ineffective fixes or the introduction of novel faults. We can capture data to help us assess the short- and long-term reliability by monitoring the time between failures. For example, we can track execution time, noting how much time passes between successive failures.

Table 11.1 displays this type of data, expressing the successive execution times, in seconds, between failures of a command-and-control system during in-house testing using a simulation of the real operational environment (Musa 1979). This data set is unusual, in that Musa took great care in its collection. In particular, it was possible to obtain the actual execution time, rather than merely calendar time (the relevance of which was described in Chapter 5). As we read across the columns and down the rows, our cursory glance detects improvement in reliability in the long run: later periods of failure-free working tend to be significantly longer than earlier ones. Figure 11.6 plots these failure times in sequence, and the improvement trend is clearly visible.

However, the individual times vary greatly, and quite short times are observed even near the end of the data set. Indeed, there are several zero observations recorded, denoting that the system failed again immediately after the previous problem was fixed. It is possible that the short inter-failure times are due to inadequate fixes, so that the same problem persists, or the fix attempt has introduced a new and severe problem. Musa claimed that the zero times are merely short execution times rounded, as are all these data, to the nearest second.

TABLE 11.1 Execution Times in Seconds between Successive Failures

3	30	113	81	115	9	2	91	112	15
138	50	77	24	108	88	670	120	26	114
325	55	242	68	422	180	10	1146	600	15
36	4	0	8	227	65	176	58	457	300
97	263	452	255	197	193	6	79	816	1351
148	21	233	134	357	193	236	31	369	748
0	232	330	365	1222	543	10	16	529	379
44	129	810	290	300	529	281	160	828	1011
445	296	1755	1064	1783	860	983	707	33	868
724	2323	2930	1461	843	12	261	1800	865	1435
30	143	108	0	3110	1247	943	700	875	245
729	1897	447	386	446	122	990	948	1082	22
75	482	5509	100	10	1071	371	790	6150	3321
1045	648	5485	1160	1864	4116				

Note: Read left to right in rows.

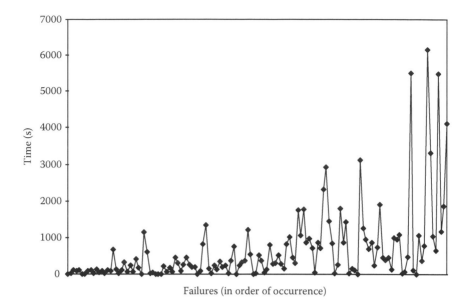

FIGURE 11.6   Plot of failure times in order of occurrence for the Musa dataset.

We capture this data sequentially, and at any time, we can ask several questions about it:

- How reliable is the software now?

- Is it sufficiently reliable that we can cease testing and ship it?

- How reliable will it be after we spend a given amount of further testing effort?

- How long are we likely to have to wait until the reliability target is achieved?

Reliability measures address each of these questions. The remainder of this section explains how we can answer the first question.

Inherent in the data of Table 11.1 is a natural uncertainty about the sequence of numbers: even if we had complete knowledge of all the faults in the software, we would not be able to state with certainty when next it would fail. This result may seem counterintuitive, but it rests on the fact that we do not know with certainty what inputs will be supplied to the software and in what order; so, we cannot predict which fault will be triggered next (and result in failure). In other words, at a given point in time, the time to the next failure is uncertain: it is a *random variable*. But the randomness means that we can apply the reliability theory and functions introduced in Section 11.1.

Assume that we have seen $i - 1$ failures, with recorded inter-failure times

$$t_1, t_2, \ldots, t_{i-1}$$

as illustrated by Musa's data. Fixes have been attempted to correct the underlying faults, and the program has just been set running. Denote by T_i the random variable that represents the yet-to-be-observed next time to failure.* To answer our first question, the current reliability must be expressed in terms of probability statements about T_i. As we saw earlier, we can describe this probability by knowing *one* of the pdf (f_i), the distribution function (F_i), or the reliability function (R_i); we have seen that once we know one of them, then we can easily compute the others. We can express the reliability function as

$$R_i(t) = P(T_i > t) = 1 - F_i(t)$$

Recall that $R_i(t)$ is the probability that the program will survive for a time t before failing next, and $F_i(t)$ is the distribution function of the random variable T_i. We can now compute several measures of the current reliability. For example, the mean time to failure is

$$E(T_i) = \int t f_i(t) dt$$

and the median time to failure is the value of m_i satisfying $F_i(m_i) = 1/2$. We can use similar measures to compute the reliability at some specified time in the future.

These theoretical measures answer our question, but in reality, we are far from done. The actual functions f_i and F_i (for each i) are unknown to us; so, we cannot compute numerical reliability measures. Instead, we must use the observed failure times together with an understanding of the nature of the failure process to obtain *estimates* $\hat{F}_i(t_i)$ of the unknown

---

* Note the convention that random variables are uppercase, while observations of random variables (i.e., data) are lowercase.

distribution functions $F_i(t_i)$. In other words, we are not computing an exact time for the next failure; we are using past history to help us make a *prediction* of the failure time. This point is quite important; so, we emphasize it: all attempts to measure reliability, however expressed, are examples of *prediction*. Even the simplest questions about *current* reliability relate to a random variable, T_i, that we shall observe *in the future*. Thus, we are not *assessing* a product's reliability; rather, software reliability measurement is a *prediction* problem, and we use the data we have available (namely t_1, t_2,..., t_{i-1}) to make accurate predictions about the future T_i, T_{i+1},...

Chapter 2 tells us that to solve a prediction problem, we must define a *prediction system*. Consequently, we need:

1. *A prediction model* that gives a complete probability specification of the stochastic process (such as the functions $F_i(T_i)$ and an assumption of independence of successive times).

2. *An inference procedure* for the unknown parameters of the model based on realizations of t_1, t_2,..., t_{i-1}.

3. *A prediction procedure* that combines the model and inference procedure to make predictions about future failure behavior.

EXAMPLE 11.9

We can construct a crude prediction system in the following way:

1. *The model.* If we assume that failures occur purely randomly, then Example 11.2 tells us that the model is exponential. Thus, for each *i*, we can express the distribution function as

$$F_i(t_i) = 1 - e^{-\lambda_i t_i}$$

2. *Inference procedure.* There is one unknown parameter for each *i*, namely λ_i. We have seen in Example 11.6 that, for this model, the mean time to the next failure is $1/\lambda_i$. A naive inference procedure for computing λ_i is to calculate the average of the two previously observed values of t_i. That is, we estimate that

$$\frac{1}{\lambda_i} = \frac{t_{i-2} + t_{i-1}}{2}$$

and solve for λ_i

$$\lambda_i = \frac{2}{t_{i-2} + t_{i-1}}$$

3. *Prediction procedure.* We calculate the *mean time to ith failure* by substituting our *predicted* value of λ_i in the model. The mean time to failure is $1/\lambda_i$; so we have the average of the two previously observed failure times. Alternatively, we can predict *the median time to ith failure*, which we know from Example 11.7 is equal to $1/\lambda_i \log 2$.

   We can apply this prediction system to the data in Table 11.1. When $i = 3$, we have observed $t_1 = 1$ and $t_2 = 30$. So, we estimate the mean of the time to failure T_3 to be $31/2 = 15.5$. We continue this procedure for each successive observation, t_i, so that we have:
   a. For $i = 4$, we find that $t_2 = 30$ and $t_3 = 113$; so, we estimate T_4 to be 71.5
   b. For $i = 5$, we have $t_3 = 113$ and $t_4 = 81$; so, we estimate T_5 to be 97
   c. and so on

The results of this prediction procedure are depicted in Figure 11.7. Many other, more sophisticated procedures could be used for the prediction. For example, perhaps, our predictions would be more accurate if, instead of using just the two previously observed values of t_i, we use the average of the 10 previously observed t_i. A plot for this variation, and for using the previous 20 observed t_i, is also shown in Figure 11.7.

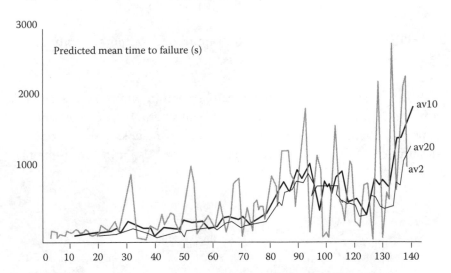

FIGURE 11.7 Plots from various crude predictions using data from Table 11.1. The *x*-axis shows the failure number, and the *y*-axis is the predicted mean time to failure (in seconds) after a given failure occurs.

For predicting the *median* time to the next failure, our procedures are similar. For this distribution, the median is $1/\lambda_i \log 2$ and the mean is $1/\lambda_i$; so, the procedure is the same, except that all the results above are multiplied by $\log 2$ (i.e., by a factor of about 0.7).

Many prediction systems have been proposed, some of which use models and procedures far more sophisticated than Example 11.9. We as users must decide which ones are best for our needs. In the next section, we review several of the most popular models, each of which is parametric (in the sense that it is a function of a set of input parameters). Then, we can turn to questions of accuracy, as accuracy is critical to the success of reliability prediction.

11.3 PARAMETRIC RELIABILITY GROWTH MODELS

Suppose we are modeling the reliability of our program according to the assumptions of the previous sections, namely that the program is operating in a real or simulated user environment, and that we keep trying to fix faults after failures occur. We make two further assumptions about our program:

1. Executing the program involves selecting inputs from some space I (the totality of all possible inputs).

2. The program transforms the inputs into outputs (comprising a space O).

This transformation is schematically shown in Figure 11.8, where P is a program transforming the inputs of I into outputs in O. For a typical program, the input space is extremely large; in most cases, a complete description of the input space is not available. Also, different users may

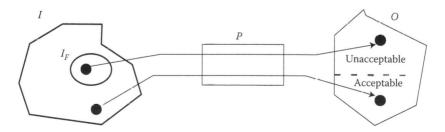

FIGURE 11.8   Basic model of program execution.

have different purposes in using the program, or they may have different habits; so, the probability that one user will select a given input may be different from the probability for another user. For example, in a word-processing program, one user may use a menu button to run a spellcheck, while a second may use a keyboard combination. Thus, we make a simplifying assumption that a single user is running and using the program.

The output space consists of two types of outputs: those that are acceptable (the program runs as it should), and those that are not (a failure occurs). In other words, a program fails when an input is selected that cannot be transformed into an acceptable output. The totality of inputs leading to unacceptable outputs is labeled I_F in Figure 11.8. In practice, a failure is detected when the output obtained by processing a particular input is compared with the output that ought to have been produced according to the program's specification. Detection of failures is, of course, not a trivial task, and we do not address this problem here.

Given these assumptions, there are two sources of uncertainty in the failure behavior:

1. *Uncertainty about the operational environment.* Even if we were to know I_F completely, we cannot know when next we will encounter it.

2. *Uncertainty about the effect of fault removal.* We do not know whether a particular attempt to fix a fault has been successful. And even if the fix is successful, we do not know how much improvement has taken place in the failure rate. That is, we do not know whether the fault that has been removed causes a large or small increase in the resulting reliability.

Good reliability models must address both types of uncertainty. Modeling type-1 uncertainty is easiest, as it seems reasonable to assume that the set I_F is encountered purely randomly during program execution. That is, the time to the next failure (and so the inter-failure times) has, conditionally, an exponential distribution. As we saw in Example 11.9, if T_1, T_2,... are the successive inter-failure times, then we have a complete description of the stochastic process if we know the rates λ_1, λ_2,...

Thus, the most difficult problem is to model type-2 uncertainty: the way in which the value of λ changes as debugging proceeds. The popular software reliability models can be characterized by the way they handle this uncertainty. We present the details of each model, but we acknowledge that the mathematics can be daunting; complete understanding of the details is not necessary for comparing and contrasting the models. However, the details are useful for implementing and tailoring the models.

11.3.1 The Jelinski–Moranda Model

The Jelinski–Moranda model (denoted JM in subsequent figures) is the earliest and probably the best-known reliability model (Jelinski and Moranda 1972). It assumes that, for each i,

$$F_i(t_i) = 1 - e^{-\lambda_i t_i}$$

with

$$\lambda_i = (N - i + 1)\, \phi$$

Here, N is the initial number of faults, and ϕ is the contribution of each fault to the overall failure rate. Thus, the underlying model is the exponential model, so that the type-1 uncertainty is random and exponential. There is no type-2 uncertainty in this model; it assumes that fault detection and correction begin when a program contains N faults, and that fixes are perfect (in that they correct the fault causing the failure, and they introduce no new faults). The model also assumes that all faults have the same rate. Since we know from Example 11.8 that the hazard rate for the exponential distribution is λ, it follows that the graph of the JM hazard rate looks like the step function in Figure 11.9. In other words, between the $(i - 1)$th and ith failure, the hazard rate is $(N - i + 1)\, \phi$.

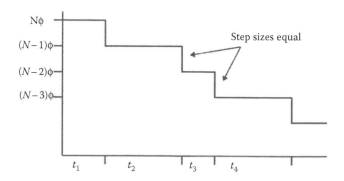

FIGURE 11.9 JM model hazard rate (y-axis) plotted against time (x-axis).

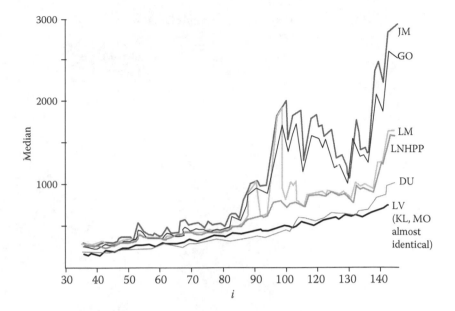

FIGURE 11.10   Data analyzed using several reliability growth models. The current median time to the next failure is plotted on the *y*-axis against failure number on the *x*-axis. We use the model abbreviations introduced in the previous sections. Two additional models are considered here: KL (Keiller–Littlewood) and MO (Musa–Okumoto).

The inference procedure for JM is called *maximum likelihood estimation*; its details need not concern us here, but a simple overview is provided in Fenton and Neil (2012) while a description in the specific context of reliability may be found in textbooks such as Rausand and Hoyland (2004), Modarres et al. (2010), and Birolini (2007). In fact, maximum likelihood is the inference procedure for all the models we present. For a given set of failure data, this procedure produces estimates of N_i and ϕ_i. Then, t_i is predicted by substituting these estimates in the model. (We shall see the JM median time to failure predictions in Figure 11.10, based on the data of Table 11.1.)

EXAMPLE 11.10

We can examine the reliability behavior described by the JM model to determine whether it is a realistic portrayal. Consider the successive inter-failure times where $N = 15$ and $\phi = 0.003$. Table 11.2 shows both the mean time to the *i*th failure and also a simulated set of failure times (produced using random numbers in the model). In the simulated data, the nature of the

TABLE 11.2   Successive Failure Times for JM when
$N = 15$ and $\phi = 0.003$

i	Mean Time to ith Failure	Simulated Time to ith Failure
1	22	11
2	24	41
3	26	13
4	28	4
5	30	30
6	33	77
7	37	11
8	42	64
9	48	54
10	56	34
11	67	183
12	83	83
13	111	17
14	167	190
15	333	436

exponential distribution produces high variability, but generally, there is reliability growth. Notice that as i approaches 15, the failure times become large. Since the model assumes there are *no* faults remaining after $i = 15$, the mean time to the 16th failure is said to be infinite.

There are three related criticisms of this model.

1. The sequence of rates is considered by the model to be purely deterministic. This assumption may not be realistic.

2. The model assumes that all faults equally contribute to the hazard rate. The Adams example in Chapter 10 provides empirical evidence that faults vary dramatically in their contribution to program unreliability.

3. We show that the reliability predictions obtained from the model are poor; they are usually too optimistic.

11.3.2 Other Models Based on JM

Several models are variations of JM. *Shooman's model* is identical (Shooman 1983). The *Musa model* (one of the most widely used) has JM as a foundation but builds some novel features on top (Musa 1975). It was the first model to insist on using execution time to capture inter-failure times. However, it also includes a model of calendar time, so that we can make project management estimates, such as the time until target reliability is achieved. By tying reliability to project management, Musa encouraged the widespread use of reliability modeling in many environments, particularly telecommunications.

11.3.3 The Littlewood Model

The *Littlewood model* (denoted LM in subsequent figures) attempts to be a more realistic finite fault model than JM by treating the hazard rates (corresponding to the different faults) as independent random variables. Thus, while JM is depicted with equal steps, Littlewood has steps of differing size. In fact, these rates are assumed to have a γ-distribution with parameters (α, β). Unlike the JM model, Littlewood introduces two sources of uncertainty for the rates; thus, we say that the model is *doubly stochastic*. In this model, there is a tendency for the larger-rate faults (i.e., the faults with larger "steps," and so larger contribution to unreliability) to be encountered (and so removed) earlier than the smaller ones, but this sequence is itself random. The model therefore represents the diminishing returns in improved reliability that comes from additional testing.

Both the JM and Littlewood models are in a general class called *exponential-order statistic models* (Miller 1986). In this type of model, the faults can be seen as competing risks: at any point in time, any of the remaining faults can cause a failure, but the chance that it will be a particular fault is determined by the hazard rate for that fault. It can be shown that the times, X_j, at which the faults show themselves (measured from when observation of the particular program began) are independent, identically distributed random variables. For the JM model, this distribution is exponential with parameter ϕ. For the Littlewood model, this distribution has a *Pareto distribution*:

$$P(X_j < x) = 1 - \left(\frac{\beta}{\beta + x} \right)^a$$

11.3.4 The Littlewood–Verrall Model

The Littlewood–Verrall model (denoted LV in subsequent figures) is a simple one that, like the Littlewood model, captures the doubly stochastic nature of the conceptual model of the failure process (Littlewood and

Verrall 1973). Here, we make the usual assumption that the inter-failure times, T_i, are conditionally independent exponentials with pdfs given by

$$\text{pdf}(t_i | \Lambda_i = \lambda_i) = \lambda_i e^{-\lambda_i t_i}$$

The λ_i are assumed to be independent γ-variables:

$$\text{pdf}(\lambda_i) = \frac{\psi(i)^\alpha \lambda_i^{\alpha-1} e^{-\psi(i)\lambda}}{\Gamma(\alpha)}$$

In this model, reliability growth (or decay) is controlled by the sequence $\psi(i)$. If this sequence is an increasing function of i, it can be shown that the λ_is are stochastically decreasing and hence, the T_is are stochastically increasing. The user chooses the ψ family; usually, $\psi(i) = \beta_1 + \beta_2 i$ works quite well. The sign of β_2 then determines whether there is growth or decay in reliability; therefore, the data themselves are allowed to determine whether the reliability is increasing, decreasing, or constant via the estimate of β_2.

11.3.5 Nonhomogeneous Poisson Process Models

Consider again the Musa data in Table 11.1. It consists of a sequence of "point events," where each point has no "memory." In other words, the future of the process is statistically independent of the past. We call this behavior *nonstationary*, and it can be modeled by what is called a *nonhomogeneous Poisson process* (NHPP) (Raus and Hoyland 2004). The behavior of the process is in a sense completely random, determined entirely by the rate at which failures are occurring. A minor drawback to this approach is that most such processes have rate functions that change continuously in time. This behavior may not be realistic for software reliability; it can be argued that, for software, the only changes that take place are the jumps in reliability occurring when a fix is made. However, one way of constructing an NHPP is to assume that N (the total number of initial faults) is Poisson distributed; we can use this process as the probability specification in models such as JM and Littlewood (Miller 1986).

The *Goel–Okumoto model* (denoted GO in subsequent figures) is an NHPP variant of JM (Goel and Okumoto 1979). Similarly, the *Littlewood NHPP model* (denoted LNHPP in subsequent figures) is an NHPP variant of the original Littlewood model. Numerous other rate functions have been proposed for NHPP models, including the *Duane model (DU)*, originally devised for hardware reliability growth arising from burn-in testing (eliminating faulty components in complex systems by forcing them to fail early in testing) (Duane 1964).

11.3.6 General Comments on the Models

We have introduced only a few of the many models proposed over the years. All of the above are parametric models, in the sense that they are defined by the values of several parameters. For example, the JM model is defined by the values of ϕ and N. As we have noted, using a model involves a two-step process: selecting a model, and then estimating the unknown parameter values from available data. You can think of this estimation as computing the likely values of the parameters, and in fact, you calculate the *maximum likelihood estimates*. Fortunately, practitioners can use available software to perform the extensive calculations such as the R open-source statistical package (http://www.r-project.org/). A full Bayesian approach (along with supporting software) is described (Neil et al. 2010).

For the rest of this chapter, we shall assume that you have access to a tool or spreadsheet to perform the calculations necessary to use the various reliability models. The next step in choosing an appropriate reliability model is evaluating model accuracy. We assist your evaluation by describing a formal procedure for determining which models perform best on a given set of data. In fact, our technique is useful regardless of the models' assumptions.

11.4 PREDICTIVE ACCURACY

Experience reveals great variation in the accuracy of software reliability prediction models (Abdel-Ghaly et al. 1986). Certainly, no single method can be trusted to give accurate results in all circumstances. In fact, accuracy is likely to vary from one data set to another, and from one type of prediction to another.

Figure 11.10 displays the results of applying several models to the data of Table 11.1; it clearly illustrates the variability of prediction. Each model is used to generate 100 successive estimates of current reliability, expressed as the median time to the next failure. Although all models agree that reliability is increasing, there is considerable disagreement about what has actually been achieved, particularly at later stages of testing. Not only are some models more optimistic in their estimates of reliability than others, but also, some are also more "noisy," giving highly fluctuating predictions as testing proceeds.

Results such as these are typical, and they can be disappointing for potential users of the techniques. Users want trustworthy predictions; so, we must question predictive accuracy and devise means of detecting inaccurate results.

There are two main ways in which reliability predictions can be inaccurate:

- Predictions are *biased* when they exhibit a reasonably consistent departure from the truth (i.e., the true reliability). We will see that the most optimistic of the predictions in Figure 11.10, JM's, are indeed truly optimistic in comparison with the true median, and the most pessimistic, LV's, is truly pessimistic.

- Predictions are *noisy* when successive predictions of, say, the median times to the next failure fluctuate in magnitude more than is the case for the true medians.

To see the difference between bias and noise, consider reliability estimates with fluctuations centered approximately on the truth. That is, the positive and negative errors have about the same magnitude. This consistency of departure means that the estimates are not biased. However, there is a great deal of noise; so, no single prediction can be trusted, since it may exhibit a large error.

It is common for poor predictions to exhibit both bias and unwarranted noise. The difficulty we face in detecting bias and noise is that, of course, we do not know the true reliabilities against which the predictions are to be judged. There is no evidence merely from Figure 11.10 to indicate which, if any, of the series of predictions is accurate. In particular, we must bear in mind that Figure 11.10 only shows a simple summary (the medians) of the successive reliability predictions; usually, we are interested in bias or noise in the complete sequence of predictive distribution functions $\hat{F}_i(t_i)$.

To address this problem, we introduce some formal techniques for analyzing the accuracy of reliability predictions. The techniques emulate how users would actually behave (similar to the descriptions in Examples 10.9 and 10.10). Typically, users make a prediction, observe the actual outcome of the predicted event, and repeat this process several times to obtain a sequence of predictive and actual data. Finally, they evaluate this evidence to determine whether the actual observations differ significantly from what had been predicted.

11.4.1 Dealing with Bias: The u-Plot

We consider first the problem of bias. To detect bias in the medians (as plotted in Figure 11.10), we can count the number of times the observed t_i is smaller than the (earlier) predicted median. If, for n observations, this count differs significantly from $n/2$, then we are likely to have bias. For example, in the data plotted in Figure 11.10, the JM model has 66 of 100 instances where the actual observations are smaller than the predicted median. The predicted medians are consistently too large, confirming that JM is too optimistic in its portrayal of the software reliability growth, as expressed by the medians. If we consider only the last 50 predictions in the data, we find the effect to be even more dramatic: 39 out of the 50 observations are smaller than the corresponding predicted medians. Thus, the clear increase in reliability predicted by JM and displayed in Figure 11.10 (about half-way through the plot) is misleading. Yet, this result is quite surprising. Intuitively, we would expect a model to produce more accurate answers later (when more data are available) than earlier (with less data). Similarly, the medians of the LV model show that the model is actually too pessimistic in its median predictions. On the other hand, there is no evidence here to suggest that the medians of LNHPP are biased.

Of course, lack of bias in the median does not guarantee lack of bias in other reliability measures. What we really seek is evidence of consistent differences between the sequences of functions $\hat{F}_i(t_i)$ (the predictions) and $F_i(t_i)$ (the actual values). To tackle this problem, we introduce the notion of the *u-plot*.

To construct a *u-plot* for a reliability model, we begin by computing the sequence of numbers $\{u_i\}$ given by

$$u_i = \hat{F}(t_i)$$

Each element of the sequence is simply the estimate of $P(T_i < t_i)$ (the probability that the observed inter-failure time is less than the previously predicted probability). In other words, we are estimating the likelihood that the actual observation was less than what we had predicted it would be.

EXAMPLE 11.11

In Example 11.9, we defined a crude prediction system in which the mean time to the next failure (based on the exponential model) was the average of the two previously observed failure times. If we apply this prediction system

TABLE 11.3   Generating u_i Values for the Crude Prediction System of Example 10.9 (Based on the Musa Data)

i	t_i	Predicted Mean Time to ith Failure	$\hat{\lambda}_i$	u_i
1	3			
2	30	16.5	0.061	0.84
3	113	71.5	0.014	0.79
4	81	97	0.010	0.57
5	115	98	0.010	0.69
6	9	62	0.016	0.14
7	2	5.5	0.182	0.30
8	91	46.5	0.022	0.86
9	112	101.5	0.010	0.67
10	15	63.5	0.016	0.21

to the successive failure time data of the Musa data, we generate the predictions shown in the third column of Table 11.3.

In this case, the estimated distribution function is

$$\hat{F}_i(t_i) = 1 - e^{-\hat{\lambda}_i t_i}$$

where

$$\hat{\lambda}_i = \frac{1}{\text{predicted mean time to } i\text{th failure}}$$

From this information, we can compute the sequence of u_is. For example,

$$u_2 = \hat{F}_2(t_2) = \hat{F}_2(30) = 1 - e^{-\frac{1}{16.5}30} = 0.84$$

Similarly, we compute the rest of the u_is as shown in the final column of Table 11.3.

Next, we construct the u-plot, shown in Figure 11.11, by placing the values of the u_is along the horizontal axis (which runs from 0 to 1) and then drawing the step function where each step has height $1/(n + 1)$ (assuming there are n u_is).

A set of predictions is said to be *perfect* if

$$\hat{F}_i(t_i) = F_i(t_i)$$

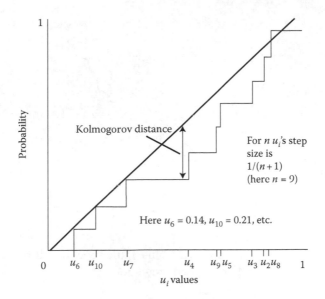

FIGURE 11.11   The u-plot (computed for the first 9 values of Example 11.11).

for all i. If we have perfect predictions, then the set of u_i will look like realizations of independent random variables with a uniform distribution on the interval (0,1) (Dawid and Vouk 1999). Thus, we draw the uniform distribution (which is the line of unit slope) on the u-plot and compare the two. Any significant difference between them indicates a deviation between predictions and observations. We measure the degree of deviation by using the *Kolmogorov distance* (the maximum vertical distance), as shown in Figure 11.11.

We leave as an exercise the drawing of the u-plot for the sequence of u_is calculated in Example 11.11.

EXAMPLE 11.12

Figure 11.12 shows two u-plots for the most extreme reliability models of Figure 11.10: JM and LV, using the data of Table 11.1. To see whether the two plots differ significantly from the line of unit slope, we measure the Kolmogorov distance (the greatest vertical distance) of a plot from the line of unit slope. For JM, this distance is 0.190, which is statistically significant at the 1% level; for LV, the distance is 0.144, which is significant at the 5% level. Thus, this dataset provides very strong evidence against the accuracy of the JM predictions, and quite strong evidence against those from LV.

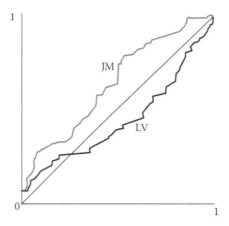

FIGURE 11.12   Jelinski-Moranda (JM) and Littlewood-Verrall (LV) u-plots for 100 predictions of the one-step-ahead predictions.

The plots show us something more important than the simple evidence of inaccuracy; they tell us about the detailed nature of the prediction errors. The plot for JM is always above the line of unit slope, indicating that there are too many small u values. That is, the model tends to underestimate the probability of failure before t_i (the later, actual, observed failure time) and so is too optimistic. Conversely, there is evidence that LV is largely below the line of unit slope and so is too pessimistic in its predictions. These conclusions certainly agree with Figure 11.10, where we saw that JM was relatively more optimistic than LV, and the earlier analysis solely in terms of the medians. We now have evidence that JM is *objectively* optimistic when compared with the truth, and this optimism exists in all measures of current reliability (i.e., not only in the median, but also in other percentiles of the time to failure distribution). A user might reasonably conclude from this analysis that the truth lies somewhere between LV and JM predictions. For example, in Figure 11.10, we might expect the LNHPP predictions to be better than LV and JM. In fact, a u-plot of these predictions is very close to uniform (the Kolmogorov distance, 0.098, is not statistically significant), which generalizes the result we have already seen for the median predictions.

11.4.2 Dealing with Noise

As we have seen, the other major source of inaccuracy in predictions is *noise*. In fact, it is quite easy to obtain predictions that are close to being unbiased but are useless because of their excessive noise.

EXAMPLE 11.13

If we seek an unbiased estimate of the median of T_i, we can use the median value of the preceding observed three inter-event times t_{i-1}, t_{i-2}, and t_{i-3}. For example, consider the data in Table 11.1. Using the three previous inter-event times, we predict the median of T_4 to be 30, the median of T_5 to be 81, the median of T_6 to be 113, etc. There is only a slight pessimistic bias here if the true reliability of the software is changing slowly. However, the individual median estimates obtained this way are likely to be very far from the true values, and they will fluctuate wildly even when the true median sequence is changing smoothly.

Informally, we can think of noise as the fluctuation we find in predictions. We must consider whether the underlying *true* reliability is also fluctuating in a similar way. If the true reliability is not fluctuating, then any noise in the predictions is *unwarranted*. On the other hand, if the true reliability is really fluctuating, then any accurate prediction should also fluctuate. For example, the true reliability of a system or program may in fact suffer reversals of fortune, with frequent decreases corresponding to the introduction of new faults during attempts to fix the old ones. In this case, what appears to be noise in the prediction is actually there for good reason, and we want the predictions to track these reversals; if they do so, the predictions exhibit *absolute noise*.

It is possible to devise a measure of absolute noise in a sequence of predictions. Then, using such a measure, we might expect the median predictions of JM in Figure 11.10 to be noisier than those of LV, which exhibits quite smooth growth. Similarly, we might find that the simple-minded medians proposed in Example 11.13 (based on only three data points) would be even noisier. However, looking at absolute noise can be misleading. It is really unwarranted noise that affects the quality of prediction.

So far, no one has been able to devise a test that is sensitive solely to such unwarranted noise. Instead, we use a more general tool, responsive to inaccuracies of prediction of all types, including unwarranted noise and bias.

11.4.3 Prequential Likelihood Function

We have seen that different prediction models can generate vastly different predictions for the same set of data. We can characterize the differences in terms of noise and bias, but so far, we have seen no way of selecting the best model for our needs. However, help is at hand. We can use the *prequential likelihood function* to compare several competing sets of predictions on the same data source, and to select the one that produces the "globally most accurate" predictions. Details of the mathematical theory behind prequential likelihood can be found in Dawid (1984); Dawid and Vouk (1999); the following is an informal and intuitive interpretation.[*]

Assume, as before, that we want to estimate $F_i(t)$, the distribution of T_i, on the basis of the observed $t_1, t_2, \ldots, t_{i-1}$. Using a prediction system, A, we make predictions for a range of values of i, say from $i = m$ to $i = n$. After each prediction, we eventually observe the actual t_i. The *prequential likelihood function* for these predictions, coming from model A, is defined as

$$\text{PL} = \prod_{i=m}^{i=n} \hat{f}_i(t_i)$$

where $\hat{f}_i(t_i)$ is the estimate of the pdf of T_i. Notice how this estimating technique works. We begin by estimating the pdf of the random variable T_i, using the observations of previous times between failures. Then, when the actual t_i is observed, we substitute the actual observations in the pdf. This procedure is similar to the way in which we formed the u-plot from the distribution function.

EXAMPLE 11.14

In Example 11.9, we built a crude prediction system by using an exponential model in which we estimated the mean time to the next failure by simply computing the average of the two previously observed values. In this situation, the estimated pdf is described by the equation

$$\hat{f}_i(t_i) = \lambda_i \hat{e}^{-\hat{\lambda}_i t_i}$$

where $1/\hat{\lambda}_i$ is the average of t_{i-2} and t_{i-1}. Using the data of Table 11.1 and a spreadsheet package, we can easily compute the sequence of prequential likelihood functions. Table 11.4 shows the results of these calculations for i ranging from 1 to 13. In the table, the right-hand column contains the prequential likelihood values. For example, when $i = 5$,

---

[*] The word "prequential," commonly used in the statistical community, was introduced by Phil Dawid, who defined the prequential likelihood technique. It is a merger of the two words, "predicting" and "sequential."

TABLE 11.4   Computing the Sequence of Prequential Likelihood Functions for
the Crude Prediction System of Example 11.9

i	t_i	T_i	$\hat{\lambda}_i$	$\hat{f}_i(t_i)$	PL
3	113	16.5	0.060606	0.0000643	6.43E−05
4	81	71.5	0.013986	0.0045050	2.9E−07
5	115	97	0.010309	0.0031502	9.13E−10
6	9	98	0.010204	0.0093087	8.5E−12
7	2	62	0.016129	0.0156170	1.33EE−13
8	91	5.5	0.181818	0.0000000	1.57EE−21
9	112	46.5	0.021505	0.0019342	3.04EE−24
10	15	101.5	0.009852	0.0084987	2.59EE−26
11	138	63.5	0.015748	0.0017923	4.64EE−29
12	50	76.5	0.013072	0.0067996	3.15EE−31
13	77	94	0.010638	0.0046894	1.48EE−33

$$PL = \prod_{i=3}^{i=5} \hat{f}_i(t_i) = \hat{f}_3(t_3) \times \hat{f}_4(t_4) \times \hat{f}_5(t_5)$$

Similarly, we can compute the sequence of prequential likelihood values of other prediction systems. For example, an exercise at the end of this chapter asks you to calculate PL for the prediction system where the estimate of the mean time to the next failure is the average of the *three* previously observed values.

To see how prequential likelihood allows us to compare models, suppose we have two candidate sets of predictions, from prediction systems A and B. For each prediction system, we compute the respective prequential likelihood functions, denoting them as PL_A and PL_B. The major theoretical result in Dawid (1984) states that if

$$\frac{PL_A}{PL_B} \to \infty \quad \text{as } n \to \infty$$

then model A discredits model B, in the sense that model B can be rejected. In a looser interpretation of this result, we can say that if PL_A/PL_B increases consistently as n increases, then A is producing more accurate predictions than B.

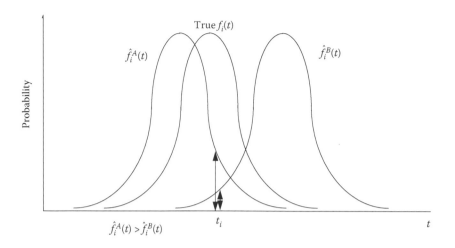

FIGURE 11.13   How prequential likelihood works.

Figure 11.13 is an intuitive justification for this looser result. Here, we examine a single prediction of T_i. We show the true pdf, $f_i(t)$, and two predictions of it. Clearly, the prediction system for A is closer to the "truth" and is thus more accurate. When we eventually see the actual t_i, it is most likely to lie in the central body of the true pdf $f_i(t)$, from which it is sampled. The body of this pdf covers the tails of both the pdf functions. The further away that a prediction is from the truth, the smaller will be the height of this predictive pdf at this point. In other words,

$$\hat{f}_i^B(t)$$

will be small compared to

$$\hat{f}_i^A(t)$$

(which in turn is smaller than the true value). For a consistently poor set of predictions, this effect will tend to occur for each of the terms making up the product in the PL equation. Thus, if model B is giving consistently worse results than A, the ratio

$$\frac{\mathrm{PL}_A}{\mathrm{PL}_B}$$

will increase.

This technique is responsive to the various ways in which predictions can be inaccurate. For instance, if a set of predictions was consistently optimistic (a special case of what we have called bias), the predictive pdf- would tend to be displaced to the right in comparison with the true pdf. If predictions were noisy but unbiased, the sequence of successive predictive pdfs would be randomly displaced to the left and right. In these cases, and for other kinds of departure from accuracy, the prequential likelihood will detect when one series of predictions is inferior to another.

EXAMPLE 11.15

Table 11.5 shows a prequential likelihood comparison of the predictions made by the LNHPP and JM on the data of Table 11.1. Here, n is the number of predictions on which the prequential likelihood ratio (PLR) is based. Thus, when $n = 10$, the PLR involves predictions of $T_{35}, \ldots, T_{44}$; when $n = 15$, it involves $T_{35}, \ldots, T_{49}$, etc. From this analysis, there is little difference between the models for about the first 60 predictions (i.e., until about T_{95}), but there- after, there is strong evidence of the superiority of LNHPP to JM. This com- parison confirms the earlier analysis, since it was precisely after this late stage that the JM predictions become too optimistic.

Of course, the fact that a particular model has been producing superior predictions in the past is no guarantee that it will continue to do so in the future. However, experience suggests that such reversals on a given data source are quite rare. Certainly, if method A has been producing better predictions than method B in the recent past, then it seems sensible to prefer A for current predictions.

TABLE 11.5   Prequential Likelihood Analysis of LNHPP
and JM Predictions for the Data of Table 11.1

n	Prequential Likelihood Ratio LNHPP:JM
10	1.28
20	2.21
30	2.54
40	4.55
50	2.14
60	4.15
70	66.0
80	1516
90	8647
100	6727

11.4.4 Choosing the Best Model

We have presented several ideas for contrasting models, including examining basic assumptions, degree of bias and noise, and prequential likelihood. These techniques may help you to decide whether one model is less plausible than another, but they do not point to a definite best model. The behavior of models greatly depends on the data to which they are applied, and we have seen how several models can produce dramatically different results even for the same dataset. In fact, the accuracy of models also varies from one data set to another (Abdel-Ghaly et al. 1986; Littlewood 1988). And sometimes, *no* parametric prediction system can produce reasonably accurate results!

EXAMPLE 11.16

Table 11.6 shows another of Musa's datasets. Using this data, we can plot the median time to failure for several different prediction systems, shown in Figure 11.14. There is clear and profound disagreement among the predictions: all but two sets are in close agreement with one another, but differ by an order of magnitude from the remaining two (LV and Keillor–Littlewood (KL)), which are close to one another. Advocates of majority voting might believe that the truth probably lies closer to the more optimistic sets of predictions, but in fact, this is not so. Prequential likelihood analysis shows that the pair of more pessimistic prediction systems is performing significantly better than the others. However, the *u*-plot analysis in Figure 11.15 shows that all are extremely poor: six are grossly optimistic, and two are very pessimistic.

The situation in Example 11.16 presents us with a serious problem: we can try all available prediction techniques, but we can demonstrate that none of them produces trustworthy reliability measures. Indeed, it is possible for all of them to be very inaccurate. In the next section, we examine a recalibration technique that can rescue us in many cases.

11.5 RECALIBRATION OF SOFTWARE RELIABILITY GROWTH PREDICTIONS

So far, we have been using the models to make predictions about reliability as we observe and record failure data for a given system. Each time a new failure is observed, we have applied the models in the same way, regardless of the known inaccuracies that we have discussed. Now, we turn to improving predictions as we examine the current data. In other words, if we can learn about the nature of the inaccuracies of reliability prediction at

TABLE 11.6 Musa SS3 Data

1,07,400	17,220	180	32,880	960	26,100	44,160	3,33,720	17820
40,860	18,780	960	79,860		240	120	1800	480
780	37,260	2100	72,060	2,58,704	480	21,900	4,78,620	80760
1200	80,700	6,88,860	2220	7,58,880	1,66,620	8280	9,51,354	1320
14,700	3420	2520	1,62,480	5,20,320	96,720	4,18,200	4,34,760	543780
8820	4,88,280	480	540	2220	1080	1,37,340	91,860	22800
22,920	4,73,340	3,54,901	3,69,480	3,80,220	8,48,640	120	3416	74160
2,62,500	8,79,300	360	8160	180	2,37,920	120	70,800	12960
300	120	5,58,540	1,88,040	56,280	420	4,14,464	2,40,780	206640
4740	10,140	300	4140	4,72,080	300	87,600	48,240	41940
5,76,612	71,820	83,100	900	2,40,300	73,740	1,69,800	1	302280
3360	2340	82,260	5,59,920	780	10,740	180	4,30,860	166740
600	3,76,140	5100	5,49,540	540	900	5,21,252	420	518640
1020	4140	480	180	600	53,760	82,440	180	273000
59,880	840	7140	76,320	1,48,680	2,37,840	4560	1920	16860
77,040	74,760	7,38,180	1,47,000	76,680	70,800	66,180	27,540	55020
120	2,96,796	90,180	7,24,560	1,67,100	1,06,200	480	1,17,360	6480
60	97,860	3,98,580	3,91,380	180	180	240	540	336900
2,64,480	8,47,080	26,460	3,49,320	4080	64,680	840	540	589980

continued

TABLE 11.6 (continued) Musa SS3 Data

3,32,280	94,140	2,40,060	2700	900	1080	11,580	2160	192720
87,840	84,360	3,78,120	58,500	83,880	1,58,640	660	3180	1560
3180	5700	2,26,560	9840	69,060	68,880	65,460	4,02,900	75480
3,80,220	7,04,968	5,05,680	54,420	3,19,020	95,220	5100	6240	49440
420	6,67,320	120	7200	68,940	26,820	4,48,620	3,39,420	480
1,042,680	7,79,580	8040	11,58,240	9,07,140	58,500	3,83,940	20,39,460	522240
66,000	43,500	2040	600	2,26,320	3,27,600	2,01,300	2,26,980	553440
1020	960	5,12,760	8,19,240	8,01,660	1,60,380	71,640	3,63,990	9090
2,27,970	17,190	5,97,900	6,89,400	11,520	23,850	75,870	1,23,030	26010
75,240	68,130	8,11,050	4,98,360	6,23,280	3330	7290	47,160	1328400
1,09,800	3,43,890	16,15,860	14,940	6,80,760	26,220	3,76,110	1,81,890	64320
4,68,180	15,68,580	3,33,720	180	810	3,22,110	21,960	3,63,600	

Note: Execution time to failure in seconds. Read from left to right.

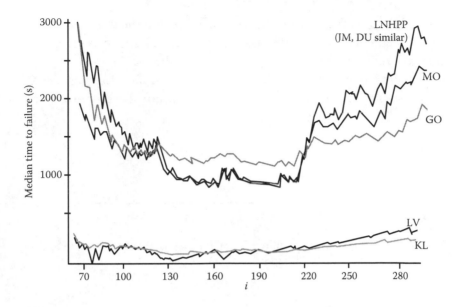

FIGURE 11.14 Raw predictive median time to failure of T_{106} through T_{278} for eight models, using the data of Table 11.6. The x-axis displays the failure number.

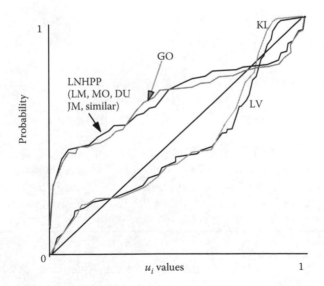

FIGURE 11.15 u-Plots for raw predictions of T_{106} through T_{278} for eight models, using the data of Table 11.16.

the early stages of observation, then we can use this knowledge to improve future predictions *on the same data source*. We call this process *recalibration*, and our recalibration technique involves the *u*-plot.

Consider a prediction $\hat{F}_i(t_i)$ of the random variable T_i, when the true (unknown) distribution is $F_i(t_i)$. Let the relationship between these be represented by the function G_i where

$$F_i(t) = G_i[\hat{F}_i(t)]$$

EXAMPLE 11.17

Suppose that the true distribution $F_i(t_i)$ is the one based on the JM model in Example 10.10, with $N = 15$ and $\phi = 0.003$. In this case,

$$F_i(t_i) = 1 - e^{-(15-i+1)0.003t_i} = 1 - e^{(i-16)0.003t_i}$$

Suppose that our estimate has $N = 17$ and $\phi = 0.003$. Then

$$\hat{F}_i(t_i) = 1 - e^{(i-18)0.003t_i} = 1 - e^{(i-16)0.003t_i}e^{(-2)0.003t_i} = 1 - e^{(-2)0.003t_i}F_i(t_i)$$

These results tell us that we have

$$F_i(t_i) = e^{0.006t_i}(1 - \hat{F}_i(t_i))$$

Thus, the function G_i is defined by

$$G_i(x) = e^{0.006t_i}(1 - x)$$

By knowing G_i, we can recover the true distribution of T_i from the inaccurate predictor, $\hat{F}_i(t_i)$. In fact, if the sequence $<G_i>$ was completely stationary (i.e., $G_i = G$ for all i), then G would characterize the notion of consistent bias discussed earlier. In this case, we could also estimate G from past predictions and use it to improve the accuracy of future predictions.

Of course, in practice, it is unlikely that the sequence $<G_i>$ will be completely stationary. But, in many cases, the sequence changes slowly as i increases; so, in some sense, it is "approximately stationary." This characteristic is the key to our recalibration approach.

We want to approximate G_i with an estimate G_i^* and so form a new prediction

$$\hat{F}^*_{i\,i}(t_i) = G_i^*[\hat{F}_i(t_i)]$$

If we look at how G_i is defined, we observe that it is the distribution function of $U_i = \hat{F}_i(T_i)$. Therefore, we can base our estimate G_i^* on the u-plot, calculated from predictions that have been made prior to T_i. The new prediction recalibrates the raw model output, $\hat{F}_i(t_i)$, in light of our knowledge of the accuracy of past predictions for the data source under study. The new procedure is a truly predictive one, "learning" from past errors.

The simplest form for G_i^* is the u-plot with steps joined up to form a polygon. For technical reasons, it is usually best to smooth this polygon, as shown in Figure 11.16 (Brocklehurst et al. 1990).

Thus, we can describe a complete procedure for forming a recalibrated prediction for the next time to failure, T_i:

1. Check that the error in the previous predictions is approximately stationary. (See (Abdel-Ghaly et al. 1986) for a plotting technique, the *y-plot*, which detects nonstationarity.)

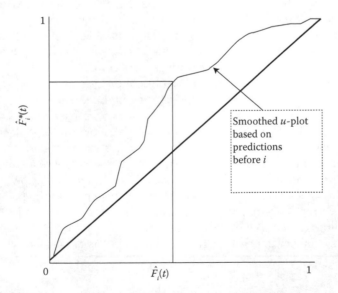

FIGURE 11.16 Using u-plot analysis of past predictions to improve future predictions.

2. Obtain the u-plot for predictions made before T_i, that is, based on t_1, $t_2, \ldots, t_{i-1}$. Join up the steps to form a polygon, and smooth it to form G_i^*, as shown in Figure 11.16.

3. Use the basic prediction system to make a "raw" prediction, $\hat{F}_i(t_i)$.

4. Recalibrate the raw prediction using $\hat{F}_i^*(t_i) = G_i^*[\hat{F}_i(t_i)]$.

The procedure can be repeated at each stage, so that the functions G_i^* used for recalibration are based on more information about past errors as i increases. This technique is not computationally daunting; by far, the greatest computational effort is needed for the statistical inference procedures used to obtain the raw model predictions. Most importantly, this procedure produces a genuine prediction system in the sense described earlier: at each stage, we are using only past observations to make predictions about unobserved future failure behavior.

EXAMPLE 11.18

Consider the u-plots of Figure 11.15. The most notable feature of each prediction system is its extreme optimism or pessimism. However, the extreme nature is not the simple effect of a single characteristic of the system. For example, in JM, the behavior of the plot at each extremity suggests too many very small u values and too many very large ones. For LV, there seem to be too many fairly large u values and too few u values near 1. Thus, although the statements about optimism and pessimism are correct at first glance, a more detailed look reveals that the u-plots present precise information about the incorrect shapes of the complete predictive distributions.

The recalibration technique works dramatically well here for all eight models discussed so far. All eight raw u-plots have Kolmogorov distances that are significant well beyond the 1% level, which is the most extreme value in the tables of this statistic. After recalibration, all the distances are more than halved, and none is significant at this high level. Figure 11.17 shows the dramatic improvement resulting from recalibrating the u-plots in comparison with the raw predictions of Figure 11.15. The differences in the detailed median predictions can be seen by comparing Figures 11.14 and 11.18. We have much closer agreement among the recalibrated models than among the raw ones.

Example 11.18 provides evidence that prediction systems that were in disagreement have been brought into closer agreement by the recalibration technique. Much more important, however, is that we have objective

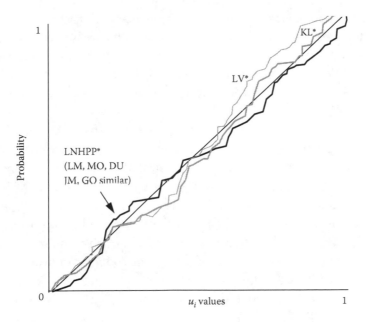

FIGURE 11.17 *u*-Plots for eight models using data of Table 11.6 after recalibration. Note the improvement over raw *u*-plots of Figure 11.15.

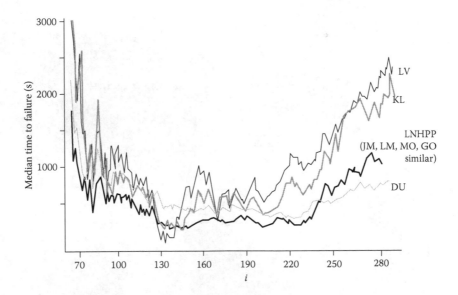

FIGURE 11.18 Medians of recalibrated predictions from eight models, using data of Table 11.6; note how these are now in close agreement compared with the great disagreement shown in the raw predictions of Figure 11.15.

evidence (by comparing the u-plot with the u^*-plot) that the recalibrated predictions are less biased than the raw ones. The prequential likelihood analysis confirms that the recalibration is working dramatically well in this case. We see the effects of recalibration in Figure 11.19, which shows the evolution of the PLRs for the recalibrated predictions against raw model predictions.

There is, for example, overwhelming evidence that the PLR, LV*:LV, is increasing rapidly; it has reached more than e^{40} during these predictions. Therefore, we can be very confident that the LV* predictions here are more accurate than the LV ones. A comparison of JM* and JM is even more dramatic: the PLR reaches e^{90} over the range of predictions shown. This result is due in part to the fact that raw JM predictions are significantly less accurate than those of raw LV (although both are bad from u-plot evidence). Thus, JM starts off with more room for improvement. In fact, after recalibration, the two predictors LV* and JM* have comparable accuracy on the prequential likelihood evidence, with slight evidence of superiority for JM*.

The recalibration technique described works well in most situations. We can evaluate its performance directly: it creates a genuine prediction

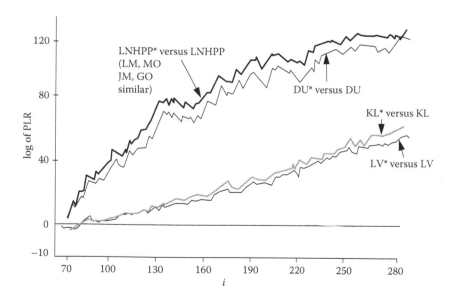

FIGURE 11.19 Prequential likelihood analysis of recalibrated versus raw predictions, using the data of Table 11.6. Note that, since this is a plot of log of the PLR, the results give strong evidence for the superiority of the recalibrated results.

system when it operates on a raw model; so, we can apply the usual analytic techniques to examine predictive accuracy.

11.6 IMPORTANCE OF THE OPERATIONAL ENVIRONMENT

We have seen how to investigate the accuracy of predictions of software reliability growth emanating from several models on a single data set. Of course, there is no guarantee of future accuracy from a model that was accurate in the past. Intuitively, we assume that the model will be accurate again if its conditions of use are the same as in the past. However, Example 11.19 below shows us the difficulty of assuring this similarity.

EXAMPLE 11.19

Suppose we have developed a software system, and we seek to predict its reliability in operation. Usually, our prediction will be based on observations of failures occurring during testing. Unfortunately, our testing environment may not reproduce the typical use of the system in actual operation. Thus, our reliability prediction may be an inaccurate reflection of the reliability as seen by the user.

The problem of realism is even more difficult when there are different modes of system use, different user experience levels, and different environments in which the system will be placed. For example, a novice user of a word-processor system is not likely to use the same shortcuts and sophisticated techniques as an experienced user; so, the failure profiles for each are likely to be quite different.

John Musa recognized these differences and pioneered an approach to address this problem by devising a scheme to anticipate typical user interaction with the system. The typical use is captured in an *operational profile*, a description of likely user input over time. Ideally, the operational profile is a probability distribution of inputs. The testing strategy is based on the operational profile, and test data reflect the probability distribution.

EXAMPLE 11.20

An operational profile is often created by dividing the input space into a number of distinct classes, and assigning to each class a probability that an input from that class will be selected. For example, suppose a program allows you to run one of three different menu options: f, g, and h. We determine from tests with users that option *f* is selected twice as often as *g* or *h* (which is selected equally often). We can assign a probability of 0.5 to *f*, 0.25 to *g*, and 0.25 to *h*. Then, our testing strategy is to select inputs randomly so that the probability of an input being an option on *f* is 0.5, on *g* is 0.25, and on *h* is 0.25.

Such a test strategy is called *statistical testing*, and it has at least two benefits:

1. Testing concentrates on the parts of the system most likely to be used, and hence should result in a system that the user finds more reliable.

2. Using the techniques described in this chapter, we have confidence that reliability predictions based on the test results will give us an accurate prediction of reliability as seen by the user.

However, it is not easy to do statistical testing properly. There is no simple or repeatable way of defining operational profiles. However, Mills' *cleanroom* technique incorporates statistical testing. The results of cleanroom provide a body of empirical evidence to show that statistical testing can be applied practically and effectively (Linger 1994; Stavely 1998).

11.7 WIDER ASPECTS OF SOFTWARE RELIABILITY

We have presented the software reliability growth problem in some detail, as it has the most complete solution. However, there are many more aspects of software reliability to be considered. In particular, measuring and predicting reliability cannot be separated from the methods for achieving it.

In Chapter 7, Section 7.3, where we described a causal modeling approach to defect prediction, our model included three variables:

1. *Residual defects*

2. *Operational usage*

3. *Defects found during operation*

The third was "conditioned" on the first two. In other words, what we had there was a causal/explanatory method of assessing software reliability (the number of defects found in operation is dependent not just on the number of residual defects in the system but also on the amount of operational usage it is subject to). In fact, the probability table associated with the third variable can be considered as a reliability prediction model of the kind described in this chapter.

The model in Section 7.3 also incorporates the (causal) process by which defects were introduced into the software and the testing processes that potentially led to their discovery and removal. Hence, the model there provides a more holistic view of software reliability assessment. Extensions of this approach to reliability modeling can be found in Neil et al. (2010); Fenton and Neil (2012).

The role of testing in the causal model is critical. Testing addresses the dual role of achievement and assessment. We have discussed the difficulty of constructing testing strategies to emulate operational environments, allowing estimates of user-perceived reliability. In addition, some practitioners argue that statistical testing is an inefficient means of removing faults and hence of achieving reliability. More conventional test strategies, it is claimed, allow testers to use their knowledge of the likely types of faults present to remove them more efficiently. However, the conventional techniques do not produce data suitable for software reliability measurement. There is thus an apparent conflict between testing as a means of achieving reliability, and testing as a means of evaluation. We must resolve the conflict by developing new testing strategies.

Ideally, we want to test in an environment that supports fault finding, so that we can increase the likelihood of achieving reliability, and then use the collected data to estimate and predict operational reliability. In classical hardware reliability, this is accomplished by *stress testing*. A component is tested in an environment that is more stressful, in a measured way, than that in which it will operate. For example, the component may be subjected to high temperature, more vibration, or greater pressure than normal. The survival times of stressed components can then be used to estimate the survival times of items in operational use. Unfortunately, the notion of stress is not well understood for software, and stress is not likely to be described as a simple scalar quantity, as it is with hardware. However, some attempts have been made. Several software researchers have characterized the environment in terms of module call frequencies and sojourn times (Littlewood 1979; Cheung 1980). However, such architecture-based methods remain hard to effectively use on real systems primarily because of difficulties in collecting needed failure data (Koziolek et al. 2010).

As we increase our use of software in *safety-critical systems*, we commonly require measurement of the achievement and assurance of very high software reliability.

EXAMPLE 11.21

Rouquet and Traverse report that the reliability requirement for the fly-by-wire computer system of the Airbus A320 is a failure rate of 10^{-9}/h, because loss of this function cannot be tolerated (Rouquet and Traverse 1986). Since this reliability constraint is the system requirement, the software requirement must be even more restrictive.

Example 11.21 describes *ultra-high reliability*, where the system can tolerate at most one failure in 10^9 h. Since this constraint translates to over 100,000 years of operation, we cannot possibly run the system and observe the failure times to measure reliability. Thus, the reliability growth techniques described in this chapter are of little help here.

Figure 11.20 illustrates this problem of assuring ultra-high reliability. The figure depicts an analysis of the failure data from a system in operational use, for which software and hardware design changes were being introduced as a result of the failures. Here, the current ROCOF is computed at various times in the history, using the LV model. The dotted line

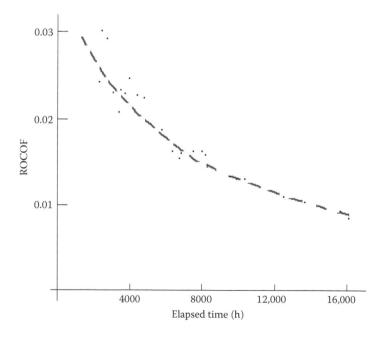

FIGURE 11.20   Successive estimates of the rate of occurrence of failures for the data of Table 11.6. The broken line here is fitted by the eye.

is fitted manually to give a visual impression of what seems to be a very clear law of diminishing returns: later improvements in the MTTF are brought about through proportionally longer testing. It is by no means obvious how the details of the future reliability growth of this system will look. For example, it is not clear to what the curve is asymptotic: will the ROCOF approach zero, or is there an irreducible level of residual unreliability reached when the effects of correct fault removal are balanced by those of new fault insertion? Clearly, even if we were sure that the system would achieve a particular very high reliability figure, the time needed to demonstrate and measure it would be very high.

In fact, even if a program is tested for a very long time and never fails, we still do not have the level of assurance we need.

EXAMPLE 11.22

Littlewood has shown that if a program has worked failure free for x hours, there is about a 50:50 chance that it will survive the next x hours before failing. To have the kind of confidence apparently needed for the A320 would require a failure-free performance of the software for several billion hours (Littlewood 1991). It is easy to see, therefore, that even in the unlikely event that the system had achieved such reliability, we could not assure ourselves of that achievement in an acceptable time.

Assuring software to very high levels of reliability is one of the most difficult, yet important challenges confronting the software industry (Littlewood and Strigini 1993). Mathematical verification techniques may eventually become available for larger control programs than is currently possible, but they cannot address the fundamental and inevitable problem of faults in the specification. Moreover, despite some developments in automated proof support, much verification still depends on humans, meaning that there is no such thing as a foolproof "verified" system. Research addressing ultra-high reliability has involved three techniques: *design diversity, testability,* and *program self-checking.*

Design diversity for fault tolerance: Design diversity has been advocated as a means of achieving high reliability in a cost-effective manner (Avizienis and Kelly 1984). The technique involves building the same system in several independent ways, each involving a different design, to decrease the likelihood that all versions will fail at the same time. Several systems have been built using such techniques (including the Airbus A320) (Rouquet and

Traverse 1986). Unfortunately, evidence suggests that independently developed software versions will not fail independently, and so will not deliver dramatic increases in reliability over single versions. For example, Knight and Leveson report an experiment in which 27 versions were developed independently but contained a high incidence of common faults (Knight and Leveson 1986). Eckhardt and Lee present a theoretical scenario, based on the notion of varying the difficulty of different inputs, which supports these empirical findings (Eckhardt and Lee 1985). These problems of dependence make design diversity an untrustworthy technique for achieving ultra-high reliability. More importantly, they make the assurance problem essentially impossible. Even if we could build a system with such ultra-high reliability, we could not validate our achievement. We cannot apply simple hardware-like theories of redundancy with independent failures; we must try to estimate the dependence between versions. This problem has been shown to be as difficult as the problem of simply testing the complete fault-tolerant system as a black box, and so is essentially impossible (Miller 1986).

Testability: Voas and his colleagues extend the traditional reliability-modeling approach to tackle the ultra-high reliability problem (Hamlet and Voas 1993; Voas and Miller 1995). The authors introduce a notion of *testability* as the probability that the software will fail if it is faulty. They propose methods for improving program testability as a way of improving reliability. The simple idea behind this approach is that if a highly testable program has revealed no failures during testing, then we have increased confidence that the software is truly fault free. However, there are potential dangers. By making a program highly testable, you may increase the probability that the program is fault free, but you also increase the probability that failures will occur if faults do remain. Using this notion of testability, a fault-tolerant system will have low testability. Bertolini and Strigini address this and other concerns about testability (Bertolini and Strigini 1995). Hays and Hayes find that Voas's approach to testability analysis can be effective in predicting the location of program faults (Hays and Hayes 2012). We describe notions of testability based on counting test requirements, and program design structures that make it easier or harder to conduct testing in Chapter 10.

Program self-testing: Blum, Luby, and Rubinfield describe a theory of program self-testing, in which programs can be shown to be correct with a probability arbitrarily close to 1. The drawback of their approach is that it is restricted to a narrow class of mathematical-type problems (Blum et al.

1993), or limited to self-testing of particular algebraic, graph, or function properties (Ron 2010).

In the absence of proven methods of assuring ultra-high software reliability levels, it is surprising to find that systems are being built whose safe functioning relies upon their achievement.

11.8 SUMMARY

Software reliability is a key concern of many users and developers of software. Since reliability is defined in terms of failures, it is impossible to measure before development is complete. However, if we carefully collect data on inter-failure times, we can make accurate predictions of software reliability. There are several software reliability growth models to aid the estimation, but none can be guaranteed to produce accurate predictions on all data sets in all environments. On the other hand, we can objectively analyze the predictive accuracy of the models and select those that are working best for the given data. Moreover, we can then get improved predictions using a technique known as recalibration. In this way, we make predictions in whose accuracy we can be confident, providing that the software's future operational environment is similar to the one in which the failure data were collected. If we must predict reliability for a system that has not yet been released to a user, then we must simulate the target operational environment in our testing. Since there are no current methods that can feasibly assure software systems with ultra-high reliability requirements, the techniques described in this chapter should be restricted to systems with relatively modest reliability requirements. However, most software requires modest reliability; so, the techniques in this chapter are useful.

Some of the techniques described in this chapter are computationally intense, but tools are available to automate these procedures. Moreover, the types of prediction systems described in this chapter can be incorporated into more holistic causal models of software reliability as described in Chapter 7.

EXERCISES

1. Consider the functions $f(t)$, $F(t)$, and $R(t)$ defined in this chapter. Redefine each in terms of the others.

2. For the pdf of Example 11.4, show that the median time to failure is 5 h.

3. For the pdf of Example 11.5, compute the median time to failure when $\lambda = 4$.

4. Draw the u-plot for the sequence of u_is calculated in Example 11.11. On the same chart, plot the true values of the $F_i(t_i)$. (You can calculate these values using Table 11.2, which gives the true means $1/\lambda_i$ of the distribution.) Compare the true values with the uniform distribution.

5. Imagine that your job is to evaluate the reliability of the software produced by your company. This task involves establishing an in-house testing service, whereby test cases would be generated to provide input to the reliability growth models described in this chapter. How would you accomplish this when the software in question is in each case, discuss whether you can justify the test procedure you are advocating as the only testing to be carried out on the particular product, bearing in mind that there is a fixed budget from which all testing must be funded.

   a. A word-processing system.

   b. An air-traffic control system.

   c. A car engine management system.

   d. A controller for a domestic washing machine.

6. Using the data of Table 11.1 and one of the most accurate sets of predictions from the median plots of Figure 11.10, obtain an approximate plot of achieved median against total elapsed test time. Comment on the shape of this graph. You have been asked to continue testing until you are confident that the median time to failure is 1000 h (remember that the raw data are in seconds); comment on the feasibility of this approach.

7. Look at the crude prediction system described in Example 11.9 (the exponential model in which we estimate the mean time to the next failure by simply computing the average of the two previously observed values). Using a spreadsheet, compute successive mean and median time to the next failure for the data of Table 11.16 (up to about $i = 20$). Analyze informally the accuracy of this prediction system on these data (if you wish, you could do this formally by carrying out u-plot and prequential likelihood analyses), and try to address the

issues of bias and noise as they are defined in this chapter. Does the situation improve if you change the prediction system by

a. Basing your prediction of the mean on the average of the three previously observed t_is ?

b. Basing your prediction of the mean on just the one previously observed t_i?

8. Example 11.14 shows how to compute the sequence of prequential likelihood values for the prediction system of Example 11.9 using the data in Table 11.1. Call this prediction system A. Repeat the computations for the prediction system B where the estimate of the mean time to the next failure is the average of the *three* previously observed values. Now, you can compute both PL(A)/PL(B) and PL(B)/PL(A) for a number of values of i and see what happens as i increases. Does this give you any evidence of which of A or B is superior?

9. When faults are found and fixed in a program, we observe reliability growth. Specifically, we see a sequence of times between successive failures $t_1, t_2,..., t_n$ that tend to increase in magnitude. Explain carefully the underlying failure process, identifying the sources of uncertainty that require us to adopt a probabilistic approach when we wish to predict the future failure behavior of the program.

10. Many different probability models have been proposed to represent the growth in reliability as fault detection and correction progress. Unfortunately, it is not possible to select one of these *a priori* and be confident that it will give accurate predictions. Imagine that you are responsible for making reliability predictions from the failure data coming from a particular program under test. What could you learn from the u-plots and prequential likelihood functions about the accuracy of the predictions coming from the different models? How could you use the u-plot to improve the accuracy of predictions?

11. Figures 11.17 through 11.19 show, successively, one-step-ahead median predictions, u-plots and prequential likelihood functions for several models operating upon some software failure data obtained in the testing of a telephone switch. State, very briefly, what you would conclude from these figures. (*Hint:* There is no need to give a detailed analysis of each model's performance individually.) Suggest how the analysis might be continued.

12. You are part of a team that is to build a critical system, the safety of which depends on the correct functioning of its software. Your responsibility will be to assess the system, and in particular the software, to decide whether it is sufficiently safe to use. In discussion with your colleagues responsible for building the system, several sources of evidence arise that might be relevant to your eventual assessment of the software. Comment briefly on the strengths and weaknesses of the following:

    a. Failure data from debugging in operational testing

    b. Observation of long periods of failure-free working

    c. Design diversity, such as three versions with 2-out-of-3 voting at run time

    d. Formal verification

    e. Quality of the development process

13. Why is reliability an external attribute of software? List three internal software product attributes that could affect reliability.

14. Suppose you can remove 50% of all faults resident in an operational software system. What corresponding improvements would you expect in the reliability of the system?

15. Consider a design or programming methodology with which you are familiar. List the ways in which this methodology influences the documents produced, and propose ways in which they can increase external quality. How can you determine that an actual increase had been achieved?

FURTHER READING

The extensive theory of hardware reliability (including how to compute maximum likelihood estimates, and nonhomogeneous Poisson processes) is described in the following textbooks.

Birolini A., *Reliability Engineering: Theory and Practice,* 5th edition. Springer, Berlin, 2007.
Modarres M., Kaminskiy M., and Krivtsov V., *Reliability Engineering and Risk Analysis: A Practical Guide,* 2nd edition. CRC Press, Boca Raton, Florida, 2010.

Rausand M. and Hoyland A., *System Reliability Theory: Models, Statistical Methods, and Applications*, 2nd edition. John Wiley & Sons, Inc., Hoboken, New Jersey, 2004.

Neil et al. provide an extensive description of how to incorporate causal factors effectively into reliability modeling in the following texts:

Fenton N.E. and Neil M., *Risk Assessment and Decision Analysis with Bayesian Networks*. 2012, CRC Press, Boca Raton, Florida, ISBN: 9781439809105.

Neil M., Marquez D., and Fenton N.E., Improved reliability modeling using Bayesian networks and dynamic discretization, *Reliability Engineering and System Safety*, 95(4), 412–425, 2010.

Musa and Baur et al. provide practical guidance on using software reliability modeling techniques.

Baur E., Zhang X., and Kimber D.A., *Practical System Reliability*, Wiley IEEE Press, Hoboken, New Jersey, 2009.

Musa J., *Software Reliability Engineering: More Reliable Software, Faster and Cheaper*, 2nd edition. Tata McGraw-Hill Education, New York, 2004

For full details of the method used for analyzing the predictive accuracy of models, see these papers by Littlewood and his colleagues.

Abdel-Ghaly A.A., Chan P.Y., and Littlewood B., Evaluation of competing software reliability predictions, *IEEE Transactions on Software Engineering*, SE-12 (9), 950–967, 1986.

Brocklehurst S. and Littlewood B., New ways to get accurate software reliability modeling, *IEEE Software*, 9(4), 34–42, July 1992.

Littlewood B., Predicting software reliability, *Philosophical Transactions of the Royal Society of London A*, 327, 513–527, 1989.

Brocklehurst and her research team describe the details of the recalibration techniques for improving predictions.

Brocklehurst S., Chan P.Y., Littlewood B., and Snell J., Recalibrating software reliability models, *IEEE Transactions on Software Engineering*, SE-16(4), 458–470, 1990.

Kapur, Pham, Anand, and Yadav develop a reliability modeling approach that uses a realistic model of the fault removal process.

Kapur P.K., Pham H., Anand S., and Yadav K., A unified approach for developing software reliability growth models in the presence of imperfect debugging and error generation, *IEEE Transactions on Reliability*, 60(1), 331–340, March 2011.

Zhu M. and Smidts develop a method to include software reliability analyses into the probabilistic risk assessment technique often used to evaluate safety-critical systems.

Zhu D., Mosleh A., and Smidts C., A framework to integrate software behavior into dynamic probabilistic risk assessment, *Reliability Engineering and System Safety*, 92, 1733–1755, 2007.

Crespo, Jino, Pasquini, and Maldonado develop a binomial reliability model that is based on the achievement of various data-flow test coverage criteria.

Crespo N.C., Jino M., Pasquini P., and Maldonado J.C., A binomial software reliability model based on coverage of structural testing criteria, *Empirical Software Engineering* 13, 185–209, 2008.

Littlewood and Strigini provide an excellent discussion of the limitations of assuring software systems with ultra-high reliability requirements.

Littlewood B. and Strigini L., Validation of ultra-high dependability for software-based systems, *Communications of the ACM*, 36(11), 1993.

Appendix: Solutions to Selected Exercises

2. Answers to the questions:

   a. Nominal, ordinal, interval, ratio, and absolute are the five-scale types.

   b. The scale type for the complexity measure is ordinal, because there is a clear notion of order, but no meaningful notion of "equal difference" in complexity between, say, a trivial and simple module and a simple and moderate module. A meaningful measure of average for any ordinal scale data is the median. The mode is also meaningful, but the mean is not.

3. The answer is given in the following table:

Scale Type	Software Measure	Entity	Entity Type	Attribute
Nominal	Classification C, C++, Java, ...	Compiler	Resource	Language
Ratio	Lines of code	Source code	Product	Size
Absolute	Number of defects found	Unit testing	Process	Defects found

14. A statement about measurement is meaningful if its truth value is invariant of the particular measurement scale being used.

The statement about Windows is meaningful (although probably not true) assuming that by size we mean something like the amount of code in the program. To justify its meaningfulness, we must consider how to measure this notion of size. We could use lines of code, thousands of lines of code, number of executable statements, number of characters, or number of bytes, for example. In each case, the statement's truth value is unchanged; that is, if a Windows program is about four times as big as the Linux program when we measure in lines of code, then it will also be about four times bigger when we measure in number of characters. We can say this because the notion of size is measurable on a ratio scale.

The rating scale for usability is (at best) ordinal, so there is no way to say that the difference in usability from 1 to 2 is the same as from 2 to 3. Therefore, if "average" refers to the mean, then the statement is not meaningful; changing to a different ordinal scale does not necessarily preserve the order of the mean values. However, if "average" refers to the median, then the statement is meaningful.

17. We must assume that the complexity ranking is ordinal, in the sense that the individual categories represent (from left to right) notions of increasing complexity.

a. You can use the median of the M values. This choice is meaningful, because the measure M is an ordinal scale measure. Mode is also acceptable.

b. Consider the following statement A: "The average complexity of modules in system X is greater than the average complexity of modules in system Y." We must show that A is not meaningful when "average" is interpreted as mean. To do this, we show that the truth value of the statement is not invariant of the particular measure used. Thus, consider the following two valid measures M and M':

	Trivial	Simple	Moderate	Complex	Very Complex	Incomprehensible
M	1	2	3	4	5	6
M'	1	2	3	4	5	10

Suppose that the complexities of the modules in X are given by:

x_1 trivial, x_2 and x_3 simple, x_4 moderate, x_5 incomprehensible while the complexities of the modules in Y are given by:

y_1 simple, $y_2, y_3,$ and y_4 moderate, $y_5, y_6,$ and y_7 complex.

Under M, the mean of the X values is 2.6, while the mean of the Y values is 3.1; so statement A is false under the measure M. However, under M', the mean of the X values is 3.6, while the mean of the Y values is 3.1. Thus, statement A is true under M'. From the definition of meaningfulness, it follows that the mean is not a meaningful average measure for this (ordinal scale) data.

c. Your answer might include a crude criterion based on some metric value, such as McCabe's cyclomatic complexity, v. For example, if $v(x) < 2$, we assign x as trivial; if $1 < v(x) < 5$, we assign x as simple, etc. Finally, if $v(x) > 100$, we assign x as incomprehensible. A more sophisticated criterion might be based on some measurable notion that closely matches our intuition about complexity. For example, the complexity of software modules is (intuitively) closely related to the difficulty of maintaining the modules. If maintenance data were available, such as mean time to repair (MTTR) faults for each module, then we could base our criterion on it. For example, if $MTTR(x) < 30$ min, then we assign x as trivial; if 30 min $< MTTR(x) < 90$ min, we assign x as simple, etc., until if $MTTR(x) > 5$ days, we assign x as incomprehensible.

22. The only way to report "number of bugs found" is to count number of bugs found. The answer to the second part depends very much on how you define program correctness. If you believe that a program is either correct or not, then correctness is measurable on a nominal scale with two values, such as (yes, no) or (1, 0). Or you may decide that correctness is measurable on a ratio scale, where the notion of zero incorrectness equals absolute correctness. (In this case, the number of bugs found could be a ratio-scale measure of program correctness.) In either of these cases, the number of bugs found cannot be an absolute scale measure for the attribute. Moreover, if "correctness" is taken to be synonymous with reliability, then counting bugs found may not be a measure of correctness at all, since we may not be finding the bugs that cause actual failures.

23. Answers to questions about the proposed *application domain (AD)*:

    a. Clearly AD is a nominal measure as there is no notion of ordering; it is only a classification.

    b. The central tendency for ordinal measures is computed only as the mode, which is 1 (WWW browsers) for the given data.

    c. The statement "my compiler has three times the AD as your browser" is not meaningful.

24. Consider a proposed measure of *program adaptability:*

    a. There are lots of possibilities here. It could be time required to make a change, or better; yet, time required to make specific types of changes. Some structural measures can be OK, but then the empirical relations (below) are implied, and will not always be satisfied. The empirical relations should be of the nature of "it is more difficult to make some specific change to program A than program B".

    b. If the measure is time, then it is a ratio scale.

    c. It satisfies the representation condition if the measure is consistent with all empirical relations in *i*.

    d. Advantages and disadvantages vary.

CHAPTER 3: A GOAL-BASED FRAMEWORK FOR SOFTWARE MEASUREMENT

5. GQM means Goal-Question-Metric, and it is based on the notion that measurement activities should always be preceded by identifying clear goals for them. To determine whether you have met a particular goal, you ask questions whose answers will tell you if the goal has been met. Then, you generate from each question the attributes you must measure in order to answer the questions.

   Goal-oriented measurement is common sense, but there are many situations where some measurement activity can be beneficial even though the goals are not clearly defined. This situation is especially true where a small number of metrics address several different goals, such as monitoring quality changes while predicting testing effort, for example. Also, GQM introduces the problem of determining who sets the goals. High-level managers may be bad at identifying goals, or they may have goals for which no metrics can be practically computed at the engineering level. Because of this mismatch between goals and practicality, some people have proposed the need for bottom-up measurement, where the engineers collect metrics data that are useful and practical.

A typical GQM tree may resemble this one:

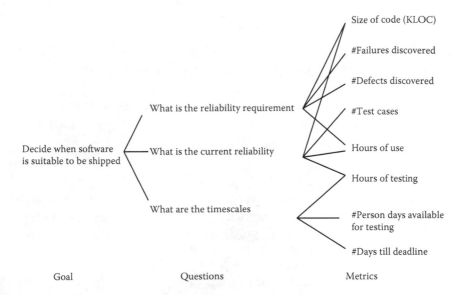

This tree depicts only direct measures. Questions such as "What is the reliability requirement?" are answered in terms of derived measures derived from the direct ones, and the tree can be extended to include both types. For example, the reliability requirement might be expressed as an average number of failures per 100 hours of use, or a number of defects per KLOC. Similarly, current reliability might be measured by number of defects per test case. It might also be useful if failures and defects were partitioned into two categories, "serious" (i.e.,, really affecting reliability) and "not serious," since many anomalies found during testing may be benign.

7. A typical GQM tree on improving maintainability of software may resemble the following figure:

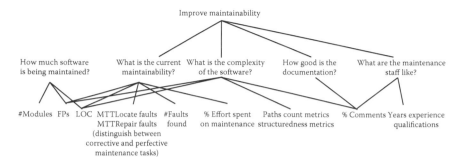

CHAPTER 5: SOFTWARE METRICS DATA COLLECTION

1. An error is a mistake made by a human designer. A fault is the encoding of an error in the software. A failure is the manifestation of a fault during software execution.

4. Answers:

   a. Note first that only *project*-wide issues can be addressed because of the data's coarse granularity. Moreover, only goals and questions relating to project-wide comparisons are reasonable. Some possibilities are:

      i. General project-wide trends or comparisons in *productivity*, measured as the ratio of effort and lines of code. An example goal might be: "Improve project productivity," and a question might be: "What is current best, average, and worst productivity?" The well-known problems with this particular productivity measure should be cited here (e.g., lines of code do not really measure utility or even size of output; there are problems with the definition of the lines of code measure itself, etc.).

      ii. General project-wide trends or comparisons in *quality*, measured as the ratio of faults with lines of code. An example goal might be: "Improve product quality," and a question might be: "What is current best, average, and worst quality?" Note problems with the lines of code measure again, but the real limitation is the use of faults. Total number of faults recorded may be more an indication of differences in recording or testing practice than of genuine quality. It does not reflect the quality of the end product.

iii. Provided you have data from a reasonable number of similar projects, it is possible to use the measures to improve project forecasting (i.e., effort estimation). You achieve this goal by using regression analysis on effort and lines of code data from the past projects.

iv. Provided you have data from a reasonable number of similar projects, you can use the measures for outlier analysis. For example, your goal may be to identify those projects having unusual productivity or quality values. However, all the limitations noted above still apply. To address this goal, you can use boxplots, two-dimensional scatter diagrams, etc.

b. You must collect data at the module level, rather than only at the project level. In particular, size, effort, and fault data should be available for each module. For each module, you must also identify the specific quality-assurance techniques that were used. (Ideally, you would know the extent of such use, measured by effort.) It would be helpful if the effort data per module could be partitioned into effort by activity, such as design, coding, testing, etc.

To get a more accurate perspective on whether a particular method has been effective, we need an external quality measure for each module. The number of faults discovered during testing is weak (and totally dependent on the level of testing and honesty of the tester). It would be better to obtain data on operational failures, where the failure can be traced back to individual modules.

To overcome the problems of lines of code as a poor size measure, it may be useful to collect other size data for each module. Static analysis tools or metrics packages offer inexpensive ways to collect structural metrics, which at least will indicate differences in structural complexity between modules (a difference not highlighted by lines of code). It may even be useful (although not as inexpensive) to use function point analysis.

5. Answers:

a. There are several important aspects of dependability:

    i. *Safety* must be the main concern, since lives are endangered if the system malfunctions.

    ii. *Availability* is important, since it contributes to safety. Long or frequent "outages" will require personal supervision of patients by nursing staff, and drugs will have to be administered by hand.

    iii. *Reliability* is similarly important. Safety will be adversely affected by frequent failure, and frequent false alarms will cause unnecessary work for the medical staff, and may be to complacency, which could mean that a genuine warning is ignored. This and the following attributes may be considered to be of equal importance.

    iv. *Recoverability* is important since it contributes to high availability.

    v. *Maintainability* also contributes to high availability, and (in the case of removal of design faults) to growth in safety and reliability.

    vi. *Usability* is important from the point of view of the nursing staff, mainly because it contributes to safety by reducing the probability of human error, for example, in missing an indication that a patient requires attention.

It is important to recognize that death or deterioration in a patient's condition is worse than those resulting only in inconvenience to staff or financial loss to the hospital. A case may be made for maintainability or recoverability being more important than availability or reliability. Security is of relatively minor concern, since the system is unlikely to be open to potential external interference. Features such as "fault-tolerance" assist reliability and safety, but are not themselves dependability attributes.

b. Any of the following are acceptable:

    i. *Location*: Identity of the installation and piece of equipment on which the incident was observed.

    ii. *Timing(1)*: When did the incident occur in real time?

iii. *Timing(2):* When did the incident occur in relation to system operational time, or software execution time?

iv. *Mode(1):* Type of symptoms observed.

v. *Mode(2):* Conditions of use of system.

vi. *Effect:* Type of consequence. This attribute will partly determine whether the incident affects safety, availability, or reliability.

vii. *Mechanism:* How did the incident come about?

viii. *Cause(1):* Types of trigger, such as physical hardware failure, operating conditions, malicious action, user error, erroneous report, "unexplained." This attribute will determine whether the incident is classified as relevant to usability or reliability.

ix. *Cause(2):* Type of source, such as physical hardware fault, unintentional design fault, intentional design fault, usability problem, "no cause found."

x. *Cause(3):* Identity of source (mode of failure and component in the case of physical failure, identity of fault in the case of design failure).

xi. *Severity(1):* Categorization of how serious the effect of the incident was. This will determine whether the incident affects safety.

xii. *Severity(2):* Cost to user (nursing staff or hospital) in terms of time lost, inconvenience, or extra effort incurred.

xiii. *Cost:* How much effort and other resources were expended by the vendor in order to diagnose and respond to the report of the incident?

xiv. *Count:* How many incidents occurred in given time interval?

c. These are examples of types of data that might be needed:

i. Records of design faults and design modifications. Operating time of each part of the system, and (ideally) execution.

ii. Time of each software module.

iii. The operational profile of the various parts of the system must also be recorded, since this will affect the levels of safety and reliability.

iv. To measure availability, recoverability, or maintainability, the time to restore service and the effort to diagnose and repair faults must be recorded.

v. To measure operating time and mode of operation accurately, it will be necessary to instrument the system, and write the data automatically to a file.

It is important to distinguish clearly between operating time and real time (calendar, or wall-clock, time), and to note that the operating time-base for the BED computer and for PAN computers (which execute many copies of the same code simultaneously) are different.

d. Five possible modes are the following:

i. PAN sends signal to operate drug pump when not required. Affects safety. *Effect:* Pump administers overdose. *Severity:* Critical/major (depending on the drug).

ii. BED software crashes. Affects availability, possibly safety. *Effect:* Loss of patient status information to nurse. *Severity:* Critical/major (depending on length of outage).

iii. PAN software does not detect that a sensor has failed. *Effect:* Relevant vital sign not monitored. *Severity:* Major.

iv. BED software indicates emergency although PAN has not reported deviation of vital signs. Affects reliability. *Effect:* Nurse visits patient unnecessarily. *Severity:* Minor.

v. "Wake up" signal not sent to junior doctor sufficiently often. Affects reliability, possibly safety. *Effect:* Delay in response of doctor to emergency. *Severity:* Major/minor.

6. Answers:

a. For each failure, it will be necessary to record:

i. *Time:* Both calendar time of the failure and total execution time of the software up to the failure (or alternatively, the total execution time on all instances of the software during a given calendar time period) are required.

ii. *Location:* The particular instance of the software that failed, identified by the installation on which it is running. In this case, the control room will operate one (or more) instance(s) of the CAM software, and each taxi will operate an instance of the COT software.

iii. *Mode:* What symptoms were observed? That is, what was the system seen to do that was not part of its required functions?

iv. *Effect:* What were the consequences of the failure for the driver of the taxi and for the whole of the SCAMS operation?

v. *Severity:* How serious were these consequences? (This may be expressed either as a classification into major/minor/negligible, or in terms of the cost of lost business, or of the time that CAM or COT were out of action.)

vi. *Reporter:* Person making the report (taxi driver, control room operator, customer).

The following are five examples of software failure modes:

- *CAM software crashes*
    - *Time:* Date and time. Hours and minutes CAM was running until failure.
    - *Location:* Control room.
    - *Mode:* Loss of all screen displays, keyboard unresponsive, all COT systems lose contact with control room.
    - *Effect:* Loss of central control and guidance for all vehicles until CAM is to be rebooted and reestablishes contact with COT systems.
    - *Severity:* Major. Disruption of service for all drivers and passengers, with considerable inconvenience and loss of business.
    - *Reporter:* Control room operator.

- *COT software crashes*

  - *Time:* Date and time. Hours and minutes of all COT systems have been running until failure. (Alternatively, if that is not possible, total running hours on all COT systems on that date.)

  - *Location:* Number of the taxi whose COT failed, and where it was at the time. (The latter might be useful if the COT failure was in response to certain geographical data being input.)

  - *Mode:* Driver's display blank, driver's control console unresponsive, loss of communication of CAM with that vehicle.

  - *Effect:* Loss of central control and guidance for the affected vehicle until the driver could restart the COT.

  - *Severity:* Minor. Disruption of service for one driver and (possibly) passenger. Some loss of business (for instance, if driver could not pick up a passenger at the agreed place and time).

  - *Reporter*: Driver.

- *CAM sends incorrect directions to COT*

  - *Time:* Date and time. Hours and minutes CAM was running until failure.

  - *Mode:* Directions displayed on driver's screen are inconsistent with the actual position of the vehicle as observed by the driver.

  - *Effect*: Driver confused, and possibly gets lost.

  - *Severity:* Minor, unless the same CAM failure affects many vehicles.

  - *Reporter:* Driver or control room operator (depending on who first notices the incompatibility).

- *COT reports incorrect location to CAM*

  - *Time:* Date and time. Operating time of all COT systems.

  - *Mode:* Apparent discontinuity in vehicle movement as tracked by CAM.

- *Effect*: CAM cannot send correct directions to COT. Driver confused or lost.

- *Severity:* Minor. Affects one vehicle only.

- *Reporter:* Driver or control room operator.

- *Journey time is calculated by CAM to be shorter than it should be*

  - *Time:* Date and time. Hours and minutes CAM was running until failure.

  - *Mode:* ETA displayed by CAM is found to be inaccurate when compared to driver's actual reported journey time.

  - *Effect:* Driver arrives to pick up passenger later than arranged.

  - *Severity:* Minor, unless the inaccuracy affects many journeys.

  - *Reporter:* Driver.

b. There are several important dependability attributes.

Safety: The only critical failure mode is the failure of the silent alarm. Target: 1 failure per thousand demands (maximum).

Reliability: Targets will differ for different modes of failure:

  i. For crash of COT software: 1 per 1000 operating hours.

  ii. For crash of CAM software: 1 per 10,000 operating hours.

  iii. For wrong directions: 1 per 5000 CAM operating hours, etc.

Recoverability: Target is to have the software recover after crash or error in an acceptable time. Different targets will apply to CAM and COT:

  i. CAM: reload and resume service after crash within 1 min on average.

  ii. COT: reload and resume service after crash within 10 s on average.

  iii. Other targets may be set for recovery from lesser error conditions; for example, detection of CAM that it has given wrong directions, and alerting driver.

Availability: Target is to have the whole system up and running for an acceptable proportion of real time. Subsidiary targets may be set for individual subsystems; for example,

i. CAM: 99.8% up time.

ii. COT: 98% up time.

Maintainability: Target can be set for diagnosing and fixing a software fault.

c. Reliability, usability, safety, and security all relate to our expectation that certain types of events will *not* occur in operation. They can be measured by "the probability that the system will deliver a required service, under given conditions of use, for a given time interval, without relevant incident."

Reliability relates to departures of any kind from the required service. Usability relates to difficulties experienced by a human user attempting to operate the system, and roughly equates to "user-friendliness." Safety is concerned only with incidents that can result in death, injury, or other catastrophic damage. Security relates to incidents that result in unauthorized disclosure or alteration of sensitive information, or in denial of service to legitimate users by unauthorized persons.

It is therefore necessary to record all incidents, and measure their location, time of occurrence, the execution time over which they occur, mode, effect, and severity. Once again, it is necessary to record the execution time for CAM and COT software separately. The COT execution time needs to be totaled over all vehicles, and this will probably mean that only "failure count" data are available for COT, whereas "time to failure" data can be obtained fairly easily for CAM.

Once an incident has been shown to be a genuine failure and diagnosed as being due to a software fault, the identity of the fault responsible should be recorded. The type of fault can be used to establish to which attribute a given incident is relevant. The fault identity and cross-reference from each record of a failure in which it manifested itself can be used to extract "execution time up to first manifestation of each fault" (for CAM), or "count of faults manifest and total execution time in a given period" (for COT).

Measurement of maintainability requires the recording of the effort to diagnose and repair each fault. Recoverability requires records of time to restore service after each failure. A measure of availability can be derived from reliability and recoverability.

CHAPTER 6: ANALYZING SOFTWARE MEASUREMENT DATA

10. Answers:

a. The standard quality measure in this context is defect density, calculated as number of defects divided by thousands of lines of code. The defect density for each system is given in the table below:

Module	Defect Density
A	0.88
B	0
C	190
D	0.7
E	0.46
F	0.57
G	0.92
H	0.4
I	1.125

b. Box plots reveal that the defect metric has two outliers: one high (system C, with 95 defects) and one low (system B, with no defects). The lines of code metric has no outliers. The defect density metric has one outlier (over 15 times bigger than any other), namely system C at 19 defects/KLOC.

c. Clearly, subsystem C is the problem area. Despite the fact that it is exceptionally small (just 4KLOC), it contains an abnormally high number of defects. You would be well advised to investigate this subsystem further to determine why. Is it exceptionally complex? Is it the part of the system most used? Has it the worst programmers or tools? There is also something unusual about subsystem B. Although large, this subsystem had no defects. Are there positive lessons to learn from it (for example, was it developed with the best programmers or tools?), or is it simply that this subsystem was not used?

d. A major weakness is that there is no notion of usage; subsystem C may be used much more than others, and system B may be never used. There are also no notions of severity of defects nor of complexity of subsystems. (KLOC is just a crude measure of size.) There is a blurring of the key distinction between fault and failure. Why should a failure always be due to a single defect in a single subsystem?

e. It would be extremely useful to get some measure of usage for each subsystem, since this would enable us to use the existing data to produce genuine reliability assessments of the different subsystems. Since system users are already logging failures as they occur, it should not be too difficult to get them to log the way they use the system. We should at least be able to get some estimate of the relative calendar time spent using the different subsystems. It would not be too difficult to insert some monitoring code into the system so that this can be done automatically. Another addition to the data-collection should be the measurement of actual time to locate and fix faults. This would reveal important maintainability information. Given the size of the system and the number of defects per year, maintainability is a major concern.

11. The outliers for CFP are modules R and P. The outliers for faults are P, Q, and R. So, the outliers for CFP and faults are more or less the same, whereas LOC has none. This result suggests that a high CFP is a more reliable indicator of a faulty module than high LOC. Thus, it would be a wise testing strategy to compute CFP for each module in a system and devote more time to testing, or redesigning, those that are outliers. Contrary to what many will say, there may be no point at all in redesigning modules R and P to lower their CFP, as the faults have already occurred!

12. The outliers of MOD and FD are the same, namely systems A, D, and L. Systems A and D are outliers of MOD because they have exceptionally low average module size, while system L is an outlier because it has exceptionally high average module size. Each of these three systems also had abnormally high fault density. This suggests that those systems in which average module size is either very high or very low are likely to be the most fault-prone. On the basis of this data, the outliers of MOD are excellent predictors of the high outliers of

FD. A reasonable quality assurance procedure would therefore be to restrict module sizes (on average) to be between 16 and 88 LOC. This procedure should avoid systems whose fault-proneness is explained by module size. A stricter quality procedure might focus on values between the upper and lower quartiles. In other words, we should restrict module sizes (on average) to be between 43 and 61 LOC. From the data, we can also conclude that size of a system alone (measured by KLOC) gives no indication of its fault proneness, nor of its likely average module size. For example, the only outlier for KLOC—the exceptionally large system R—does not have either an unusual average module size or an unusually high fault density.

CHAPTER 8: MEASURING INTERNAL PRODUCT ATTRIBUTES: SIZE

7. Function points are a measure of the functionality of a software system. The unadjusted function count UFC is derived from counting system inputs, outputs, enquiries, and files. A technical complexity factor, F, is then computed for the system, and the function point count is FP = UFC*F.

The main applications of FPs are:

a. Sizing for purposes of effort/cost estimation (provided that you have data about previous projects that relates the number of FPs in a system to the actual cost/effort).

b. Sizing for purposes of normalization. Thus, FPs are used to compute quality density (defects/FP), productivity (person-months/FP), etc.

Comparing FPs with LOC:

a. Unlike LOC, FPs can be extracted early in software life-cycle (from requirements definition or specification) and so can be used in simple cost-estimation models where size is the key parameter.

b. FPs, being a measure of functionality, are more closely related to utility than LOC.

c. FPs are language-independent.

d. FPs can be used as a basis for contracts at the requirements phase.

However:

a. FPs are difficult to compute, and different people may count FPs differently.

b. Unlike LOC, FPs cannot be automatically extracted.

c. There is some empirical evidence to suggest that FPs are not very good for predicting effort. Empirical evidence also suggests that FPs are unnecessarily complex.

8. Answers:

a. Function points are supposed to measure the amount of functionality in a software "product," where product can mean any document from which the functional specification can be extracted: the code itself, the detailed design, or the specification. Function points are defined in a language-independent manner, so the number of function points should not depend on the particular product representation. Function points are also commonly interpreted as a measure of size.

Drawbacks:

i. The main drawback of function points is the difficulty in computing them. You must have at least a very detailed specification. This task is not easily automated or even repeatable. Different people will generally arrive at a different FP count for the same specification, although the existence of standards helps minimize the variance.

ii. The definition of function points was heavily influenced by the assumption that the number should be a good predictor of effort; in this sense, the function point measure is trying to capture more than just functionality. Thus, FPs are not very well-defined from the measurement theory perspective.

iii. FPs have been shown to be unnecessarily complicated. In particular, the TCF appears to add nothing in terms of measuring functionality, nor does it help to improve the predictive accuracy when FPs are used for effort prediction.

b. Most common answers will be:

    i. FPs can be used as the main input variable in an effort prediction system. Traditionally, LOC (or some similar code-based metric) has been used. The advantage of FPs is that they can be computed directly from the specification, so you need not predict LOC early in development.

    ii. Function points can be used in any application where you need to normalize by size. So, if you measure productivity traditionally by LOC/effort, it would be advantageous to use FPs instead of LOC, since FPs are more obviously related to the utility of output, and they are independent of the language used. You could also use FPs in this same equation to measure designer (or even specifier) productivity and not just coder productivity.

    iii. You could use FPs to measure defect density as defects/FP, rather than defects/LOC.

c. There is no single correct answer here. However, good answers should contain similar information.

    We identify the following items and their associated "complexity," weighting them as follows (using the table for UFC weighting factors):

External Inputs		
Coursework marks	Simple	3
Exam marks	Simple	3
Menu-selection: course choice	Simple	3
Menu-selection: operation choice	Simple	3
External Inquiries		
Average	Simple	3
Letter grade	Average	4
External Outputs		
List of student marks, etc.	Average	5
External Files		
None.		
Internal File		
Course file database: This contains the course list, student lists, and all known marks and grades.	Complex	10

Then UFC = 35

To compute the technical complexity factor (TCF), we consider the 14 listed factors in the Albrecht model. We assume that all but the following are irrelevant:

F6	Online data entry
F7	Operational ease
F9	Online update
F13	Multiple sites

Each of these is rated essential and hence get a weighting of 5.
Therefore, TCF = 0.65 + 0.01 * (4 * 5) = 0.85
Thus, FP = UFC * TCF = 30.

CHAPTER 9: MEASURING INTERNAL PRODUCT ATTRIBUTES: STRUCTURE

15. We look at each of the properties:

    a. Property 1—Nonnegativity: Does not hold. Any module or system with fan-out more than double the fan-in will have negative Cnew or CnewSyst.

    b. Property 2—Null value: Holds, as a very simple one module system can have no fan-in or fan-out.

    c. Property 3—Symmetry: Holds, as long as fan-in and fan-out links can be recognized.

    d. Property 4—Module monotonicity: Does not hold. A system with modules with very low Cnew values (very high fan-out) could cause CnewSyst to be less than the sum of two selected modules. This is a consequence of Property 1 being violated.

    e. Property 5—Disjoint module additivity: Does not hold. This is because individual Cnew and CnewSyst factors are computed differently. To compute CnewSyst, the total size of the system multiplies all fan-in and fan-out factors. The CnewSyst can vary dramatically from the sum of the Cnew values for individual modules, especially when the sizes of the systems vary greatly.

17. Coupling is a property of pairs of modules, while cohesion is a property of individual modules. Coupling between modules is the extent of interdependence between modules, whereas cohesion of a module is the extent to which the elements of the module have a common purpose, that is, are part of the same function.

It is generally believed that if a design is made up of a number of related modules, then there should be (as far as possible) a low level of coupling; this way, errors made in any one module should affect a minimum number of others. Also, low coupling should help keep independent the implementation of the modules. On the other hand, high cohesion of each module may be desirable for conceptual simplicity and ease of testing. The lowest level of coupling can be achieved by having a single module for the whole system, but such a module will have very low cohesion. Analogously, we could ensure the highest level of cohesion of each module at the expense of very high coupling, namely where each individual statement corresponds to a module. Therefore, we have to find an optimal balance of low coupling and high cohesion.

18. The software entity is source code (or the flowgraph representation of source code). The attribute is, strictly speaking, the number of linearly independent paths through the code, or the number of decisions plus one. The attribute is internal.

19. The statement coverage strategy is a form of white box testing. The tester must select inputs so that, when the program is executed, enough paths are executed for each statement of the program to lie on at least one path.

22. A procedure is D-structured, formally, if its decomposition tree (shown below) contains only primes of the form $P_n, D_0, D_1, D_2,$ or $D_3,$ which is true in this case:

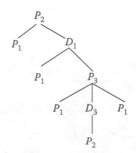

23. The key thing to note about this algorithm is that its underlying flowgraph is the double-exit loop shown below.

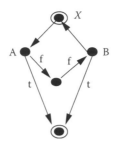

In the strict interpretation of structured programming, only certain single-exit loops are allowed as building blocks. The two-exit loop is a prime structure that cannot be structured in terms of other such loops. Thus, in the strict sense, the algorithm is unstructured. But this example highlights the limitations of the strict view. Any attempt to rewrite the algorithm in "structured form" requires the introduction of new dummy variables which mar the simple and intuitive structure of the original algorithm. More liberal views of structured programming allow primes such as the two-exit loop, which is very natural in any control environment.

CHAPTER 10: MEASURING EXTERNAL PRODUCT ATTRIBUTES

2. The measure is quite useful for developers but is almost useless for users or potential purchasers. For developers who measure both LOC and faults found (in the code) in a consistent manner, the measure will indicate:

a. Broad differences in quality among different modules, systems, teams

b. Trends that can aid quality control efforts

c. Potential troublespots in the system

d. When the developers have reached the point of diminishing returns on testing

However, the measure is of limited usefulness to the user, as it

a. Gives no *real* indication of reliability (since faults may not be good predictor of failures, as shown in the Adams data)

b. Gives no indication at all of *usability*

c. Gives no indication of the severity of the faults

d. Cannot be used to compare products from different producers; the producers may count LOC differently, and they may have different definitions or classifications of faults

e. May say more about the rigor of the testing process (or of the tester) than about the quality of the code; clearly, the testing strategy influences how many faults are found

f. Invites abuse by programmers who may artificially increase the length of a program in order to be seen to be producing higher-quality code

g. Is irrelevant for assessing quality before coding has begun

h. Cannot be used comparatively with earlier products whose system size is not LOC

i. Does not take account of reused code

3. To compare projects with a given measure, it must be either normalized using size or type of project or independent of size/type of project. None of the measures here is normalized. Therefore, the only ones of value for comparative purposes are:

- Mean time to failure, for all projects at beta-test phase (a good measure of system reliability)

- Mean time to repair reported defects (a reasonable measure of maintainability)

- Maximum cyclomatic number (a very crude measure of system structuredness, and much less useful than the previous two measures)

- Average number of function points produced per month of programmer effort (a crude measure of programmer productivity; it would be dangerous to use this if the projects involved vastly different applications).

Each of the other measures, although useful for tracking and quality control purposes within a project, cannot be used sensibly for cross-project comparisons unless normalized.

- Total number of user-reported failures: For this to be used as a comparative measure of quality, it should be normalized against both system size and usage time. The former could be measured by total number of function points in the system (presumably already collected) or a simple measure like LOC. The latter would be much more difficult to measure, since it can be done only by the system users.

- Total number of defects found during system testing: At the very least, this measure should be normalized against size (as above). It may also need normalization against amount of testing.

- Total number of changes made during development: If this measure is normalized against size (preferably measured by function points), it may be a measure either of system volatility (if the changes are being suggested by customers) or quality (if the changes are being made as a result of discovering faults).

- Total project overspend/underspend: This measure should be normalized against actual expenditures to yield a measure of the accuracy of the cost predictions.

CHAPTER 11: SOFTWARE RELIABILITY: MEASUREMENT AND PREDICTION

8. Both prequential likelihood ratios actually converge at around $i = 10$ for this particular dataset. This result suggests that neither one is better than the other. (In fact, neither is particularly good.)

9. The totality of all possible inputs forms an input space. A fault can be thought of as a collection of points in the input space, each of which, when executed, results in output that is regarded as failed. (The decision to label an output failed will depend upon a comparison of what was required with what was produced.) Execution of the program involves the successive execution of a sequence of inputs. This sequence will be a random walk in the input space—random because the selection of future inputs is unpredictable and determined by the outside world. Such a trajectory through the input space will fall over the fault regions and so cause failures randomly.

There is a further source of randomness (or uncertainty) when a fix is carried out. If the fix is successful, we are uncertain of the magnitude of the effect of its removal on the unreliability. There will be a tendency for high rate faults to be encountered, and removed, earlier than low rate ones. In addition there is the possibility that a fix is not perfect. In this case, the effect on the program may be to increase the unreliability by an unknown amount.

10. The u-plot essentially detects consistent bias in predictions of reliability: the magnitude of the maximum departure from unit slope can be used to test whether the bias is statistically significant. The precise shape of the u-plot gives information about the nature of the errors. For example, a plot that is consistently above the line of unit slope tells us that the predictions are too optimistic.

The prequential likelihood ratio allows us to compare the accuracies of prediction of model A with model B on the same data. If prequential likelihood ratio is consistently increasing as data vector increases in dimension, the model in numerator can be said to be more accurate than the model in the denominator. The prequential likelihood ratio is sensitive to all kinds of departure from the truth, and not just bias, but it is only a comparative analysis.

The smoothed u-plot of previous predictions can be used to modify the current prediction as follows:

$$F_i{}^* (t) = G_{i-1}\{F_i(t)\}$$

where $F_i(t)$ is a raw estimate from a particular model of $P(T_i < t)$, G_{i-1} is the smoothed u-plot based on predictions of … t_{i-2}, t_{i-1}, and $F_i^*(t)$ is the modified—recalibrated—prediction of $P(T_i < t)$.

11. The results clump into two groups: Six model predictions in one group, two in the other. There is reasonable agreement of the median predictions within each group, but big differences between groups. From the u-plot, all predictions are bad: 6 too optimistic, 2 too pessimistic. From the prequential likelihood ratio, the two pessimistic ones are not as bad as the six optimistic ones. None of the predictions can be trusted as they are; they should be recalibrated.

12. Answers:

    a. *Debugging data:* Use a reliability growth model, but you can get only quite modest reliability because of the law of diminishing returns. You must also worry about the efficacy of fixes, since models tend to assume fixes are correct and do not introduce new faults; this is not a conservative assumption in the case of safety-critical systems.

b. *Failure-free working:* This is quite weak evidence of reliability. Roughly, we can expect there to be a 50:50 chance of working failure-free for a time *t* when we have seen a time *t* of failure-free working.

c. *Diversity:* There are experimental and theoretical reasons to doubt that the versions will fail independently. There is some evidence that this approach does deliver improvements over single version, but the improvement is hard to quantify. Evaluating a particular diverse system is equivalent to treating it as a black box, and thus the previous two paragraphs apply.

d. *Verification:* This technique addresses consistency with formal specification, but does not always address the issue of whether this is complete and accurate representation of the informal engineering requirements. There are practical difficulties with verification for any except small programs, because of resource constraints.

e. *Process:* There is very weak evidence to show a link between process and product.

14. Unless you know a lot about the faults you have removed, there is very little you can say about the improvements to reliability. It might be tempting to assume a proportional improvement in reliability, but on the basis of the Adams data, you could actually remove 95% of all faults and yet see no perceptible reliability improvements; this is because a very small proportion of faults cause almost all the common failures.

Bibliography

Abdel-Ghaly A.A., Chan P.Y., and Littlewood B., Evaluation of competing software reliability predictions, *IEEE Transactions on Software Engineering*, SE-12(9), 950–967, 1986.

Abran A., *Software Metrics and Software Metrology*, John Wiley & Sons, Hoboken, New Jersey, 2010.

Abran A. and Robillard P.N., Function points: A study of their measurement processes and scale transformations, *Journal of Systems and Software*, 25(2), 171–184, 1994.

Adams E., Optimizing preventive service of software products, *IBM Journal of Research and Development*, 28(1), 2–14, 1984.

Albrecht A.J., Measuring application development, *Proceedings of IBM Applications Development Joint SHARE/GUIDE Symposium*, Monterey, California, pp. 83–92, 1979.

Alhazmi O.H., Malaiya Y., and Ray I., Measuring, analyzing and predicting security vulnerabilities in software systems. *Computers and Security*, 26(3), 219–228, May 2007.

Ammann P. and Offutt J., *Introduction to Software Testing*, Cambridge University Press, New York, 2008.

Anda B. Comparing effort estimates based on use case points with expert estimates, *Empirical Assessment in Software Engineering (EASE 2002)*, Keele, UK, 2002.

Anda B., Dreiem H., Sjoberg D.I.K., and Jorgensen M., Estimating software development effort based on use cases—Experiences from industry, *Proceedings of UML 2001—The Unified Modeling Language. Modeling Languages, Concepts, and Tools*, pp. 487–502, 2001.

Antoniol G., Fiutem R., and Lokan C., Object-oriented function points: An empirical evaluation, *Empirical Software Engineering*, 8(3), 225–254, 2003.

Antoniol G., Lokan C., Caldiera G., and Fiutem R., A function-point like measure for object oriented software, *Empirical Software Engineering*, 4(3), 263–287, September 1999.

Arisholm E. and Sjoberg D.I.K., Evaluating the effect of a delegated versus centralized control style on the maintainability of object-oriented software, *IEEE Transactions on Software Engineering*, 30(8), 521–534, August 2004.

Bache R. and Neil M., Introducing metrics into industry: A perspective on GQM, in *Software Quality Assurance and Metrics: A Worldwide Perspective* (Eds:

Fenton N.E., Whitty R.W., and Iizuka Y.), International Thomson Press, London, pp. 59–68, 1995.

Bachmann A. and Bernstein A., Software process data quality and characteristics—A historical view on open and closed source projects, *Proceedings of the Joint ERCIM Workshop on Software Evolution and International Workshop on Principles of Software Evolution (IWPSE-Evol'09)*, Amsterdam, Netherlands, August 24–25, 2009, pp. 119–128, 2009.

Baker A.L., Bieman J.M., Fenton N.E., Gustafson D., and Melton A., A philosophy for software measurement, *Journal of Systems and Software*, 12, 277–281, July 1990.

Banker R., Kauffman R., and Kumar R., An empirical test of object-based output measurement metrics in a computer-aided software engineering (CASE) environment, *Journal of Management Information Systems*, 8(3), 127–150, 1994a.

Banker R., Chang H., and Kemerer C.F., Evidence on economies of scale in software development, *Information and Software Technology*, 36(5), 275–282, 1994b.

Barford N.C., *Experimental Measurements: Precision, Error and Truth*, 2nd Edition. Addison-Wesley, Reading, Massachusetts, 1985.

Barnard J. and Price A., Managing code inspection information, *IEEE Software*, 11(2), 59–69, March 1994.

Basili V.R. and Rombach H.D., The TAME project: Towards improvement-oriented software environments, *IEEE Transactions on Software Engineering*, 14(6), 758–773, 1988.

Basili V.R., Selby R.W., and D.H. Hutchens, Experimentation in software engineering, *IEEE Transactions on Software Engineering*, 12(7), 733–743, July 1986.

Basili V.R. and Weiss D., A methodology for collecting valid software engineering data, *IEEE Transactions on Software Engineering*, SE-10(6), 728–738, 1984.

Basili V.R., Heidrich J., Lindvall M., Münch J., Seaman C, Regardie M., and Trendowicz, A., Determining the impact of business strategies using principles from goal-oriented measurement, *Proceedings of the 2009 3rd International Symposium on Empirical Software Engineering and Measurement (ESEM)*, Lake Buena Vista, Florida, pp. 390–400, 2009.

Baudry B. and Sunye G., Improving the testability of UML class diagrams, *Proceedings of the First International Workshop on Testability Assessment, 2004 (IWoTA)*, pp. 70–80, November 2004.

Baudry B., Le Traon Y., and Sunyé G., Testability analysis of a UML class diagram, *Software Metrics, 2002. Proceedings of the Eighth IEEE Symposium*, Ottawa, Canada, IEEE, pp. 54–63, 2002.

Baudry B., Traon Y., Sunye G., and Jezequel J.M., Towards a safe use of design patterns to improve OO software testability, *Proceedings of the 12th International Symposium on Software Reliability Engineering (ISSRE 2001)*, Hong Kong, China, pp. 324–329, 2001.

Baur E., Zhang X., and Kimber D.A., *Practical System Reliability*, Wiley IEEE Press, Hoboken, New Jersey, 2009.

Belton V., A comparison of the analytic hierarchy process and a simple multi-attribute utility function, *European Journal of Operational Research*, 26, 7–21, 1986.

Bennett P.A., Software development for the channel tunnel: A summary, *High Integrity Systems*, 1(2), 213–220, 1994.

Berg van den K.G. and Broek van den P.M., Static analysis of functional programs, *Information and Software Technology*, 37(4), 213–224, 1995.

Berger B., Data-centric quantitative computer security risk assessment, *SANS Institute InfoSec Reading Room*, August 2003, http://www.sans.org/reading-room, August 7, 2014.

Bertolino A. and Marre M., How many paths are needed for branch testing?, *Journal of Systems and Software*, 35(2), 95–106, 1995.

Bertolino A. and Strigini L., On the use of testability measures for dependability assessment, SHIP project, document T34/v0.3, 1995.

Bevan N., Measuring usability as quality of use, *Software Quality Journal*, 4(2), 115–130, 1995.

Bieman J. and Kang B.-L. Cohesion and reuse in an object-oriented system, *Proceedings of the ACM Symposium on Software Reusability (SSR'95)*, Seattle, Washington, pp. 259–262, April 1995.

Bieman J. and Karunanithi S. Measurement of language supported reuse in object oriented and object based software, *The Journal of Systems and Software*, 28(9), 271–293, 1995.

Bieman J. and Zhao J.X., Reuse through inheritance: A quantitative study of C++ software, *Proceedings of the ACM Symposium on Software Reusability (SSR'95)*, Seattle, Washington, pp. 47–52, April 1995.

Bieman J.M. and Debnath N.C., An analysis of software structure using a generalized program graph, *Proceedings of the IEEE-CS 9th International Computer Software and Applications Conference (COMPSAC 85)*, Chicago, Illinois, pp. 254–259, 1985.

Bieman J. and Kang B.-K., Measuring design level cohesion, *IEEE Transactions on Software Engineering*, 24(2), 111–124, February 1998.

Bieman J.M. and Ott L.M., Measuring functional cohesion, *IEEE Transactions on Software Engineering*, 20(8), 644–657, 1994.

Bieman J.M. and Schultz J., Estimating the number of test cases required to satisfy the all-du-paths testing criterion, *Proceedings of ACM TAV3 Conference*, Key West, Florida, ACM SIGSOFT Notes, pp. 179–186, 1989.

Bieman J., Andrews A., and Yang H., Understanding change-proneness in OO software through visualization, *Proceedings of the International Workshop on Program Comprehension (IWPC 2003)*, Portland, Oregon, pp. 44–53, 2003a.

Bieman J., Fenton N.E., Gustafson D., Melton A., and Whitty R., Moving from philosophy to practice in software measurement, in *Formal Aspects of Software Measurement* (Eds: Denvir T., Herman R., and Whitty R.), Chapman & Hall, London, 1992.

Bieman J., Jain D., and Yang H., Design patterns, design structure, and program changes: An industrial case study, *Proceedings of the International Conference on Software Maintenance (ICSM 2001)*, Florence, Italy, November 2001.

Bieman J.M., Baker A.L., Clites P.N., Gustafson D.A., and Melton A., A standard representation of imperative language programs for data collection and software measures specification, *Journal of Systems and Software*, 8(1), 13–37, January 1988.

Bieman J.M. and Schultz J.L., An empirical evaluation (and specification) of the all-du-paths testing criterion, *Software Engineering Journal*, 7(1), 43–51, 1992.

Bieman J., Straw G., Wang H., Munger P.W., and Alexander R., Design patterns and change proneness: An examination of five evolving systems, *Proceedings of the Ninth International Software Metrics Symposium (Metrics 2003)*, Sydney, Australia, pp. 40–49, 2003b.

Birolini A., *Reliability Engineering: Theory and Practice*, 5th Edition. Springer, Berlin, 2007.

Blum M., Luby M., and Rubinfield R., Self-testing/correcting with applications to numerical problems, *Journal of Computer and Systems Sciences*, 47, 549–595, 1993.

Boehm B.W., *Software Engineering Economics*, Prentice-Hall, Englewood Cliffs, New Jersey, 1981.

Boehm B.W., Brown J.R., Kaspar J.R., Lipow M., and MacCleod G., *Characteristics of Software Quality*, Elsevier Science Ltd., Amsterdam, North Holland, 1978.

Boehm B.W., Abts C., Brown A.W., Chulani S., Clark K.C., Horowitz E., Madachy R, Reifer D.J., and Steece B., *Software Cost Estimation with Cocomo II*, Prentice-Hall, Upper Saddle River, New Jersey, 2000.

Böhm B.W. and Jacopini G., Flow diagrams, Turing machines and languages with only two formation rules, *Communications of the ACM*, 9(5), 366–371, 1966.

Brewer E.A., Lessons from giant-scale services, *IEEE Internet Computing*, July–August, 46–55, 2001.

Briand L.C., Morasca S., and Basili V.R., Property-based software engineering measurement, *IEEE Transactions on Software Engineering*, 22(1), 68–86, 1996.

Briand L.C., Daly J.W., and Wüst J.C., A unified framework for cohesion measurement in object-oriented systems, *Empirical Software Engineering*, 3(1), 65–117, 1998.

Briand L.C., Wüst J., Ikonomovski V., and Lounis H., Investigating quality factors in object-oriented designs: An industrial case study, *Proceedings of the International Conference on Software Engineering (ICSE)*, ACM, Los Angeles, California, pp. 345–354, 1999a.

Briand L.C., Daly J.W., and Wüst J.C., A unified framework for coupling measurement in object-oriented systems, *IEEE Transactions on Software Engineering*, 25(1), 91–121, 1999b.

Brilliant S., Knight J.C., and Leveson N., Analysis of faults in an n-version software experiment, *IEEE Transactions on Software Engineering*, 16(2), 1990.

Brocklehurst S., Chan P.Y., Littlewood B., and Snell J., Recalibrating software reliability models, *IEEE Transactions on Software Engineering*, SE-16(4), 458–470, 1990.

Brocklehurst S. and Littlewood B., New ways to get accurate software reliability modeling, *IEEE Software*, 9(4), 34–42, July 1992.

Broek van den P.M. and Berg van den K.G., Generalised approach to software structure metrics, *Software Engineering Journal*, 10(2), 61–68, 1995.

Brooks F.P., *The Mythical Man-Month: Essays on Software Engineering*, 2nd Edition. Addison-Wesley, Reading, Massachusetts, 1995.

Brown W.H., Malveau R.C., McCormick III H.W., and Mowbray T.J., *AntiPatterns: Refactoring Software, Architectures and Projects in Crisis*, John Wiley & Sons, Hoboken, New Jersey, 1998.

Bush M. and Fenton N.E., Software measurement: A conceptual framework, *Journal of Systems and Software*, 12, 223–231, July 1990.

Butler S.A., Security attribute evaluation method: A cost-benefit approach, *Proceedings of the International Conference on Software Engineering (ICSE)*, Raleigh, North Carolina, pp. 232–240, 2002.

Butler R.W. and Finelli G.B., The infeasibility of quantifying the reliability of life-critical real-time software, *IEEE Transactions on Software Engineering*, 19(3), 3–12, 1993.

Butler S.A. and Fischbeck P., Multi-attribute risk assessment, *Proceedings of the 2nd Symposium on Requirements Engineering for Information Security*, Raleigh, North Carolina, 2002.

Campbell D.T. and Stanley J., *Experimental and Quasi-Experimental Designs for Research*, Rand McNally, Chicago, 1966.

Campbell N.R., *Physics: The Elements*, Cambridge University Press, Cambridge, Massachusetts, 1920. Reprinted as *Foundations of Science: The Philosophy of Theory and Experiment*, Dover, New York, 1957.

Casscells W., Schoenberger A., and Graboys T.B., Interpretation by physicians of clinical laboratory results, *New England Journal of Medicine*, 299, 999–1001, 1978.

Caulcutt R., *Statistics in Research and Development*, Chapman & Hall, London, England, 1991.

Causevic A., Sundmark D., and Punnekkat S., Impact of test design technique knowledge on test driven development: A controlled experiment, in *Agile Processes in Software Engineering and Extreme Programming*, Springer, Berlin, pp. 138–152, 2012.

Chatfield C., *Statistics for Technology: A Course in Applied Statistics*, 3rd Edition (Revised). Chapman & Hall, London, 1998.

Chen K., Schach S., Yu L., Offutt J., and Heller G., Open-source change logs, *Empirical Software Engineering*, 9, 197–210, 2004.

Cherniavsky J.C. and Smith C.H., On Weyuker's axioms for software complexity measures, *IEEE Transactions on Software Engineering*, 17(6), 636–638, 1991.

Cheung R.C., A user-oriented software reliability model, *IEEE Transactions on Software Engineering*, SE-6(3), 118–125, 1980.

Chidamber S.R. and Kemerer C.F., A metrics suite for object oriented design, *IEEE Transactions on Software Engineering*, 20(6), 476–498, 1994.

Chillarege R., Bhandari I.S., Chaar J.K., Halliday M.J. Moebus D.S., Ray B.K. and Wong M.-Y., Orthogonal defect classification: A concept for in-process measurements, *IEEE Transactions on Software Engineering*, 18(11), 943–956, November 1992.

Churcher N.I. and Shepperd M.J., Comments on 'A metrics suite for object oriented design', *IEEE Transactions on Software Engineering*, 21(3), 263–265, 1995.

CMMI Product Team, *CMMI for Development, Version 1.3* (CMU/SEI-2010-TR-033). Software Engineering Institute, Carnegie Mellon University, 2010. http://www.sei.cmu.edu/library/abstracts/reports/10tr033.cfm, February 25, 2011.

Cochran W.G., *Sampling Techniques*, 2nd Edition. John Wiley & Sons, New York, 1963.

Conte S.D., Dunsmore H.D., and Shen V.Y., *Software Engineering Metrics and Models*, Benjamin-Cummings, Menlo Park, California, 1986.

Cook T.D., Campbell D.T., and Day A., *Quasi-Experimentation: Design and Analysis Issues for Field Settings*, Houghton-Mifflin, Boston, Massachusetts, 1979.

Coverity, 2011. Coverity Scan: 2011 Open Source Integrity Report. Published by Coverity, Inc., San Fransisco, California, http://www.coverity.com/library/pdf/coverity-scan-2011-open-source-integrity-report.pdf, August 8, 2014.

Courtney R.E. and Gustafson D.A., Shotgun correlations in software measures, *Software Engineering Journal*, 8(1), 5–13, 1993.

Crespo N.C., Jino M., Pasquini P., and Maldonado J.C., A binomial software reliability model based on coverage of structural testing criteria, *Empirical Software Engineering*, 13(2), 185–209, 2008.

Curtis B., Measurement and experimentation in software engineering, *Proceedings of the IEEE*, 68(9), 1144–1157, September 1980.

DaCosta D., Dahn C., Mancoridis S., and Prevelakis V., Characterizing the security vulnerability likelihood of software functions, *Proceedings of the 19th International Conference on Software Maintenance*, Amsterdam, Netherlands, pp. 266–274, September 2003.

Davis A., Overmyer S., Jordan K., Caruso J., Dandashi F., Dinh A., Kincaid G., Ledeboer G., Reynolds P., Sitaram P., Ta A., and Theofanos M., Identifying and measuring quality in a software requirements specification, *Proceedings of the First International Software Metrics Symposium*, Baltimore, Maryland, IEEE Computer Society Press, pp. 141–152, 1993.

Dawid A.P., Statistical theory: The prequential approach, *Journal of the Royal Statistical Society*, A147, 278–292, 1984.

Dawid A.P. and Vouk V.G., Prequential probability: Principles and properties, *Bernoulli*, 5(1), 125–162, 1999.

De Young G.E. and Kampen G.R., Program factors as predictors of program readability, *Proceedings of the Computer Software and Applications Conference (COMPSAC)*, IEEE Computer Society Press, Chicago, Illinois, pp. 668–673, 1979.

DeMillo R.A. and Lipton R.J., Software project forecasting, in *Software Metrics*, (Eds: Perlis A.J., Sayward F.G. and Shaw M.), MIT Press, Cambridge, Massachusetts, pp. 77–89, 1981.

DeMillo R.A., Lipton R.J, and Sayward F.G, Hints on test data selection: Help for the practicing programmer, IEEE Computer, 11(4), 34–41, 1978.

Denvir T., Herman R., and Whitty R.W. (Eds), *Formal Aspects of Software Measurement*, Springer Verlag, Heidelberger, Germany, 1992.

Dobson A.J., *An Introduction to Generalized Linear Models*, 3rd Edition. Chapman & Hall, London, 2008.

Draper N. and Smith H., *Applied Regression Analysis*, 3rd Edition. John Wiley & Sons, New York, 1998.

Drapkin T. and Forsyth R., *The Punters Revenge*, Chapman & Hall, London, 1987.

Duane J.T., Learning curve approach to reliability monitoring, *IEEE TransAerospace*, 2, 563–566, 1964.

Eckhardt D.E. and Lee L.D., A theoretical basis for the analysis of multi-version software subject to coincident errors, *IEEE Transactions on Software Engineering*, SE-11(12), 1511–1517, 1985.

Ellis B., *Basic Concepts of Measurement*, Cambridge University Press, Oxford, England, 1966.

Falmagne J.-C. and Narens L., Scales and meaningfulness of quantitative laws, *Synthese*, 55, 287–325, 1983.

Fenton, N.E., The structural complexity of flowgraphs. In Y. Alavi., G. Chartrand, L. Lesniak, and Lick (Eds.), *Graph Theory and its applications to Algorithms and Computer Science*, Kalamazoo, Wiley, New York, pp. 273–282, 1985.

Fenton N.E., Software measurement: Why a formal approach?, in *Formal Aspects of Software Measurement* (Eds: Denvir T., Herman R., and Whitty R.W.), Springer Verlag, Heidelberger, Germany, pp. 3–27, 1992a.

Fenton N.E., When a software measure is not a measure, *Software Engineering Journal*, 7(5), 357–362, 1992b.

Fenton N.E., How effective are software engineering methods?, *Journal of Systems and Software*, 20, 93–100, 1993a.

Fenton N.E., The effectiveness of software engineering methods, *Proceedings of AQuIS '93 (Second International Conference on Achieving Quality in Software)*, Venice, Italy, pp. 295–305, 1993b.

Fenton N.E., Software measurement: A necessary scientific basis, *IEEE Transactions on Software Engineering*, 20(3), 199–206, 1994.

Fenton N.E. and Hill G., *Systems Construction and Analysis: A Mathematical and Logical Approach*, McGraw-Hill, New York, 1992.

Fenton N.E. and Kaposi A.A., Metrics and software structure, *Journal of Information and Software Technology*, 29, 301–320, July 1987.

Fenton N.E. and Kitchenham B.A., Validating software measures, *Journal of Software Testing, Verification and Reliability*, 1(2), 27–42, 1991.

Fenton N.E. and Littlewood B. (Eds), *Software Reliability and Metrics*, Elsevier, 1991 (Edited Proceedings of the Centre for Software Reliability Conference, Garmisch-Partenkirchen, Germany, September 12–14, 1990).

Fenton N.E. and Melton A., Deriving structurally based software measures, *Journal of Systems and Software*, 12, 177–187, 1990.

Fenton N.E. and Neil M., A critique of software defect prediction models, *IEEE Transactions on Software Engineering*, 25(5), 675–689, 1999.

Fenton N.E. and Neil, M., *Risk Assessment and Decision Analysis with Bayesian Networks*, CRC Press, Boca Raton, Florida, ISBN: 9781439809105, ISBN 10: 1439809100, 2012.

Fenton N.E. and Ohlsson N. Quantitative analysis of faults and failures in a complex software system, *IEEE Transactions on Software Engineering*, 26(8), 797–814, 2000.

Fenton N.E. and Whitty R.W., Axiomatic approach to software metrication through program decomposition, *Computer Journal*, 29(4), 329–339, 1986.

Fenton N.E. and Whitty R.W., Program structures: Some new characterizations, *Journal of Computer and System Sciences*, 43(3), 467–483, 1991.

Fenton N., Pfleeger S.L., and Glass R.L., Science and substance: A challenge to software engineers, *IEEE Software*, 11(4), 86–95, July 1994.

Fenton N.E., Littlewood B., and Page S., Evaluating software engineering standards and methods, in *Software Engineering: A European Perspective* (Eds: Thayer R. and McGettrick A.D.), IEEE Computer Society Press, Los Alamitos, California, pp. 463–470, 1993.

Fenton N.E., Marsh W., Cates P., Forey S., and Tailor M., Making resource decisions for software projects, *Proceedings of the 26th International Conference on Software Engineering (ICSE2004)*, Edinburgh, United Kingdom, IEEE Computer Society, pp. 397–406, 2004.

Fenton N.E., Neil M., and Gallan J., Using ranked nodes to model qualitative judgments in Bayesian networks, *IEEE Transactions on Knowledge and Data Engineering*, 19(10), 1420–1432, 2007a.

Fenton N. E., Neil M., and Marquez D., Using Bayesian networks to predict software defects and reliability, *Proceedings of the Institution of Mechanical Engineers, Part O, Journal of Risk and Reliability*, 222(O4), 701–712, 2008a.

Fenton N.E., Neil M., Marsh W., Hearty P., Marquez D., Krause P., and Mishra, R., Predicting software defects in varying development lifecycles using Bayesian nets, *Information & Software Technology*, 49, 32–43, 2007b.

Fenton N.E., Neil M., Marsh W., Hearty P., Radlinski L., and Krause P., On the effectiveness of early life cycle defect prediction with Bayesian Nets, *Empirical Software Engineering*, 13, 499–537, 2008b.

Fenton N.E., Neil M., and Krause P., Software measurement: Uncertainty and causal modelling, *IEEE Software*, 10(4), 116–122, 2002.

Fenton N.E., Whitty R.W., and Kaposi A.A., A generalized mathematical theory of structured programming, *Theoretical Computer Science*, 36, 145–171, 1985.

Figueira, J., Salvatore G., Matthias E., *Multiple Criteria Decision Analysis: State of the Art Surveys*, Springer Science + Business Media, Inc., New York, ISBN 0-387-23081-5, 2005.

Finkelstein L., Representation by symbol systems as an extension of the concept of measurement, *Kybernetes*, 4, 215–223, 1975.

Finkelstein L., What is not measurable, make measurable, *Measurement and Control*, 15, 25–32, 1982.

Finkelstein L., A review of the fundamental concepts of measurement, *Measurement*, 2(1), 25–34, 1984.

Fisher R.A., *Statistical Methods for Research Workers*, Genesis Publishing, London, 1925.

Fowler M. *Refactoring: Improving the Design of Existing Code*, Addison-Wesley, Reading, MA, 1999.

Frank M.V., Choosing among safety improvement strategies: A discussion with examples of risk assessment and multi-criteria decision approaches for NASA, *Reliability Engineering and System Safety*, 49, 311–324, 1995.

Fredreiksen H.D. and Mathiassen L., Information-centric assessment of software metrics practices, *IEEE Transactions on Engineering Management*, 52(3), 350–362, August 2005.

Fuggetta A., Lavazza L., Morasca S., Cinti S., Oldano G., and Orazi E., Applying GQM in an industrial software factory, *ACM Transactions on Software Engineering and Methodology*, 7(4), 411–448, October 1998.

Gamma E., Helm R., Johnson R., and Vlissides J., *Design Patterns: Elements of Reusable Object-Oriented Software*. Addison-Wesley, Boston, 1994.

Garey M.R. and Johnson D.S., *Computers and Intractability: A Guide to the Theory of NP-Completeness*, W.H. Freeman, San Francisco, 1979.

Gilb T., *Software Metrics*, Chartwell-Bratt, Cambridge, Massachusetts, 1976.

Gilb T., *Principles of Software Engineering Management*, 2nd Edition. Addison-Wesley, Reading Massachusetts, 1988.

Gill G.K. and Kemerer C.F., Cyclomatic complexity density and software maintenance productivity, *IEEE Transactions on Software Engineering*, 17(12), 1284–1288, 1991.

GNU 2013, *GNU Coding Standards*, Free Software Foundation, http://www.gnu.org/prep/standards, August 20, 2014.

Godfrey M.W. and Tu Q., Evolution in open source software: A case study, *Proceedings of the International Conference on Software Maintenance (ICSM)*, San Jose, California, pp. 131–142, 2000.

Goel A.L., Software reliability models: Assumptions, limitations and applicability, *IEEE Transactions on Software Engineering*, 11(12), 1411–1423, 1985.

Goel A.L. and Okumoto K., Time-dependent error-detection rate model for software reliability and other performance measures, *IEEE Transactions on Reliability*, R-28, 206–211, 1979.

González-Barahona J.M., Pérez M.A.O., de las Heras Quirós P., González J.C. and Olivera V.M., Counting potatoes: The size of Debian 2.2, *Upgrade*, II(6), 60–66, December 2001.

Grady R.B., Successfully applying software metrics, *IEEE Computer*, 27, 18–25, September 1994.

Grady R.B. and Caswell D., *Software Metrics: Establishing a Company-Wide Program*, Prentice-Hall, Englewood Cliffs, New Jersey, 1987.

Gras J.-J., End-to-end defect modeling, *IEEE Software*, 21(5), 98–100, 2004.

Green J. and d'Oliveira M., *Units 16 & 21 Methodology Handbook (Part 2)*, Open University, Milton Keynes, England, 1990.

Gueheneuc Y.-G. and Antoniol G., DeMIMA: A multilayered approach for design pattern identification, *IEEE Transactions on Software Engineering*, 34(5), 667–684, 2008.

Gunning R., *The Technique of Clear Writing*, McGraw-Hill, New York, 1968.

Haigh M., Software quality, non-functional software requirements and IT-business alignment, *Software Quality Journal*, 18(3), 361–385, 2010.

Halstead M., *Elements of Software Science*, Elsevier, North Holland, New York, 1977.

Hamer P. and Frewin G., Halstead's software science: A critical examination, *Proceedings of the 6th International Conference on Software Engineering*, Tokyo, Japan, pp. 197–206, 1982.

Hamlet D. and Voas J., Faults on its sleeve: Amplifying software reliability testing, *Proceedings of the ISSTA '93*, Boston, Massachusetts, pp. 89–98, 1993.

Harel D., *Algorithmics*, 3rd Edition. Addison-Wesley, Reading, Massachusetts, 2004.

Harrison W., A flexible method for maintaining software metrics data: A universal metrics repository, *The Journal of Systems and Software*, 72(2), 225–234, 2004.

Hatton L. and Safer C., *Developing Software for High-Integrity and Safety-Critical Systems*, McGraw-Hill, New York, 1995.

Hausen H.-L., Yet another model of software quality and productivity, in *Measurement for Software Control and Assurance* (Ed: B.A. Kichenham and B. Littlewood), Elsevier, London and New York, 1989.

Hays M. and Hayes J., The effect of testability on fault proneness: A case study of the Apache HTTP server, *Proceedings of the International Symposium on Software Reliability Engineering Workshops (ISSREW)*, Dallas, Texas, pp. 153–158, 2012.

Hecht M.S., *Flow Analysis of Computer Programs*, Elsevier, New York, 1977.

Henry S. and Kafura D., Software structure metrics based on information flow, *IEEE Transactions on Software Engineering*, SE-7(5), 510–518, 1981.

Heston K.M and Phifer W., The multiple quality models paradox: How much 'best practice' is just enough? *Journal of Software Maintenance and Evolution: Research and Practice*, 23(8), 517–531, 2011.

Hetzel W.C., *Making Software Measurement Work: Building an Effective Software Measurement Program*, QED Publishing Group, Wellesley, Massachusetts, 1993.

Hoaglin D.C., Mosteller F., and Tukey J.W., *Understanding Exploratory Data Analysis*, John Wiley & Sons, New York, 2000.

Holzinger A., Usability engineering methods for software developers, *Communications of the ACM*, 48(1), 71–74, January 2005.

Hornbaek K., Current practice in measuring usability: Challenges to usability studies and research, *International Journal of Human-Computer Studies*, 64(2), 79–102, 2006.

Hubbard D.W., *How to Measure Anything: Finding the Value of Intangibles in Business*, 2nd Edition. John Wiley & Sons, Hoboken, New Jersey, 2010.

Huber J.T., A comparison of IBM's orthogonal defect classification to Hewlett Packard's defect origins, types and modes, *Proceedings of the International Conference on Applications of Software Measurement*, San Jose, California, 1–17, 2000.

Humphrey W.S., *Managing the Software Process*, Addison-Wesley, Reading, Massachusetts, 1989.

Humphrey W.S., *A Discipline for Software Engineering*, Addison-Wesley, Reading, Massachusetts, 1995.

Humphrey W.S., *PSP(sm): A Self Improvement Process for Software Engineers*, Addison-Wesley Professional, Upper Saddle River, New Jersey, 2005.

IEEE Standard 610.12-1990, *Glossary of Software Engineering Terminology*, IEEE Computer Society Press, New York, 1990.

IEEE Draft Standard 1044-1993, *Draft Standard Classification for Software Anomalies*, IEEE Computer Society Press, New York, 1993.

IEEE Standard 1044-2009, *Standard Classification for Software Anomalies*, IEEE Computer Society Press, New York, 2009.

IEEE Standard 1061, *Software Quality Metrics Methodology*, IEEE Computer Society Press, New York, 2009.

Ince D.C. and Hekmatpour S., An approach to automated software design based on product metrics, *Software Engineering Journal*, 3, 53–56, March 1988.

Inglis J., Standard software quality metrics, *AT&T Technical Journal*, 65(2), 113–118, 1985.

International Standards Organisation, Software engineering—Product quality—Part 1: Quality model, SS-ISO/IEC 9126-1, 2003.

International Standards Organisation, Software engineering—Product quality—Part 2: External metrics, SS-ISO/IEC 9126-2, 2003.

International Standards Organisation, Software engineering—Product quality—Part 3: Internal metrics, SS-ISO/IEC 9126-3, 2003.

International Standards Organisation, Software engineering—Product quality—Part 4: Quality in use metrics, SS-ISO/IEC 9126-4, 2004.

International Standards Organization, Systems and software engineering—systems and software quality requirements and evaluation (SQUARE)—systems and software quality models, ISO/IEC 25010:2011(E), 2011.

International Standards Organization, Systems and software engineering—systems and software quality requirements and evaluation (SQUARE)—systems and software quality models, ISO/IEC 25040:2011(E), 2011.

ISQAA, *Metrics Handbook*, Information Systems Quality Assurance Association, London, 1989.

Izurieta C. and Bieman J., The evolution of FreeBSD and Linux, *Proceedings of the ACM/IEEE International Symposium on Empirical Software Engineering (ISESE 2006)*, Rio de Janeiro, Brazil, 204–211, 2006.

Izurieta C. and Bieman J., Testing consequences of grime buildup in object oriented design patterns, *Proceedings of the International Conference Software Testing, Verification, and Reliability (ICST)*, Lillehammer, Norway, 171–179, 2008.

Izurieta C. and Bieman J., A multiple case study of design pattern decay, grime, and rot in evolving software systems, *Software Quality Journal*, 21(2), 289–323, June 2013.

Jackson D., Alloy: A lightweight object modeling notation, *ACM Transactions on Software Engineering and Methodology*, 11(2), 256–290, April, 2002.

Jeffery D.R., Low G.C., and Barnes M., A comparison of function point counting techniques, *IEEE Transactions on Software Engineering*, 19(5), 529–532, 1993.

Jelinski Z. and Moranda P.B., Software reliability research, in *Statistical Computer Performance Evaluation* (Ed: Freiberger), Academic Press, New York, pp. 465–484, 1972.

Jensen F.V. and Nielsen T., *Bayesian Networks and Decision Graphs*, Springer-Verlag, New York Inc., 2007.

Jones C., *Applied Software Measurement: Global Analysis of Productivity and Quality*, 3rd Edition. McGraw-Hill, New York, 2008.

Kafura D. and Henry S., Software quality metrics based on interconnectivity, *Journal of Systems and Software*, 2, 121–131, 1981.

Kaposi A.A., Measurement theory, in *Software Engineer's Reference Book* (Ed: McDermid J.), Butterworth Heinemann, Oxford, Boston, 1991.

Kaposi A.A. and Kitchenham B.A., The architecture of systems quality, *Software Engineering Journal*, 2(1), 2–8, 1987.

Kaposi A.A. and Myers M., *Systems, Models and Measures*, Springer-Verlag, London, 1993.

Kapur P.K., Pham H., Anand S., and Yadav K., A unified approach for developing software reliability growth models in the presence of imperfect debugging and error generation, *IEEE Transactions on Reliability*, 60(1), 331–340, 2011.

Kauffman R. and Kumar R., Modeling estimation expertise in object based CASE environments, Stern School of Business Report, New York University, January 1993.

Kemerer C.F., Reliability of function points measurement: A field experiment, *Communications of the ACM*, 36, 85–97, February 1993.

Kemerer C.F. and Porter B., Improving the reliability of function points measurement: An empirical study, *IEEE Transactions on Software Engineering*, 18(10), 1011–1024, 1992.

Khoshgoftaar T.M. and Allen E.B., Applications of information theory to software engineering measurement, *Software Quality Journal*, 3(2), 105–112, 1994.

Khoshgoftaar T.M., Allen E.B., Jones W.B., and Hudepole J.P., Classification-tree models of software quality over multiple releases, *IEEE Transactions on Reliability*, 49(1), 4–11, March 2000.

Kitchenham B.A., Empirical studies of assumptions that underlie software cost-estimation models, *Information and Software Technology*, 34(4), 211–218, 1992.

Kitchenham B.A., Using function points for software cost estimation, in *Software Quality Assurance and Measurement* (Eds: Fenton N.E., Whitty R.W., and Iizuka Y.), International Thomson Computer Press, London, pp. 266–280, 1995.

Kitchenham B.A, Series on experimentation in software engineering, *ACM Software Engineering Notes*, 1996.

Kitchenham B.A. and Känsälä K., Inter-item correlations among function points, *Proceedings of the IEEE Software Metrics Symposium*, IEEE Computer Society Press, Baltimore, Maryland, pp. 11–15, 1993.

Kitchenham B.A. and Walker J.G., A quantitative approach to monitoring software development, *Software Engineering Journal*, 4(1), 2–13, 1989.

Kitchenham B.A., Kok P.A.M., and Kirakowski J., The MERMAID approach to software cost estimation, *ESPRIT '90*, Kluwer Academic Press, Brussels, pp. 296–314, 1990a.

Kitchenham B.A., Pfleeger S.L., and Fenton N.E., Towards a framework for software measurement validation, *IEEE Transactions on Software Engineering*, 21(12), 929–944, 1995.

Kitchenham B.A., Pickard L.M., and Linkman S.J., An evaluation of some design metrics, *Software Engineering Journal*, 5(1), 50–58, 1990b.

Knight J.C. and Leveson N.G. An empirical study of failure probabilities in multi-version software. *In Fault Tolerant Computing Symposium*, Vienna, Austria, vol. 16, pp. 165–170, 1986.

Knuth D.E., *The Art of Computer Programming*, Volume 3, Addison-Wesley, Reading, Massachusetts, 1973.

Knuth D.E., The errors of tex, *Software Practice and Experience*, 19(7), 607–685, 1989.

Koller D. and Pfeffer A. Object-oriented Bayesian networks, *Proceedings of the 13th Annual Conference on Uncertainty in AI (UAI)*, Providence, Rhode Island, pp. 302–313, 1997.

Koru G., Liu H., Zang, D., and El Emam K., Testing the theory of relative defect proneness for closed-source software. *Empirical Software Engineering*, 15, 577–598, 2010.

Kosaraju R.S., Analysis of structured programs, *Journal of the CSS*, 9, 232–255, 1974.

Koziolek H., Schlich B., and Bilich C., A large-scale industrial case study on architecture-based software reliability analysis, In *Proceedings of the IEEE 21st International Symposium on Software Reliability Engineering (ISSRE)*, San Jose, California, pp. 279–288, 2010.

Kpodjedo S., Ricca F., Galinier P., Gueheneuc Y.-G., and Antoniol G., Design evolution metrics for defect prediction in object oriented systems, *Empirical Software Engineering*, 16(1), 141–175, 2011.

Krantz D.H., Luce R.D., Suppes P., and Tversky A., *Foundations of Measurement*, Volume 1, Academic Press, New York, 1971.

Kraska-Miller M., *Nonparametric Statistics for Social and Behavioral Science*, CRC Press, Boca Raton, Florida, 2014.

Kusumoto S, Matukawa F., Inoue K., Hanabasa S., and Maegawa Y., Effort estimation tool based on use case points: Method, tool, and case study, *Proceedings of the 10th International Symposium on Software Metrics*, IEEE, Chicago, Illinois, pp. 292–299, 2004.

Kyburg H.E., *Theory and Measurement*, Cambridge University Press, Cambridge, England, 1984.

Lackshmanan K.B., Jayaprakesh S., and Sinha P.K., Properties of control-flow complexity measures, *IEEE Transactions on Software Engineering*, 17(12), 1289–1295, 1991.

Larman C., *Applying UML and Patterns: An Introduction to Object-Oriented Analysis and Design and Iterative Development*, 3rd Edition. Prentice-Hall, Upper Saddle River, New Jersey, 2004.

Lee W., *Experimental Design and Analysis*, W.H. Freeman and Company, San Francisco, California, 1975.

Lethbridge T.C., Sim S.E., and Singer J., *Empirical Software Engineering*, 10(3), 311–341, 2005.

Le Traon Y., Baudry B., and Jézéquel J.-M., Design by contract to improve software vigilance, *IEEE Transactions on Software Engineering*, 32(8), 571–586, 2006.

Le Traon Y., Oubdesselam F., and Robach C., Analyzing testability on data flow designs, *Proceedings of the 11th International Symposium on Software Reliability Engineering (ISSRE)*, IEEE, Denver, Colorado, pp. 162–173, 2003.

Leveson N., *Safeware: System Safety and Computers*, Addison-Wesley, Reading, Massachusetts, 1995.

Leveson N.G. and Turner C.S., An investigation of the Therac-25 accidents, *IEEE Computer*, 26, 18–41, July 1993.

Lincke, R., Lundberg, J., and Löwe, W., Comparing software metrics tools, *Proceedings of the International Symposium on Software Testing and Analysis (ISSTA)*, ACM, Seattle, Washington, pp. 131–142, 2008.

Li W. and Henry S., Object oriented metrics that predict maintainability, *Journal of Systems and Software*, 23, 111–122, 1993.

Lim W.C., Effects of reuse on quality, productivity and economics, *IEEE Software*, 11(5), 23–30, September 1994.

Linger R., Cleanroom process model, *IEEE Software*, 11(2), 50–58, March 1994.

Littlewood B., A software reliability model for modular program structure, *IEEE Transactions on Reliability*, R-28(3), 241–246, 1979.

Littlewood B., Stochastic reliability growth: A model for fault removal in computer programs and hardware designs, *IEEE Transactions on Reliability*, R-30, 313–320, 1981.

Littlewood B. (Ed.), *Software Reliability: Achievement and Assessment*, Blackwell Scientific Publications, Oxford, 1987.

Littlewood B., Forecasting software reliability, in *Software Reliability, Modelling and Identification* (Ed: Bittanti S.), Lecture Notes in Computer Science 341, Springer-Verlag, Berlin Heidelberg, pp. 141–209, 1988.

Littlewood B., Predicting software reliability, *Philosophical Transactions of the Royal Society of London*, A 327, 513–527, 1989.

Littlewood B., Limits to evaluation of software dependability, in *Software Reliability and Metrics* (Eds: Littlewood B. and Fenton N.) Elsevier, London, New York, 1991.

Littlewood B. and Miller D., *Software Reliability and Safety*, Elsevier, London, New York, 1991.

Littlewood B. and Miller D.R., Conceptual modeling of coincident failures in multiversion software, *IEEE Transactions on Software Engineering*, 15(12), 1596–1614, 1989.

Littlewood B. and Strigini L., Validation of ultra-high dependability for software-based systems, *Communications of the ACM*, 36(11), 1993.

Littlewood B. and Verrall J.L., A Bayesian reliability growth model for computer software, *Journal of the Royal Statistical Society*, C22, 332–34, 1973.

Littlewood B., Brocklehurst S., Fenton N.E., Mellor P., Page S., Wright D., and Dobson J., Towards operational measures of security, *Journal of Computer Security*, 2, 211–229, 1993.

Lokan C.J., Function points, *Advances in Computers*, 65, 297–347, 2005.

MacDonnell S.G., Rigor in software complexity measurement experimentation, *Journal of Systems and Software*, 16, 141–149, 1991.

Madsen A.L., *Bayesian Networks and Influence Diagrams*, Springer Verlag, New York, 2007.

Malevanny S., *Case Study: Software Project Cost Estimates Using COCOMO II Model*, 2005, www.codeproject.com, The Code Project, Full URL http://www.codeproject.com/Articles/9266/Software-Project-Cost-Estimates-Using-COCOMO-II-Mo, August 7, 2014.

Manadhata P.K. and Wing. J.M., An attack surface metric, *IEEE Transactions on Software Engineering*, 37(3), 371–386, May/June 2011.

Martin R.C., *Agile Software Development: Principles, Patterns, and Practices*, Prentice-Hall, Upper Saddle River, New Jersey, 2003.

Mascena, J.C.C.P., de Almeida, E.S., and de Lemos Meira, S.R., A comparative study on software reuse metrics and economic models from a traceability perspective, *Proceedings of the IEEE International Conference Information Reuse and Integration (IRI)*, Las Vegas, Nevada, pp. 72–77, 2005.

Mayer A. and Sykes S.A., Probability model for analysing complexity metrics data, *Software Engineering Journal*, 4(5), 254–258, 1989.

McCabe T, A software complexity measure, *IEEE Transactions on Software Engineering*, SE-2(4), 308–320, 1976.

McCall J.A., Richards P.K., and Walters, G.F., Factors in software quality, RADC TR-77-369, 1977. Vols I, II, III, US Rome Air Development center Reports NTIS AD/A-049 014, 015, 055, 1977.

McGrayne S.B., *The Theory That Would Not Die*, Yale University Press, New Haven, Connecticut, 2011.

Mell P., Scarfone K., and Romanosky S., CVSS A complete guide to the common vulnerability scoring system version 2.0, *Forum of Incidence Response and Security Teams (FIRST)*, June 2007, http://www.first.org/cvss/cvss-guide. html, August 7, 2014.

Mellor P., Failures, faults and changes in dependability measurement, *Information and Software Technology*, 34(10), 640–654, 1992.

Mellor P., CAD—Computer aided disaster, *High Integrity Systems Journal*, 1(2), 101–156, 1994.

Melton A. (Ed.), *Software Measurement*, International Thomson Computer Press, London, Boston, 1995.

Melton A.C., Bieman J.M., Baker A., and Gustafson D.A., Mathematical perspective of software measures research, *Software Engineering Journal*, 5(5), 246–254, 1990.

Mendonça M.G and Basili V.R., Validation of an approach for improving existing measurement frameworks, *IEEE Transactions on Software Engineering*, 26(6), 484–499, 2000.

Menzies T., Greenwald J., and Frank A. Data mining static code attributes to learn defect predictors. *IEEE Transactions on Software Engineering*, 33(1), 2–13, 2007.

Miller D.R., Exponential order statistic models of software reliability growth, *IEEE Transactions on Software Engineering*, SE-12(1), 12–24, 1986.

Mockus A. and Weiss D., Interval quality: Relating customer-perceived quality to process quality, *Proceedings of the International Conference on Software Engineering (ICSE 2008)*, Leipzig, Germany, pp. 723–732, 2008.

Modarres M., Kaminskiy M., and Krivtsov V., *Reliability Engineering and Risk Analysis: A Practical Guide*, 2nd Edition. CRC Press, Boca Raton, Florida, 2010.

Mohagheghi P. and Conradi R., Quality, productivity and economic benefits of software reuse: A review of industrial studies, *Empirical Software Engineering*, 12(5), 471–516, 2007.

Mohagheghi P. and Conradi R., An empirical investigation of software reuse benefits in a large telecom product, *ACM Transactions on Software Engineering and Methodology (TOSEM)*, 17(3), 1–31, June 2008.

Moller K.-H. and Paulish D., An empirical investigation of software fault distribution, in *Software Quality Assurance and Measurement* (Eds: Fenton N.E., Whitty R.W., and Iizuka Y.), International Thomson Computer Press, London, Boston, pp. 242–253, 1995.

Morres T.T., Developing a software size model for rule-based systems: A case study, *Expert Systems with Applications*, 21, 229–237, 2001.

Moroney M.J., *Facts from Figures*, Third and revised edition, Pelican Books, London, 1962.

Mosteller F. and Tukey J.W., *Data Analysis and Regression*, Addison-Wesley, Reading, Massachusetts, 1977.

Munger W., Bieman J., and Alexander R., Coding concerns: Do they matter?, *Proceedings of the Workshop on Empirical Studies of Software Maintenance (WESS 2002)*, Montreal, Canada, 2002.

Musa J., A theory of software reliability and its application, *IEEE Transactions on Software Engineering*, SE-1, 312–327, 1975.

Musa J., Software reliability data, Technical Report available from Data Analysis Center for Software, Rome Air Development Center, New York, USA, 1979.

Musa J., *Software Reliability Engineering: More Reliable Software Faster and Cheaper*, 2nd Edition. Tata McGraw-Hill Education, New York, 2004.

Myers G.J., *Composite Structured Design*, Van Nostrand Reinhold, New York, 1978.

Neil M.D., Multivariate assessment of software products, *Journal of Software Testing, Verification and Reliability*, 1(4), 17–37, 1992.

Neil M.D., Measurement as an alternative to bureaucracy for the achievement of software quality, *Software Quality Journal*, 3(2), 65–78, 1994.

Neil M., M. Tailor M., and D. Marquez D., Inference in hybrid Bayesian networks using dynamic discretization, *Statistics and Computing*, 17(3), 219–233, 2007.

Neil M., Marquez D., and Fenton N.E. Improved reliability modeling using Bayesian networks and dynamic discretization, *Reliability Engineering & System Safety*, 95(4), 412–425, 2010.

Neapolitan R.E., *Learning Bayesian Networks*, Upper Saddle River Pearson Prentice-Hall, Upper Saddle River, New Jersey, 2004.

NetFocus: Software Program Managers Network, number 207, Department of the Navy (US), January 1995.

Neumann P.G. (moderator), *The Risks Digest: Forum on Risks to the Public in Computers and Related Systems*, ACM Committee on Computers and Public Policy. http://catless.ncl.ac.uk/Risks/, August 8, 2014.

Nishiyama S. and Furayama T., The validity and applicability of function point analysis, pre-print, 1994.

NIST/SEMATECH e-Handbook of Statistical Methods, http://www.itl.nist.gov/div898/handbook/pmc/section5/pmc51.htm. Image of Normal distribution, visited October 17, 2011.

Oivo M. and Basili V.R., Representing software engineering models: The TAME goal oriented approach, *IEEE Transactions on Software Engineering*, 18(10), 886–898, 1992.

Onvlee J., Use of function points for estimation and contracts. In: Fenton, N. E., Iizuka, Y., and Whitty, R. W., (Eds.), *Software Quality Assurance and Metrics: A Worldwide Perspective*. International Thomson Computer Press, London, pp. 88–93, 1995.

Ostle B. and Malone L.C., *Statistics in Research*, 4th Edition. Iowa State University Press, Ames, Iowa, 1988.

Ott R.L. and Longnecker M.T., *An Introduction to Statistical Methods and Data Analysis*, 6th Edition. Ducxbury Press, Pacific Grove, California, United Kingdom, 2010.

Pai G.J. and Dugan J.B., Empirical analysis of software fault content and fault proneness using Bayesian methods, *IEEE Transactions on Software Engineering*, 33(10), 675–685, October 2007.

Park R., Software size measurement: A framework for counting source statements, CMU/SEI-92-TR-20, Software Engineering Institute Technical Report, Pittsburgh, Pennsylvania, 1992.

Parnas D.L., On the criteria to be used in decomposing systems into modules, *Communications of the ACM*, 15(12), 1052–1058, 1972.

Parnas D.L., Madey J., and Iglewski M., Precise documentation of well-structured programs, *IEEE Transactions on Software Engineering*, 20(12), 948–976, 1994.

Pearl J., *Probabilistic Reasoning in Intelligent Systems*, Morgan Kaufmann, Palo Alto, CA, 1988.

Pearl J., *Causality: Models Reasoning and Inference*, Cambridge University Press, 2000.

Perlis A.J., Sayward F.G., and Shaw M. (Eds.), *Software Metrics: An Analysis and Evaluation*, MIT Press, Cambridge, Massachusetts, 1981.

Petersen K. and Wohlin C., The effect of moving from a plan-driven to an incremental software development approach with agile processes, *Empirical Software Engineering*, 15(6), 654–693, 2010.

Pfleeger, S.L., Lessons learned in building a corporate metrics program, *IEEE Software*, 10(3), 67–74, May 1993.

Pfleeger S.L. and Atlee J.M., *Software Engineering Theory and Practice*, 3rd Edition. Pearson Education, Inc., Upper Saddle River, New Jersey, 2006.

Pfleeger S.L. and McGowan C.L., Software metrics in a process maturity framework, *Journal of Systems and Software*, 12, 255–261, 1990.

Pfleeger S.L., Fenton N.E., and Page S., Evaluating software engineering standards, *IEEE Computer*, 27, 71–79, September 1994.

Ponisio L. and Nierstrasz O., Using contextual information to assess package cohesion, Technical Report IAM-06-002, University of Bern, 2006.

Poolsappasit N., Dewri R., and Ray I., Dynamic security risk management using Bayesian attack graphs, *IEEE Transactions on Dependable and Secure Computing*, 9(1), 61–74, January-February 2012.

Prather R.E., An axiomatic theory of software complexity measure, *Computer Journal*, 27, 273–347, 1984.

Prather R.E., On hierarchical software metrics, *Software Engineering Journal*, 2(2), 42–45, 1987.

Prather R.E., Hierarchical metrics and the prime generation problem, *Software Engineering Journal*, 8(5), 246–252, 1993. (Interesting theoretical paper that describes a method for generating prime flowgraphs.)

Prather R.E. and Giulieri S.G., Decomposition of flowchart schemata, *Computer Journal*, 24(3), 258–262, 1981.

Pressman R.S., *Software Engineering: A Practitioner's Approach*, 7th Edition. McGraw-Hill, New York, 2010.

Pulford K., Kuntzmann-Combelles A., and Shirlaw S., *A Quantitative Approach to Software Management*, Addison-Wesley, Reading, Massachusetts, 1995.

Riaz M., Mendes E., and Tempero E., A systematic review of software maintainability prediction and metrics, *Proceedings of the Third International Symposium on Empirical Software Engineering and Measurement*, Lake Buena Vista, Florida, pp. 367–377, 2009.

Rapps S. and Weyuker E.J., Selecting software test data using data flow information, *IEEE Transactions on Software Engineering*, 11(4), 367–375, 1985.

Rasool G. and Mader P., Flexible design pattern detection based on feature types, *Proceedings of the Automated Software Engineering (ASE)*, Lawrence, Kansas, pp. 243–252, 2011.

Ratcliffe B. and Rollo A.L., Adapting function point analysis to Jackson System Development, *Software Engineering Journal*, 5(1), 79–84, 1990.

Rausand M. and A. Hoyland A., *System Reliability Theory: Models, Statistical Methods, and Applications*. 2nd Edition. John Wiley & Sons, Inc., Hoboken, New Jersey, 2004.

Riaz M., Mendes E., and Tempero E., A systematic review of software maintainability prediction and metrics, *Proceedings of the Third International Symposium on Empirical Software Engineering and Measurement*, Lake Buena Vista, Florida, pp. 367–377, 2009.

Rifkin S. and Cox C., Measurement in practice, SEI Technical Report SEI—CMU—91—TR—16, Software Engineering Institute, Pittsburgh, Pennsylvania, 1991.

Riley P., Towards safe and reliable software for Eurostar, *GEC Journal of Research*, 12(1), 3–12, 1995.

Roberts F.S., *Measurement Theory with Applications to Decision Making, Utility, and the Social Sciences*, Addison-Wesley, Reading, Massachusetts, 1979.

Roberts F.S., Applications of the theory of meaningfulness to psychology, *Journal of Mathematical Psychology*, 29, 311–332, 1985.

Robertson J. Microsoft Zune's New Year Crash, *The Street*, 2009, http://www.thestreet.com/story/10455712/1/microsoft-zunes-new-year-crash.html, August 7, 2014.

Ron D., Algorithmic and analysis techniques in property testing, *Foundations and Trends® in Theoretical Computer Science*, 5(2), 73–205, February 2010.

Rooijmans J., Aerts H., and van Genutchen M., Software quality in consumer electronics products, *IEEE Software*, 13(1), 55–64, January 1996.

Rook P. (Ed.), *Software Reliability Handbook*, Elsevier, North Holland, 1990.

Rouquet J.C. and Traverse P.J., Safe and reliable computing on board the Airbus and ATR aircraft, *Proceedings of the 5th IFAC Workshop on Safety of Computer Control Systems* (Ed: Quirk W.J.) Pergamon Press, Oxford, pp. 93–97, 1986.

Roy B., Decision aid and decision making, *European Journal of Operational Research*, 45, 324–331, 1990.

Rozum J.A. and Florac W.A., A DoD software measurement pilot: Applying the SEI core measures, Software Engineering Institute Technical Report CMU/SEI-94-TR-016, Pittsburgh, Pennsylvania, May 1995.

Saaty T. and Vargas L., *Models, Methods, Concepts & Applications of the Analytic Hierarchy Process*, 2nd Edition. Springer, New York, 2012.

Schneidewind N.F., Validating metrics for ensuring space shuttle flight software quality, *IEEE Computer*, 27, 50–58, August 1994.

Schneidewind N.F., Controlling and predicting the quality of space shuttle software using metrics, *Software Quality Journal*, 4(1), 49–68, 1995.

Schneidewind N.F. and Keller T.W., Applying reliability models to the space shuttle, *IEEE Software*, 9(4), 28–33, July 1992.

Schulmeyer G.G. and McManus J.I., *Handbook of Software Quality Assurance*, Van Nostrand Reinhold, New York, Boston: Artech House, 2008.

Seffah A., Donyaee M., Kline R.B., and Padda H.K., Usability measurement and metrics: A consolidated model, *Software Quality Journal*, 14(2), 159–178, 2006.

Selby R.W., Extensible integration frameworks for measurement, *IEEE Software*, 7(6), 83–84, November 1990.

Selby R.W., Enabling reuse-based software development of large-scale systems, *IEEE Transactions on Software Engineering*, 31(6), 495–510, 2005.

Shen V.Y., Conte S.D. and Dunsmore H.E., Software science revisited: A critical analysis of the theory and its empirical support, *IEEE Transactions on Software Engineering*, 9(2), 155–165, March 1983.

Shepperd M., *Software Engineering Metrics, Volume 1: Measures and Validations*, McGraw-Hill, New York, 1993.

Shepperd M.J., A critique of cyclomatic complexity as a software metric, *Software Engineering Journal*, 3(2), 30–36, 1988.

Shepperd M. and Turner R., Real time function points: An industrial validation, *Proceedings of European Software Cost Modelling Conference* (ESCOM), Bristol, England, 1993.

Shepperd M.J. and Ince D., *Derivation and Validation of Software Metrics*, Clarendon Press, Oxford, 1993.

Shooman M.L., *Software Engineering: Design, Reliability and Management*, McGraw-Hill, New York, 1983.

Siegel S. and Castellan N.J. Jr., *Nonparametrics Statistics for the Behavioral Sciences*, 2nd Edition. McGraw-Hill, New York, 1988.

Sillitti A., Russo B., Zuliani P., and Succi G., Deploying, updating, and managing tools for collecting software metrics, *Proceedings of the 2004 Workshop on Quantitative Techniques for Software Agile Process*, Newport Beach, California, ACM, pp. 1–4, 2004.

Simpson E., Bayes at Bletchley Park, *Significance* 7(2), 76–80, 2010.

Software Productivity Consortium, Software measurement guidebook, *Software Measurement Guidebook*, (Main contributors: Bassman M.J., McGarry F., and Pajesrki R.), John Gaffney, (Ed.) et al. International Thomson Computer Press, London, Boston, 1995.

Spivey J.M., *The Z Notation: A Reference Manual*, Prentice-Hall, Englewood Cliffs, New Jersey, 1993.

Sprent P., *Applied Nonparametric Statistical Methods*, 4th Edition. Chapman & Hall, London, 1989.

Stavely A.M., *Toward Zero-Defect Programming*, Addison-Wesley Longman Publishing Co., Reading, Massachusetts, Inc., 1999.

Stevens S.S., On the theory of scale types and measurement, *Science*, 103, 677–680, 1946.

Stevens W., Myers G., and Constantine L., Structured design, *IBM Systems Journal*, 13(2), 115–139, 1974.

Sydenham P.H. (Ed.), *Handbook of Measurement Science*, Volume 1, John Wiley, New York, 1982.

Symons C.R., Function point analysis: Difficulties and improvements, *IEEE Transactions on Software Engineering*, 14(1) 2–11, 1988.

Tajima D. and Matsubara T., The computer software industry in Japan, *IEEE Computer*, 14(5), 89–96, 1981.

Thuring M. and Mahlke S., Usability, aesthetics and emotions in human-technology interaction, *International Journal of Psychology*, 42(4), 253–264, 2007.

Tian J. and Zelkowitz M.V., Complexity measure evaluation and selection, *IEEE Transactions on Software Engineering*, 21(8), 641–650, 1995.

Vaisanen A., Auer A., and Korhonen J., Assessment of the safety of PLCs: Janiksenlinna water plant study, *SHIP/T/033*, VTT, Finland, 1994.

van Solingen R. and Berghout E., *The Goal/Question/Metric Method: A Practical Guide for Quality Improvement of Software Development*, McGraw-Hill, London, 1999.

van Vliet J.C., *Software Engineering: Principles and Practice*, 3rd Edition. John Wiley & Sons, New York, 2008.

Velleman P.F. and Wilkinson L., Nominal, ordinal, interval and ratio typologies are misleading, *The American Statistician*, 47(1), 65–72, February 1993.

Verner J.M. and Tate G., Estimating size and effort in fourth generation language development, *IEEE Software*, 5(4), 173–177, July 1988.

Verner, J. and G. Tate, A software size model, *IEEE Transactions on Software Engineering*, 18(4), 265–278, 1992.

Vincke P., *Multicriteria Decision Aids*, John Wiley, New York, 1992.

Voas J.M. and Miller K.W., Software testability: The new verification, *IEEE Software*, 12(3), 17–28, May 1995.

Walker, M., *The Nature of Scientific Thought*, Prentice-Hall, Inc. Englewood Cliffs, New Jersey, 1963.

Wang H., Peng F., Zhang C., and Pietschker A., Software project level estimation model framework based on Bayesian belief networks, in *Sixth International Conference on Quality Software (QSIC'06)*, Beijing, China, pp. 209–218, 2006.

Weimer W., Forrest S., Le Goues C., and Nguyen T.V., Automatic program repair with evolutionary computation, *Communications of the ACM*, 53(5), 109–116, May 2010.

Weyuker E., Can we measure software testing effectiveness? *Proceedings of the First International Software Metrics Symposium*, Baltimore, Maryland, IEEE Computer Society Press, pp. 100–107, 1993.

Weyuker E.J., Evaluating software complexity measures, *IEEE Transactions on Software Engineering*, SE-14(9), 1357–1365, 1988.

Weyuker E.J., More experience with data-flow testing, *IEEE Transactions on Software Engineering*, 19(3), 912–919, 1993.

Wheeler D., More than a gigabuck: Estimating GNU/Linux's size, Version 1.07, 2002, http://www.dwheeler.com/sloc/redhat71-v1/redhat71sloc.html, August 8, 2014.

Whitmire S., *Object-Oriented Design Measurement*, John Wiley & Sons, New York, 1997.

Whitty R.W. and Fenton N.E., An axiomatic approach to systems complexity, in L. Evans (Ed.), *Pergamon Infotech State-of-the-Art Reports: Designing for Systems Maturity*, Pergamon Infotech Ltd., Oxford, New York, pp. 113–137, 1985.

Whitty R.W., Fenton N.E., and Kaposi A.A., A rigorous approach to structural analysis and metrication of software, *IEE Software and Microsystems*, 4(1), 2–16, 1985.

Wilson R.I., *Introduction to Graph Theory*, 5th Edition. Pearson, Harlow, New York, 2010.

(Solid, standard text on graph theory. More than adequate background for Chapter 9.)

Wirth N., Program development by stepwise refinement, *ACM Computing Surveys*, 6, 247–259, 1974.

Wohlin C. and Ahlgren M., Soft factors and their impact on time to market, *Software Quality Journal*, 4(3), 189–206, 1995.

Wohlin C., Runeson P., Höst M., Ohlsson M.C., Regnell B., and Wesslén A., *Experimentation in Software Engineering: An Introduction*, Kluwer Academic Publishers, Norwell, MA, USA, 2000.

Wong W.E. and Mathur A.P., Fault detection effectiveness of mutation and data flow testing, *Software Quality Journal*, 4(1), 69–93, 1995.

Woodward M.R., Hedley D., and Hennell M.A., Experience with path analysis and testing of programs, *IEEE Transactions on Software Engineering*, 6(5), 278–286, 1980.

Yand Y. and Weber R., An ontological model of an information system, *IEEE Transactions on Software Engineering*, 16, 1282–1292, 1990.

(Was the model used by Chidamber and Kemerer for their metrics of object-oriented designs.)

Yates D.F. and Malevris N., The effort required by LCSAJ testing: An assessment via a new path generation strategy, *Software Quality Journal*, 4(3), 227–242, 1995.

Yau S.S. and Collofello J.S., Some stability measures for software maintenance, *IEEE Transactions on Software Engineering*, 6(6), 545–552, 1980.

Yau S.S. and Collofello J.S., Design stability measures for software maintenance, *IEEE Transactions on Software Engineering*, 11(9), 849–856, 1985.

Yin B.H. and Winchester J.W., The establishment and use of measures to evaluate the quality of system designs, *Proceedings of the Software Quality and Assurance Workshop*, San Diego, California, pp. 45–52, 1978.

Yourdon E. and Constantine L.L., *Structured Design*, Prentice-Hall, Englewood Cliffs, New Jersey, 1979.

Zhu D., Mosleh A., and Smidts C., A framework to integrate software behavior into dynamic probabilistic risk assessment, *Reliability Engineering and System Safety*, 92(12), 1733–1755, 2007.

Zhou Y. and Leung H., Empirical analysis of object-oriented design metrics for predicting high and low severity faults, *IEEE Transactions on Software Engineering*, 32(10), 771–789, October 2006.

Zhou T., Xu B., Shi L., Zhou Y., and Chen L., Measuring package cohesion based on context, *Proceedings of the IEEE International Workshop Semantic Computing and Systems*, Huangshan, China, pp. 127–132, 2008.

Ziliak S.T. and McCloskey D.N, *The Cult of Statistical Significance: How the Standard Error Costs Us Jobs, Justice, and Lives*, University of Michigan Press, Ann Arbor, Michigan, 2008.

Zuneboards, 2008. Cause of Zune 30 Leapyear Problem Isolated. http://www.zuneboards.com/forums/showthread.php?t = 38143, August 26, 2011.

Zuse H., *Software Complexity: Measures and Methods*. De Gruyter, Berlin, 1991.

Zuse H., Properties of software measures, *Software Quality Journal*, 194, 225–260, 1992.

Zweben S.H., Edwards S.H., Weide B.W., and Hollingsworth J.E., The effects of layering and encapsulation on software development cost and quality, *IEEE Transactions on Software Engineering*, 21(3), 200–208, March 1995.

Index